Managing Technology and Innovation

Modern technology and innovation are vital to the success of all companies, be they high-tech firms or companies seemingly unaffected by technology and innovation, established firms or business start-ups.

Managing Technology and Innovation focuses on understanding technology as a corporate resource, covering product development, design of systems and the managerial aspects of new and high technology.

Topics covered include:

- the internal organization of high-technology firms;
- the management of technology in the society;
- managing innovation;
- dilemmas and strategies.

The wide-ranging experience of the teachers and experts who have contributed to this book has resulted in an integrated, multi-disciplinary textbook that provides an introductory overview to managing technology and innovation in the twenty-first century. This text is essential reading for students of business and engineering concerned with technology and innovation management.

Robert M. Verburg is Associate Professor of Organizational Psychology at the Faculty of Technology Policy and Management at Delft University of Technology. His current research interests include the strategic management of human resources, knowledge management, coordination mechanisms of mobile virtual work and management of technology.

J. Roland Ortt is Associate Professor of Technology Management at the Faculty of Technology, Policy and Management of the Technical University Delft. His current research interests focus on describing and explaining the patterns of development and diffusion of breakthrough technologies.

Willemijn M. Dicke is Assistant Professor in the Faculty of Technology, Policy and Management, Delft University of Technology. She is leader of the Public Values Program of the Next Generation Infrastructures Foundation. She has published on the safeguarding of public values in the utility sectors, issues of water management, the shifting public–private divide under conditions of globalization and on liberalization and privatization in utility sectors.

Managing Technology and Innovation

An introduction

Edited by
Robert M. Verburg, J. Roland Ortt
and Willemijn M. Dicke

LONDON AND NEW YORK

First published 2006
by Routledge
2 Park Square, Milton Park, Abingdon, Oxon OX14 4RN

Simultaneously published in the USA and Canada
by Routledge
270 Madison Ave, New York, NY 10016

Routledge is an imprint of the Taylor & Francis Group

Typeset in Perpetua by
Florence Production Ltd, Stoodleigh, Devon
Printed and bound in Great Britain by
TJ International Ltd, Padstow, Cornwall

British Library Cataloguing in Publication Data
A catalogue record for this book is available from the British Library

Library of Congress Cataloging in Publication Data
A catalog record for this book has been requested

ISBN10: 0–415–36228–8 (hbk)
ISBN10: 0–415–36229–6 (pbk)

ISBN13: 9–78–0–415–36228–3 (hbk)
ISBN13: 9–78–0–415–36229–0 (pbk)

Contents

Figures

Tables

Boxes

About the authors

J.H. Erik Andriessen is Professor of Work and Organizational Psychology at Delft University of Technology, the Netherlands. He studied industrial psychology and obtained his Ph.D. in 1974 at the Free University in Amsterdam. He has been a professor at Delft University of Technology since 1990. He and his group are involved in international research with regard to new forms of work and organization, knowledge management and communities of practice, distributed teamwork, innovation processes and evaluation methodology.

Guus Berkhout started his career with Shell in 1964, where he held several international positions in R&D and technology transfer. In 1976 he accepted a Chair at Delft University of Technology in the field of geophysical and acoustical imaging. During 1998–2001, he was a member of the Board, being responsible for scientific research, knowledge management and intellectual property. In 2001 he also accepted a Chair in the field of innovation management. Professor Berkhout is a member of the Royal Netherlands Academy of Arts and Sciences (KNAW) and the Netherlands Academy of Engineering (NFTW), and an honorary member of the Society of Exploration Geophysicists (SEG).

Hans de Bruijn is Professor of Organization and Management at Delft University of Technology. His research is on performance management, network management, knowledge management and the process perspective of change management. Areas of application are professional organizations, decision making on large infrastructural projects and change processes in the utility sectors. Recent publications are *Performance Management in the Public Sector* (Routledge, London 2002) and *Process Management, Why Projects Management Fails in Complex Decision Making Processes* (Kluwer, Boston 2002, with others).

Mark de Bruijne (Ph.D.) studied Public Administration at Erasmus University in Rotterdam (1999) and specialized in safety management. His graduation paper focused on the influence of stress and bureau politics on the collection and processing of warning signals by the Dutch government during the crisis that developed around Dutch New-Guinea (1959–1962). His research "networked reliability" explores the consequences of institutional fragmentation with regard to the reliability of service provision in critical infrastructures. The focus of this research specifically described how operators and organizations who manage these infrastructures cope with these changes.

Deanne N. Den Hartog is Full-Professor of Organizational Behavior at the University of Amsterdam Business School in the Netherlands. She is director of the Business Studies Bachelor and Masters programs of the Business School and teaches on OB and leadership. Her research interests include cross-cultural and inspirational leadership processes, team processes and human resource management. Among other things, she studies the impact of leadership on employees' learning, affect, cooperation and innovative work behaviors.

Willemijn M. Dicke is Assistant Professor in the School of Technology, Policy and Management, Delft University of Technology, the Netherlands. She is leader of the Public Values Program of the Next Generation Infrastructures Foundation (www.nginfra.nl). She has published on the safeguarding of public values in the utility sectors, on issues of water management, on the shifting public–private divide under conditions of globalization, and on liberalization and privatization in utility sectors.

Peer Ederer is the Head of Innovation and Growth Project of Zeppelin University in Germany, Adjunct Faculty member of Rotterdam School of Management and Associate Professor of Strategy at Strategy Academy. He holds degrees from Sophia University in Japan and Harvard Business School in Boston, and earned a Ph.D. in economics. He has worked as a financial trader for Deutsche Bank, as a management consultant for McKinsey&Company and started his own technology company as an entrepreneur. He has co-authored two best-seller listed books in Germany, co-authored a path breaking study on human capital accounting and has published numerous chapters and articles on how to organize for innovation and growth.

Dap Hartmann received his Ph.D. in astronomy from Leiden University in 1994. Subsequently, he was visiting scientist at the Harvard-Smithsonian Center for Astrophysics (Cambridge, MA), at the University of Bonn, and at the Max-Planck-Institut für Radioastronomie. After returning to Holland, he founded Cygnus Consulting, providing expert advice in science and technology. Since 2003, he has been Assistant Professor in Intellectual Property Valorization and Innovation Management at Delft University of Technology.

Bernhard R. Katzy started his professional career with an apprenticeship as a car mechanic and later earned Master's degrees in electrical engineering and business management. He holds a Ph.D. in industrial management from the University of Technology (RWTH) Aachen in Germany and a second Ph.D. (habilitation) in general management and technology management from the University of St Gallen, Switzerland. He is currently a professor at the University BW Munich (D) and Leiden University (NL) and director of CeTIM – Center for Technology and Innovation Management. His research interest is entrepreneurial management of fast growing high-tech firms and the management of strategic change in the transition to the information age.

David J. Langley has been carrying out research in the Telecom sector since 1991, including a number of years at BT (British Telecom), KPN (the incumbent Dutch telecom operator) and TNO, the largest independent Dutch research institute. Originally working in the areas of end-user research and social psychology, Langley currently focuses on new ways of gaining insight into human behavior and applying that to the question of predicting and influencing the adoption and use of products and services.

Martijn Leijten is a Ph.D. researcher in Organization and Management at the Faculty of Technology, Policy and Management of Delft University of Technology, the Netherlands, and a member of the research center for Sustainable Urban Areas of Delft University of Technology.

Jakki J. Mohr (Ph.D. 1989, University of Wisconsin, Madison) is a Professor of Marketing at the University of Montana-Missoula. Prior to joining the University of Montana in the Fall of 1997, Dr Mohr was an Assistant Professor at the University of Colorado, Boulder (1989–1997). Her interests are primarily in the area of marketing of high-technology products and services, with a focus on distribution channels and governance. Her early research focused on organizational communication and learning between partners in strategic alliances and in distribution channels. Her research has received several awards, and has appeared in *Journal of Marketing*, *Strategic Management Journal*, *Journal of Public Policy and Marketing*, other marketing journals, and specialized journals at the intersection of technology and business marketing. She consults with a variety of high-tech companies, both in the US and in Europe. Before beginning her academic career, she worked in Silicon Valley for Hewlett-Packard and TeleVideo Systems.

Karel F. Mulder received an engineering degree from Twente University, and a doctorate in Business Administration from Groningen University in 1992. He teaches in the areas of industrial innovation studies, the societal challenges for innovation and the effects of innovation upon society. His current research interests are especially focused upon the challenge of long-term issues such as Sustainable Development, for innovation. Mulder presided over the Technology & Society department of the Dutch Royal Institute of Engineers from 1994 to 1999 and is currently in charge of a project to include Sustainable Development in all engineering curricula at Delft University of Technology.

C.W.M. (Ro) Naastepad obtained an M.A. (Sociology) from Leiden University, an MSc (Economics) from the University of East Anglia and her Ph.D. (Economics) from Erasmus University Rotterdam. She worked in the economics departments of Erasmus University Rotterdam and of Utrecht University and (since 1998) is a member of the Section Economics of Innovation of Delft University of Technology. Her research focuses on the longer-run interactions between aggregate demand growth, income distribution, labour relations, productivity growth and technological change in the OECD countries. Her research work is published in such journals as the *Cambridge Journal of Economics*, *Structural Change and Economic Dynamics* and *Economic Modelling*. She has co-edited two books: *The State and the Economic Process* (published in 1996) and *Globalization and Economic Development* (2001).

J. Roland Ortt studied economics and specialized in the analysis of high-tech markets. He was an Assistant Professor and received a Ph.D. at the faculty of Industrial Design Engineering of Delft University of Technology. His thesis was devoted to developing market analysis methods for a breakthrough communication technology, the video telephone. He has worked as an R&D manager at the research institute of KPN (the incumbent Dutch telecom operator). Currently he is an Associate Professor in technology management at the faculty of Technology, Policy and Management of Delft University of Technology.

Nico Pals studied biology and educational technology. He worked as Assistant Professor in biology, producer of scientific and educational films, Assistant Professor in educational media, manager Educational Research (KPN) and assistant manager Social Science Research (KPN). Currently he works as a senior scientist at TNO Telecom. His most recent R&D activities are all related to the study of human behavior: knowledge management, virtual communities and adoption and use of ICT services.

Tom Poot is Lecturer of Economics of Innovation at the University of Utrecht, faculty of Geosciences at the department of Innovation and Environmental sciences. He received his Ph.D. from Delft University of Technology at the Faculty of Technology, Policy and Management in 2004. His research reports on knowledge-intensive R&D collaboration. He has worked as a senior researcher at the foundation of Economic Research at the University of Amsterdam. There, he supervised a number of research projects. He wrote a report on knowledge-intensive R&D collaboration, the subject of his Ph.D. thesis, for the Ministry of Economic Affairs.

Emery Roe is Professor of Public Policy at Mills College. He has a Ph.D. in public policy from the University of California and writes widely on domestic and international public policy and management issues. His recent research includes critical infrastructure reliability, especially with respect to homeland security and deregulated electricity markets. He is also working on a “professional challenges” approach in policy education to managing complexity in real-time.

Paul R. Schulman is Professor of Government at Mills College. He received his Ph.D. in political science from Johns Hopkins University and his interests are in science and technology and their challenges to the public policy-making process.

Sanjit Sengupta (Ph.D. 1990, University of California, Berkeley) is Professor and Chair of the Marketing Department at San Francisco State University. Prior to joining San Francisco State in 1996, Dr Sengupta was Assistant Professor at the University of Maryland, College Park, where he received two teaching awards. He has taught executive development programs in the US, Finland and South Korea. His research interests include new product development and technological innovation, strategic alliances, sales management and international marketing. Prior to his academic career, Dr Sengupta worked in sales and marketing for Hindustan Computers Limited and CMC Limited in Mumbai, India.

Stanley F. Slater (Ph.D. 1988, University of Washington) is the Charles and Gwen Lillis Professor of Business Administration at Colorado State University. Dr Slater serves on the editorial review boards of *Journal of Marketing*, *Industrial Marketing Management* and *Journal of Strategic Marketing*. His work has appeared in *Journal of Marketing*, *Strategic Management Journal*, *Academy of Management Journal* and *Journal of Product Innovation Management*, among others. Prior to his academic career, he held positions with IBM and Adolph Coors Company. Dr Slater has consulted with Hewlett-Packard, Johns-Manville, Monsanto, United Technologies, Cigna Insurance, Qwest, Philips Electronics and Weyerhaeuser.

Uldrik E. Speerstra joined the faculty of Technology, Policy and Management of Delft University of Technology in 2001. He works for the Special Projects department. As a member of the Management of Technology project team he has developed the curriculum of the MSc programme in Management of Technology. Currently he is the coordinator of the programme.

Servaas Storm holds an M.A. (Economics) and a Ph.D. (Economics), both from Erasmus University Rotterdam. He worked in the economics department of Erasmus University Rotterdam before joining the Section Economics of Innovation of Delft University of Technology in 2001. His research interests include economic development, macro-economic policy, globalization, productivity growth and labor relations. He has published in books and in journals including the *Cambridge Journal of Economics*, *World Development*, *Journal of Development Economics*, *Journal of Policy Modeling* and *Development and Change*. With C.W.M. Naastepad he co-edited two books: *The State and the Economic Process* (1996) and *Globalization and Economic Development* (2001).

Michel J.G. van Eeten is an Associate Professor in the School of Technology, Policy and Management, Delft University of Technology, the Netherlands. He is also leader of the Critical Infra Program of the Next Generation Infrastructures Foundation (www.nginfra.nl). He has published on large technical systems, ecosystem management, high reliability theory, land use planning, transportation policy, internet governance and recasting intractable policy issues. His recent work as a practicing policy analyst includes advice to the Directorate General of Telecommunications and Post, the Ministry of Economic Affairs, KPN Mobile, Rabobank and the Civil Aviation Authority.

Patrick Van der Duin is a research fellow at Delft University of Technology, Faculty of Technology, Policy and Management, Section Technology, Strategy and Entrepreneurship. He received his Master's degree in economics from the University of Amsterdam. He was a researcher and senior advisor at KPN Research where he participated in future studies research on use of telecommunication services and products. Currently, he is finalizing his Ph.D. on the use of qualitative methods of futures research in innovation processes in large corporations.

Robert M. Verburg (Ph.D., organizational psychology, Free University Amsterdam) joined the faculty of Technology Policy and Management at Delft University of Technology in 2002 and is currently an Associate Professor of Organizational Psychology. He teaches in the areas of human resource management, organizational psychology and knowledge management. His current research interests include the strategic management of human resources, knowledge management, coordination mechanisms of mobile virtual work and management of technology. Specifically, he is interested in the impact of human resource management on knowledge sharing and innovation in organizations.

Zofia Verwater-Lukszo works as Associate Professor in the Energy and Industry Section at Delft University of Technology. She studied applied mathematics at Technical University of Łódź and philosophy at University of Łódź, Poland. Zofia received her Ph.D. from the Eindhoven University of Technology in 1996, for the thesis "A Practical Approach to Recipe Improvement and Optimization in the Batch Processing Industry." Since November 1, 1995 Zofia has worked at the Delft University of Technology at the Faculty of Technology, Policy and Management. Her research interests are management of manufacturing operations, in particular batch process operation, integrated plant management in the process industry, quality management and process control and optimization. Zofia has extensive experience in the application of mathematical modeling and optimization to industrial scenarios at the operational and control level.

Marc A. Zegveld is a part-time Associate Professor of strategic management at Delft University of Technology. His research interests are focused on the relation between business strategy and innovation and the impact of this relation on firm-level productivity. Besides his academic career Marc Zegveld is director of TVA developments BV, a strategy consultancy company based in the Netherlands. Marc Zegveld is a non-executive board member for several companies and institutions such as the "vereniging voor strategische beleidsvorming" which is the Dutch branch of the Strategic Management Society.

Preface

At the beginning of the last century, a small university town in the Netherlands, was for a few decades the coldest place on earth. In that city, the Dutch physicist Heike Kamerlingh Onnes succeeded as the first physicist on earth to liquefy helium at a temperature of –269 degrees Celsius. He also discovered the superconductivity of metals at extreme low temperatures. In 1913, Kamerlingh Onnes received the Nobel Prize for Physics for his work on the liquefaction of helium. Superconductivity is now widely used, for instance in hospitals as one of the basic technologies in modern body-scan apparatus.

Kamerlingh Onnes was not the only Noble Prize winner in the Netherlands in the first two decades of the last century. Over a period of 13 years 5 Dutch physicists received a Noble Prize, an incredible number for such a small country.

In a recently published biography of Kamerlingh Onnes, Dirk van Delft (2005) tries to find an explanation for the successes of this physicist. For that purpose he investigates three lines: the scientific talent of Kamerlingh Onnes, his organizational talent and his personality.

Van Delft comes to the surprising conclusion that the success of Kamerlingh Onnes may primarily be attributed to his exceptional organizational talent and his personality, and to a much lesser extent to his scientific skills. He was a scientific entrepreneur, a manager of technology more than a theoretical physicist.

Van Delft and other historians of science also show that the Dutch success in science at the beginning of the last century is the result of educational reforms. The founder of the Dutch constitution in 1848, Thorbecke, radically reformed the education system in 1863. He founded a new educational stream based on teaching modern languages and science. This new type of grammar school, intended for the middle class, had to offer a general educational program and aimed at the direct needs of modern society. In this way new talent could be scouted that was inaccessible before.

"We are going, my gentlemen", said Thorbecke in the senate, "to benefit this country. We are calling into existence forces and institutions that must increase the intellectual and the practical generating power of the core of this nation" (see Van Delft, 2005, p. 39).

In 1863 the new law on the grammar school was accepted by parliament and this law was effective in the Dutch educational system for a century. By 1870 the Netherlands already had 44 new schools (in Dutch HBS), most of them situated in new buildings and equipped with the latest instruments for experimental work in physics and chemistry. Some of the schools were even better equipped than universities. New talent, eager to learn, was able to get in touch with science in the early stages of their education. Noble Prize winners Van't Hoff (1901), Lorentz (1902), Zeeman (1902) and Kamerlingh Onnes (1913) were all products of this new educational system. Apparently, a new source of talent was found.

Kamerlingh Onnes was a scientific entrepreneur. He raised new funds, educated his own personnel, and even founded a new school for glass blowers and metal workers, which is still there. Indeed, he was a manager of technology "avant la letter".

In 1998, France, Germany, Italy and the UK signed the Sorbonne Declaration. It stipulated that barriers to education and learning across European nations should be removed. European countries have to cooperate and to converge the structure of their higher educational systems. A two-phase structure was proposed.

The Bologna Accord, signed in 1999 by 29 European countries, outlined some key objectives for the coming 10 years to achieve convergence. The objectives included the creation of a two-cycle system of university studies leading to recognized qualifications with relevance to the labor market at the end of each stage. The traditional European first degree splits into two components: the Bachelors degree after 3 or 4 years of study followed by the Master degree after 1 or 2 years of study. It is the explicit intention of the Bologna Accord to stimulate students with a Bachelors degree to choose between either entering the labor market or to continue their study in a Master's program. If students choose for a Master's program they have to decide whether they continue to specialize or move into a new direction.

Through the influence of the Bologna Accord, new Master's programs will be developed in Europe, programs specifically aimed at students who have just finished their Bachelor's degree. Multidisciplinary programs especially will be appropriate for this course. Students with different backgrounds in the supporting disciplines can enrol in such a program.

An example of one such multidisciplinary program is the international Master of Science program in Management of Technology at the Delft University of Technology: a two-year Master's program for students with a Bachelor's degree in Engineering.

The Master program in Management of Technology intends to educate students as technology managers, technological analysts and entrepreneurs in highly technology-based, competitive environments for a variety of technology sectors, with the ultimate objective of improving the quality of technology management. The program focuses on decision making in a technological context, technology and strategy, knowledge management, research and development management, and innovation processes. Students gain interdisciplinary knowledge and understanding, practical skills and intellectual abilities on the themes:

- managing technology
- the corporation
- design of technological systems.

The textbook *Managing Technology and Innovation: an Introduction*, is an introduction in the field of Management of Technology. Most of the authors in this book teach and do their research mainly in and around such programs as the Management of Technology (MoT) program at Delft University of Technology or the major in Technology and Innovation Management (TIM) offered by the aerospace department at University Bw Munich in Germany. This textbook is a new European sound in modern management.

Without the MoT program there would be no book on Management in Technology, without the Bologna Accord there would be no MSc program in Management of Technology. Kamerlingh Onnes and the Dutch Nobel Prize winners at the beginning of the last century show that improvement of an education system may bring outstanding results.

Let this book be a contribution to such a new adventure.

Simon Peerdeman

REFERENCE

Van Delft, D. (2005). *Heike Kamerlingh Onnes, een Biografie*. Amsterdam: Bert Bakker.

REFERENCE

Part I

General introduction

Chapter 1

Management of technology

Setting the scene

Uldrik E. Speerstra, Robert M. Verburg, J. Roland Ortt and Willemijn M. Dicke

From the Industrial Revolution to the battlefields of the First World War and from the development and mass production of modern medicine to the Internet Revolution, technology has made and unmade societies and shaped the lives of human beings. Since technology plays such a central role in almost all aspects of modern life it may come as no surprise that it also plays a decisive role in corporate development and the competitive positioning of firms. This is no longer just the case for the products and services of the traditional technology-based companies. Today, many companies ranging from financial services firms to logistics companies depend on technology in order to be successful. Technology not only plays a significant role in existing companies but is often the basis for new technology-based start-ups. Technological entrepreneurship has even become an important driver in the economies of many countries in the world, or, as Zehner (2000) points out:

> Over 95% of all scientists and engineers who ever lived is working today. Scientific and technical discoveries can leap from the laboratory to the marketplace in months instead of years. In addition, scientific and technical knowledge is likely to be found in Singapore or São Paulo as in Silicon Valley.
>
> (p. 283)

At the beginning of the twenty-first century the pace of scientific and technical knowledge production has increased in such an unprecedented way that some even speak of a "technology explosion" (Zehner, 2000). Zehner's technology explosion, the accelerating rate of technological diffusion and the globalization of technology are trends that set the agenda for senior strategic executives. And, indeed, in boardrooms and at campuses alike, CEOs and starting entrepreneurs are trying to address challenging questions such as: "How do the abundant technological opportunities affect our mission, objectives and strategies?," "What kind of technology do we need and when?," "How do we procure the technology we need?," and "How do we implement the required technology in our operations?." The optimal utilization of technology across all business functions is a way to create customer value. Companies have, or can acquire, different types of technological assets and these assets need to be managed in different contexts (Zehner, 2000). Technology thus can be seen as a corporate resource that creates specific managerial issues.

Our introduction on managing technology and innovation aims to improve the quality of technology management by presenting an introductory overview of the interdisciplinary field of management of technology. The rest of the book consists of four parts and will focus on the internal organization of technology firms, managing technology in society, the management of innovation and dilemmas technology managers, entrepreneurs and analysts have to deal with. The book is written for both students new to the field of management of technology and practitioners who would like to broaden their technological knowledge with insights from the management of complex organizations.

THE MANAGEMENT OF TECHNOLOGY FIELD

The understanding of technology as a corporate resource and the need for managerial competences in the area of technology led to the emergence of management of technology as a new discipline. Nambisan and Willemon (2003) trace its origins back to the early 1970s and Chanaron and Jolly (1999) to the mid-1980s. Nambisan and Willemon refer explicitly to management of technology education programs. In the 1970s and 1980s business, engineering and science schools developed their first educational programs in management of technology in the periphery of their mainstream business management programs. Although no apparent explanations are mentioned in the literature, the emergence of management of technology programs can be dated back to the second half of the 1980s. This period shows also a boom in general literature on management of technology (see Box 1.1).

In his article on ranking centers of management of technology research, Linton (2004) uses a list of 10 top-ranked journals for the field to identify academic researchers active in management of technology. This list is made up of one practice-oriented and nine research-oriented journals and is presented in Box 1.2. The dates of origin of the journals range from 1954 to 1989 and half of them were founded in the 1980s. In this period, the titles of the journals show a focus on technology management and innovation management and management of technology as a research object seems to have blossomed since then.

Differences exist in conceptualization of the phenomenon of management of technology. These differences are largely due to the fact that the field is relatively young and fundamentally interdisciplinary. Linton (2004) describes management of technology as an immature field that lacks a widely accepted body of literature. He mentions the differences in the curricula of the education programs as another symptom of its immaturity. The different conceptualizations are conveyed in the different definitions in the literature (see Chanaron and Jolly, 1999 and Nambisan and Willemon, 2003 for an overview). Bayraktar (1990) defines management of technology as a rational and systematic view of responding to technological opportunities and innovations, and dealing with their consequences. Dankbaar (1993) defines management of technology as management activities associated with the procurement of technology, with research, development, adaptation and accommodation of technology in the enterprise, and the exploitation of technologies for the production of goods and services. The US National Research Council holds that management of

BOX 1.1 MANAGEMENT OF TECHNOLOGY BOOKS

Frederick Betz, 1987, *Managing Technology*

K.B. Clark, R.H. Hayes, C. Lorenz, 1985, *The Uneasy Alliance, Managing the Productivity and Technology Dilemma*

Donald D. Davis, 1986, *Managing Technological Innovation*

M. Dodgson, 1989, *Technology Strategy and the Firm: Management and Public Policy*

Richard N. Foster, 1986, *Innovation, the Attacker's Advantage*

Watts S. Humphrey, 1987, *Managing for Innovation, Leading Technical People*

B.W. Mar, 1985, *Managing High Technology*

Donald Britton Miller, 1986, *Managing Professionals in Research and Development*

Philip A. Roussel, Kamal N. Saad, Tamara J. Erickson, 1991, *Third Generation R&D Managing the Link to Corporate Strategy*

Brian C. Twiss, 1988, *Business for Engineers*

Brian Twiss, Mark Goodbridge, 1989, *Managing Technology for Competitive Advantage. Integrating technological and organisational development: from strategy to action*

Brian Twiss, 1980, *Managing Technological Innovation*

BOX 1.2 MANAGEMENT OF TECHNOLOGY JOURNALS BY LINTON (2004)

IEEE Transactions on Engineering Management (1954)

Technology Forecasting and Social Change (1969)

R&D Management (1970)

Research Policy (1971)

Technovation (1981)

Journal of Engineering and Technology Management (1984)

Journal of Product Innovation Management (1984)

International Journal of Technology Management (1986)

Technological Analysis and Strategic Management (1989)

technology forms the link between engineering, science and management disciplines and addresses the issues involved in planning, development and implementation of technological capabilities in order to shape and accomplish the strategic and operational objectives of an organization (National Research Council, 1987). According to Badawy (1998) management of technology evolves around integrating technology strategy with business strategy. Badawy presents management of technology as a field of study and practice. In this field technology is studied as a corporate resource that determines both the strategic and operational capabilities of the firm.

In the above-mentioned and often-cited definitions, management of technology is described as a systematic and rational way of responding, a management activity, an activity of linking different disciplines, and as a field of study and practice. The goals of the activities may vary and range from responding to technological opportunities and innovations to shaping and accomplishing the strategic and operational objectives of an organization to maximize customer satisfaction, corporate productivity, profitability and competitiveness (Badawy, 1998). The various definitions are at best complementary and at worst mutually exclusive. Instead of focusing on elements that exclude each other, the definitions can also serve to draw attention to the contours of the immature and still developing field that lacks a body of widely accepted literature. Certain elements recur in these definitions. These elements seem to be more or less beyond historical change or at least the rhythm of change is relatively slow. The elements fall into the categories object, perspective and practice (see Table 1.1).

All authors implicitly or explicitly take the understanding of technology as a corporate resource to be the key object of study and knowledge. This object includes issues of procurement of technology and accommodation of technology and technical knowledge within companies. In the management of technology field, technology determines the strategic and operational capabilities of companies (Badawy, 1998). This, of course, means that management of technology is only relevant in contexts where technology has such a determining force. Technology in some companies, industries or markets doesn't have that impact.

The perspective is described as fundamentally interdisciplinary with roots in the sciences and engineering on the one hand and the social sciences, especially management and business sciences, on the other. The integration of these two scientific branches is central to the thinking of the authors. Badawy presents it as an integrating activity or process and Dankbaar and the National Research Council as a bundle of related integrating activities. The integrating activities can be localized on both the strategic level and the operational level. Management of technology takes the perspective of a company rather than an individual or the society as a whole.

Management of technology is practiced in different ways. In the description of the history of the field a distinction between education and research was made. Some focus on the professional practice in companies. Others focus on management of technology as a management activity for professionals and as a field of research. It can be concluded that the activities in the management of technology

Table 1.1 *The management of technology field*

Object	*Perspective*	*Practice*
• Technology as a corporate resource • Technology determines capability of company • Technical artifacts, organizational artifacts, institutional artifacts	• Interdisciplinary and integrating: combining science and engineering with the social sciences, for example business and management sciences • Company • Strategic and operational	• Professional practice • Research • Education

field are relevant to practitioners in companies, researchers and educators alike.

TRENDS

The object, perspective and practice that constitute the management of technology field are more or less beyond historical change. Exclusively focusing on these elements creates a rather static picture of the field. Management of technology as an object of knowledge, however, is not static at all but constantly changing. This change has implications for professional practice, research and education. Questions arise, such as along which lines does change take place and which forces are responsible for the dynamics in the management of technology field? Looking back one can clearly see how the practice of technology management has changed over the last decades. Zehner (2000) identified three trends, technology explosion, the accelerating rate of technological diffusion and the globalization of technology, as agenda setting for technology management. And, indeed, changes in the technology management field seem to largely coincide with a number of trends, which are listed below:

- unbundling of value chains;
- liberalization of markets;
- alliances between organizations;
- globalization;
- decentralization of organizations;
- technology development becomes more complicated and more expensive;
- specialization of organizations;
- shortening of product life cycles;
- increased attention for societal responsibility;
- increased client orientation.

Below we will explain these major trends.

Unbundling, liberalization, alliances

Industries, especially those that show large network effects, used to be organized in large companies controlling the entire value chain. State-owned monopolies for utilities, for example, have apparent advantages such as economies of scale and controllability, yet these monopolies turn out to be untenable in a rapidly changing market. The large and bureaucratic organizations lack the agility that is required in such a market (see also Chapter 13). In addition, monopolies are thought to stifle innovation and increase price levels, so governments adopt liberalization policies and sell the rights of state-owned companies or force commercial monopolists to dissolve. Liberalization and unbundling are highly related and both stimulate the emergence of alliances between different companies. It may come as no surprise that the management of new technology in an unbundled industry requires cooperation between more stakeholders and therefore differs considerably from traditional patterns of technology management in monopolies.

Globalization, decentralization, specialization, alliances

The globalization is measurable in terms of a sharp increase in intercontinental travel and the transportation of goods as well as in the similarity of consumption patterns around the world. But still, local market situations can differ considerably. Multinationals have decentralized in order to be able to adapt swiftly to these local market situations. In addition to these multinationals, small specialized companies have entered the global marketplace. The efficient and cheap means of communication and transportation have enabled small and specialized companies to market their products all over the world. Globalization forces companies to focus on core competences. As a result, some of the competences required to innovate are no longer available in a single firm (Hamel and Prahalad, 1994). The alliances that are needed to innovate require increased coordination and introduce new risks. Alliances are often aimed at increasing development speed, and decreasing the cost and risk of development. But they also increase the pressure on R&D management (Tidd *et al.*, 2001).

Technology development becomes more complicated and more expensive/specialization

Technologies that are required to develop new products have evolved (Tidd *et al.*, 2001). Therefore, the technological competences involved in developing new products have become too complex to be mastered by a single company. Many products, such as mobile telephones, cars and consumer electronics, for example, gradually incorporate more complex and different technologies once the number of features increases.

Shortening of product life cycles

The length of product life cycles generally decreases. As a result, the period in which investments can be earned back decreases and the risk of this investment increases accordingly. To compensate for this, innovation processes should become more cost-effective and the return of these processes (in terms of the percentage of innovations that is successful in the market) should increase. Shorter product life cycles also mean that time-to-market becomes more important. Innovation processes should be completed more quickly. So, shorter product life cycles have, in multiple ways, increased the pressure on innovation processes.

Increased client orientation/ increased attention for societal responsibility

Clients are generally more demanding with regard to products. Products have to be personalized, of high quality, and easy to understand. In addition, product liability has increased in many countries. As a result, technology development and subsequent technology application in new products and services becomes more client-oriented. Finally, governments and clients demand a socially responsible attitude from producers. Examples are the demand for more environmentally friendly methods of production, methods to recycle products, social welfare programs, and so on.

The description shows that the trends are highly inter-related. The causal relationship between these trends and the changing practice of technology management is hard to identify. Changes in the practice of technology management, for example, enable globalization. Globalization, in turn, reinforces these practices of technology management.

PROFESSIONAL PRACTICE

The position of management of technology as a relatively young discipline not only implies demarcations between what is management of technology and what is not, but also concerns the question who is responsible for management of technology in practice. Viewing technology as a corporate resource implies an important role for both line managers and strategic decision makers in dealing with technology issues on a day-to-day basis.

What kind of technology do we need? How do we implement the required technology in our operations? These are questions that may occupy managers of technology in R&D laboratories, pharmaceutical firms, software developers, and many other companies in the area of advanced technology.

In addition to people who manage technology directly within the context of a specific firm, there is a growing group of people that support high-technology firms as consultants. The increasing complexity of high-technology firms nowadays calls for more and more specialized knowledge and a number of companies ask for external advice on a regular basis. We expect technology management consulting to be in high demand in the coming decade as more and more companies are concerned with the management of their technology resources in order to gain competitive advantage.

Another important group dealing with management of technology are entrepreneurs. Bringing a great idea to the market and building your own company along the way appeals to many people. Students of traditional technological disciplines, such as engineering, industrial design, and the applied sciences, are often capable of generating promising product innovations and interesting new concepts. A small number proves to be capable of making the next step in the further development of the innovation and an even smaller number succeeds eventually in bringing it to the market. Nowadays, many companies, governments and universities are actively involved in stimulating people with new ideas and promising innovations to set up their own businesses. Good technology management is a key factor in bringing great ideas to the market and entrepreneurs may benefit from the various insights that are generated by the management of technology discipline.

So far, we have identified four main categories of people who deal with the different aspects of management of technology as part of their daily jobs:

1. line managers involved in the process of technological innovation;
2. strategic decision makers of high-technology firms;
3. consultants of technology management;
4. entrepreneurs.

Although not everybody involved with technology management will fit in one of these four categories the proposed categorization shows the multiple professional perspectives of the discipline. From a developmental point of view it may be worthwhile to look at what competencies are needed in order to perform well in the area of Technology Management. The categorization implies that there is a different set of competences within each job. For instance, an entrepreneur may need to be inspiring and to be able to initiate and innovate whereas a manager may benefit from competences, such as being organized and results-oriented. Aside the obvious differences between the four jobs, a common set of competences that are valuable for all who deal with the management of technology can be identified. Some examples of generic competences are:

1. An ability to draw conclusions based on a logical interpretation and the information available, for example, with regard to the interpretation of articles from engineering disciplines.
2. An ability to handle complex situations and to form a rational judgment, even when information is not complete, for example, with regard to issues around technology choices.

3 An ability to know the consequences of a plan to see its feasibility based on practical notions. This competence is, for example, needed for the management of complex technological projects or programs.
4 An ability to oversee the consequences of (technical) innovations from different interests and perspectives. This competence will, for example, prevent mistakes in product development and enable the practitioner to be effective in multi-actor settings.
5 Knowledge of the technical and non-technical aspects of science in general, and of research and development in particular.
6 An understanding of the relation between models and the theory behind them, both in simple and in more realistic situations. This competence is, for example, needed in complex decision-making processes or issues concerning safety or logistics.
7 An ability to think of creative solutions and to develop own ideas.
8 An ability to make decisions considered and decisively, and to overlook and accept the consequences of these decisions. This competence is needed for the management of complex technological projects.
9 An ability to critically reflect on one's thinking, decisions and behavior and consequently adjust one's actions.
10 An ability to determine goals and priorities in an efficient way and to indicate the needed actions, time and resources to achieve stated goals. This competence is especially crucial for the management of complex technological projects.

RESEARCH

Research in management of technology is a global phenomenon with the US as its center of gravity. Linton (2004) identifies 120 research centers in Africa, Australia, Europe and North America. He identifies and ranks these centers by looking at authorships in peer-reviewed research journals on management of technology. His exclusive focus on research journals is explained by stating that most members of the academic community focus on these journals instead of other forms of publications such as books or conference papers.

Journals being so central in the thinking of the academic community make them a good starting point for the description of the practice of research in management of technology. Although the term management of technology as a field of research was coined in the 1980s, the research on the processes involved in the management of technology is of considerably older age. This is reflected in the inaugural date of journals such as *Technology Forecasting and Social Change* (1969) and *R&D Management* (1970). These journals have a focus on methodology and the practice of technology, more than on developing new theories on management of technology. In the journals of later date, the focus was on both theory and the practice of so-called "engineering management" (e.g. *Journal of Engineering and Technology Management* and *Technological Analysis and Strategic Management*).

For all journals in the field of management of technology, theory is never approached as "l'art pour l'art," or "theory just for the sake of theory." Analysis and theory serves a goal, namely to inform practitioners, government departments, technology executives in industry, analysts and managers. The goal of the *International Journal of Technology*

Management for example, is to "help professionals working in the field, engineering and business educators and policymakers to contribute, disseminate information and to learn form each other's work." A similar explicit link between theory and practice can be found in the aim of *Technology Analysis and Strategic Management*: "linking the analysis of science and technology with the strategic needs of policy makers and management."

The journal *Technology, Policy and Management* pays attention to the exchange between policy analysts and policy makers and the interface between analytic concepts and human and organizational problem solvers. In this explicit link between theory and practice, the field of management of technology reveals a key characteristic: with the help of the insights of both the natural and the social sciences, this discipline aims to solve a problem, to design a tool, to analyze the incentives to improve the model. In the end, management of technology is about management, and not about theory. Theory is only instrumental to build towards this solution.

The focus on practical problem solving and the prescriptive aspects of research but also the multidisciplinary perspective still leave the question open what the object of study is. Research in the area of technology management is primarily aimed at technology as a corporate resource. Objects of study range from the antecedents of performance between and within high-technology companies to detailed analyses of product innovation processes. Management of technology is an interdisciplinary field that combines insights from disciplines, such as economics, HRM, innovation management, information management, management science, manufacturing, marketing, organizational psychology, strategy and other, related, disciplines.

The main topics of technology and innovation management that will be discussed in this book are:

- organizational design;
- human resource management;
- finance and accounting;
- marketing;
- strategy;
- decision-making;
- innovation management;
- R&D management;
- manufacturing;
- market and society.

As achieving technological innovation is a central aim of the management of technology in most firms, research on technology management often involves innovation as its key outcome variable. Technological innovation is an important source of competitive advantage for firms. Innovative companies show more rapid growth of output and productivity, are more export intensive, offer more attractive jobs and tend to be more profitable. However, many companies hardly innovate. Innovation involves risks and uncertainties and many innovations fail, the factors behind failure (or success) still being poorly understood. Research that analyzes the various interactions between actors and activities will be of great value to high-technology companies. Research in the area of technology and innovation management aims to achieve generalized findings on innovation problems in companies and the external technological and economic environment.

EDUCATION

Management of technology programs all over the world educate students to become practitioners in the management of technology field. Although most programs are offered at universities in the US there are programs in Australia, Canada, China, India, France, New Zealand, Norway, Spain, South Africa, Sweden, the Netherlands and the UK (see Nambisan and Willemon 2003, for an overview). The earliest date back to the 1960s, but the majority were established in the 1990s. There are full-time and part-time programs and the duration is mostly between 1.5 and 2 years. Nambisan and Willemon's global study of graduate management of technology programs shows that most programs are offered as part of an MBA program and only a few programs offer an undergraduate degree in management of technology. Most students who enter a management of technology program have a degree in engineering.

What do students learn in the management of technology programs? In a general sense, they will acquire knowledge on the management of technology field by studying technology as a corporate resource from the specific management of technology perspective (see Table 1.1). This, of course, means studying different aspects of technology management. The most important aspects will be discussed in this book. Students will also have to develop the competences needed for professional practice.

The curricula of most management of technology programs focus more on management than on technology. The background of the students plays a role in this focus. The curricula of the programs differ considerably, but there are also several topics that are frequently part of it. Nambisan and Willemon (2003) made a list of such topics, they include: technology strategy, finance/accounting, information technology, innovation management, organizational factors, new product development, technology entrepreneurship, technology marketing, manufacturing management, statistics/decision making, technology planning, quality management, technology policy, negotiation/conflict management and telecommunications.

In management of technology education programs, students acquire competences needed for practitioners. As line manager, strategic decision maker, consultant or entrepreneur they have to be able to deal with technology issues in a company context. In academic curricula three lines of learning these competences can be distinguished. First, there is the conceptual line that focuses on the transfer of disciplinary knowledge and tools. All relevant theories and bodies of knowledge of the disciplines that contribute to the field are part of this line. The second line consists of more general academic skills from research methods to presentation skills and academic writing. The third line is where the integration of all these different bodies of knowledge and skills takes place.

Especially for the management of technology programs, with their roots in both the engineering and the social sciences, the bodies of knowledge and skills are incredibly varied, as the list of common topics found in programs described above shows. So, one challenge of these programs is to integrate the different theories and bodies of knowledge. Even more difficult is the issue of how to teach students to apply the repertoire of management of technology knowledge and skills. The traditional approach in MBA programs is case-based learning. Christensen (Christensen and Jansen, 1987) defines a case as a written description of a real company facing specific business problems. Working on a case and learning to solve problems and to work multidisciplinarily helps students develop the competences they will need as future technology managers. Another way of dealing with the problem of integration is to develop courses on this subject. The management of technology program of Delft University of Technology can serve as an example. At the end of every semester there is an integration moment, a course specifically designed for learning to apply the general academic skills and the management of technology repertoire taught in the other courses of that semester. As with case-based learning a business problem is the starting point. But now, students play management games and reflect on integration and application of the different skills. This reflection is supported and structured by an electronic portfolio. Such a portfolio is a tool that enables students to plan and reflect on their career at an early stage. It also serves as a means to proof the development of relevant management of technology competences.

HOW TO READ THIS BOOK

The aim of the book is to provide an introductory overview of the field of management of technology. The book is an edited volume, in which the chapters are written by specialists in the field and by experienced lecturers and researchers in the particular subject areas of management of technology. It is written for students of management of technology programs all over the world and for students, especially those in the engineering disciplines, who take an interest in the subject and want to broaden their horizon. The themes addressed in the book are also relevant for managers and consultants who want to learn more about the variety of aspects of technology management.

The idea for a new book came up when students in the management of technology program of the Delft University of Technology kept asking for an introduction in the different aspects of the field. They asked for a book that not only addresses issues technology managers have to deal with in their day-to-day practice, but that also serves as an introduction to the different topics and disciplines that contribute to the field. We decided that for a book about an emerging field such as management of technology an approach that focuses on showing instead of telling is most suited. Telling, in the sense of presenting strict definitions and clear demarcations, would create a static picture of the field. Showing the field by descriptions of subject areas by specialists in the field, experienced lecturers and researchers will enable the reader to create his or her own picture, but the reader will also learn to learn, in other words, he or she will develop the skills needed to deal with a rapidly changing field. Another advantage of this way of structuring the book is that the different parts of chapters can be read separately.

The book consists of five parts. The general introduction to the management of technology field may serve as a starting point for meaningful further reading. The description of the management of technology field (see Table 1.1) can also serve as a frame designed to help the reader with the plotting of the different aspects of technology management as discussed in the other parts of the book. The trends that are discussed in this chapter will recur in the next parts. We have asked all the authors to elaborate on the ways one or more of these trends shape the particular aspects of technology management they discuss.

The next three parts consist of several chapters that are organized around different themes. "The internal organization of high-technology firms" focuses on companies itself. In "The management of technology in the society" the perspective shifts to the relationship between companies and their environment. In "Managing innovation," finally, the management of processes of technological innovation is introduced.

BOX 1.3 STRUCTURE OF THE BOOK

PART I General introduction

1 Management of technology: setting the scene

PART II The internal organization of high-technology firms

2 Design of technological firms
3 Human Resource Management for advanced technology
4 Cost and financial accounting in high-technology firms
5 Foundations for successful high-technology marketing

PART III The management of technology in the society

6 Managing the dynamics of technology in modern day society
7 Development and diffusion of breakthrough communication technologies
8 Forecasting the market potential of new products
9 The innovating firm in a societal context
10 Complex decision-making in multi-actor systems

PART IV Managing innovation

11 Corporate strategy and technology
12 Innovation in context: from R&D management to innovation networks
13 Operation management with system dynamics
14 Managing knowledge processes

PART V Dilemmas and strategies

15 Making the impossible possible: controlling innovation
16 When failure is not an option: managing complex technologies under intensifying interdependencies
17 Managing performance in firms
18 Management dilemmas and strategies in practice

The last part of the book, "Dilemmas and strategies," discusses some of the dilemmas technology managers face. The development of a competences-based academic curriculum usually starts with identifying core problems in professional practice. These problems and ways to deal with them will be the starting point for the development of the education program. The dilemmas serve a similar function for the book. They have added value because they help the reader to develop sensitivity with regard to complexity of the practice of technology management.

THE INTERNAL ORGANIZATION OF HIGH-TECHNOLOGY FIRMS

In order to understand the nature of technology management it is important to know how high-technology firms are organized. Instead of focusing on the market or society, our introduction into the management of technology starts with an internal view of the firm. The internal organization of a high-technology firm encompasses topics such as organizational structure, Human Resource Management, financial accounting, and high-technology marketing. In other words what does it mean to be a market-oriented and customer-focused organization?

Chapter 2 deals with the "design of technological firms" and introduces design rules for such companies. How to decide on the most relevant design objectives, and how to achieve the "ends" through selection of the appropriate "means." To this end the chapter cuts through insights from a variety of academic disciplines, especially entrepreneurship, strategic management theory, organizational theory and operations management. The chapter calls for "technopreneurs," professionals who are capable of managing the uncertainties associated with designing technological firms.

Chapter 3 aims to provide insight into the role and nature of Human Resource Management (HRM) for the success of high-technology firms. The core of Human Resource Management is the recognition of the value of employees for organizational success, which is often defined in terms of creating and sustaining competitive advantage. HRM policies and practices enable firms to build the innovative work force they need to meet the challenges of today's business world. Selection, development and socialization of employees are core activities in managing personnel. Such practices are no longer seen as administrative tasks that are the responsibility of a separate personnel department, but become part of top management's personnel strategy. Line managers are increasingly expected to play an active role in implementing HR policies and practices. In order to prepare managers of high-technology firms for their task, it is necessary to have a good understanding of human resource issues, such as selective recruitment, reward systems, training and development of employees working in advanced technology industries.

Chapter 4 deals with cost and financial accounting in high-technology firms and provides insight into the relevant financial issues high-technology firms face. Cost accounting addresses the planning and reporting of financial flows in the firm in order to make decisions on technology and the production structure. On the other hand, financial accounting addresses the reporting of the firm's performance to the outside world: the tax authorities and the shareholders. To meet the requirements of the latter group, management needs to gain knowledge on asset pricing in financial markets because this affects firm performance. Particularly in markets of novel high-technology products, such as UMTS (Universal Mobile Telephone Service) systems in telecommunication markets, investors are very vulnerable to all kinds of news that may affect the share price of the firm positively or negatively. This strongly affects the opportunities of the firm to attract financial flows necessary to invest in new technologies.

The section concludes with a chapter on the foundation for successful high-technology marketing. Companies operating in the high-tech marketplace live on the edge of revolutionary changes that transcend industry boundaries. In order to truly capture the promise that new technological developments have to offer, companies must take careful steps in the commercialization and marketing of their new innovations. Therefore, the purpose of Chapter 5 is to provide the reader with a solid foundation of the critical ingredients of successful high-technology marketing. Topics covered include delineating the scope of marketing activities in the high-tech company, understanding what it means to be market-oriented and customer-focused, and overcoming barriers to successful high-technology marketing. By the end of the chapter, readers should understand that marketing is more than a set of activities; it operates as a philosophy of making decisions in the high-tech organization.

The management of technology in the society

Part III focuses on the society or the environment of a technology-based firm. The environment of such a firm consists of all factors and actors that directly or indirectly have an impact on the performance of this firm. In all cases, a short analysis would reveal that many actors and factors are involved, including suppliers, competitors, distributors, clients, labor unions, banks, insurance companies, governments, technological developments, cultural factors, demographic developments, environmentalists, and so on. Some of these factors, such as demographic developments, seem to develop in a quite predictable trend, whereas other actors, for example governments, act on the basis of complex negotiation processes and therefore seem to behave somewhat unpredictably. These factors, predictable or not, interact in complex patterns and thereby have an effect on the performance of a high-tech firm.

Rather than attempting to describe all relevant actors and factors in the environment of a high-tech firm, the third part of the book describes five alternative perspectives on this environment. Chapter 6, "Managing the dynamics of technology in modern day society," discusses alternative theoretical explanations of technological change. On the one hand, technological determinists consider this change as an autonomous process driven by the internal logic of the technology rather than the environment. On the other hand, social constructionists consider the development of technology as an entirely socially determined process.

Chapter 7, "Development and diffusion of breakthrough communication technologies," focuses on a specific theory of technological change. It considers technological change as an evolutionary process that is driven by market actors and factors as well as by characteristics of the technology involved. Different patterns in this process and their managerial implications are discussed.

Chapter 8 "Forecasting the market potential of new products," discusses alternative approaches to infer the future market potential of new high-tech products. In addition to the approaches that try to extrapolate past experiences, notably consumer research, data analysis and expert opinions, two approaches will be discussed that are particularly suited to high-tech products. These approaches are practical approaches, such as probe and learn, and more theoretical approaches derived from biological evolutionary theories.

Chapter 9, "The innovating firm in a societal context," focuses on labor–management relations in countries and in companies and their effect on productivity and innovativeness. Empirical results show that two entirely different ways of managing labor–management relations can be distinguished and that these alternative approaches have a significant effect on labor productivity.

The final chapter in this part, Chapter 10, "Complex decision-making in multi-actor systems," describes the interaction of actors in a network and the effect of these interactions on the progress and outcome of large technological projects.

Managing innovation

The subject of Part IV, managing innovation, refers to managing the innovation process and the implementation of its results in organizations. The results of innovation processes can refer to new products, new production systems, and new organizations. In practice, minor product innovations are possible without changes in the production system or organization. However, in case of major product innovations this is hardly the case. Major product innovations consist of entirely new combinations of product attributes or incorporate new technologies. New technologies will often require a new production system. An example is provided by the transistor radio, a radio that uses a transistor instead of a vacuum tube. The transistor is a technology that enables the amplification of small electrical signals. This technology requires an entirely different system of production than its predecessor, the vacuum tube. As a result of the transistor technology radios became smaller and cheaper and required less energy. All these changes enabled the emergence of the portable radio, the marketing of which required considerable changes in radio producing and selling organizations. The case of the transistor radio also illustrates that mastering new technologies can be a vital competence for organizations. Technology management is one of the pillars of innovation management.

Chapter 11, "Corporate strategy and technology," describes four alternative perspectives on strategy. A central issue in this chapter is the way in which these perspectives perceive technology. The strategy of an organization has a large impact on R&D management practices. For example, it determines whether an organization is an imitator, follower or leader, and that, in turn, determines the importance of R&D and the kind of R&D management practices in an organization.

Chapter 12, "Innovation in context: from R&D management to innovation networks," describes subsequent mainstream practices of R&D management in organizations. Four generations of R&D management are distinguished. An example of a fourth generation approach is elaborated on.

Chapter 13, "Operation management with system dynamics," focuses on systems of production. In general, production and R&D management are different functions in an organization that have completely different perspectives on innovation. R&D management refers to innovation in terms of product innovations; production, on the other hand, refers to innovation as production process renewal. The chapter describes how systems of production can be made more flexible using a system dynamics approach.

Chapter 14, "Managing knowledge processes," focuses on a secondary process that has become crucial in technology-intensive organizations. Explicit knowledge management practices are required when knowledge flows across disciplinary and departmental boundaries in an organization. An example of these boundary-crossing flows of knowledge occur between the production and the R&D function in an organization when new products are developed (that require adaptations in the production system) or when new production machines are installed (that require adaptations in the product parts).

Dilemmas and strategies

The book concludes with a section on dilemmas that managers operating in a technology context are faced with. The idea behind this section is that the manager is confronted with many developments, both inside and outside the organization. The manager seeks to respond adequately to each and every one of these developments, whether it is globalization, the development of new technologies, new legal requirements affecting financial management or the production line, or demographical changes resulting in the need for different products.

In responding to all kinds of developments, the manager will find out that different tendencies ask for conflicting responses: an action that is an adequate answer to one development might have an adverse effect on a different part of the organization. For example, from the viewpoint of the design of technological firms, it might be an excellent idea to make the organization more transparent, to standardize procedures, and to control the entire organization. With a standardized and quantifiable output measurement system, the whole organization becomes easier to manage. For innovation, on the other hand, creativity is needed. And creativity does not combine well with total control and standardized procedures. It is precisely in the unusual that innovations develop. In other words: employees and parts within the organization need some freedom in order to be able to innovate. How can they control the uncontrollable? What should the CEO do in this case? Should he or she exert control and standardize procedures because otherwise the organization will be out of control? Or should he or she trust the R&D department to come up with something good in the end, and in doing so risk a financial disaster or risk R&D developing a product that the market does not ask for?

This last section of the book focuses on the question of how managers deal with these and other dilemmas. The dilemmas have their origins both in developments outside the organization (e.g. globalization and liberalization) and within the organization (e.g. conflicting requirements of different parts within the organization). The manager has to respond to all these developments. How does he weigh the impact of his actions on different parts of the organization?

The first three chapters in Part V describe the theory behind three major dilemmas. In Chapter 15, "Making the impossible possible," the dilemmas around innovation are described. Chapter 16, "When failure is not an option," depicts how managers can maintain reliability in circumstances that are characterized by a great deal of uncertainty. In Chapter 17, "Managing performance in firms," the dilemmas around the use of performance management are analyzed. The last chapter of Part V consists of interviews with six senior managers of typical management of technology companies. In their own words, these managers report on the dilemmas they are confronted with in their daily work and what their strategies are to deal with these dilemmas.

ACKNOWLEDGMENTS

The authors are grateful for the work and the valuable comments of all the members of the management of technology community who reviewed the chapters of this book. The authors would also like to thank our students and colleagues for the discussions and debates and for the open and inspiring atmosphere within our faculty. Finally, the authors wish to express a special word of gratitude to Simon Peerdeman, Hans Wissema and Jenny Brakels for sharing their valuable ideas on the management of technology field, education programs and competences.

REFERENCES

Badawy, M.K. (1998). Technology Management Education: Alternative Models. *California Management Review*. 40(4), 94–115.

Bayraktar, B.A. (1990). On the Concepts of Technology and Management of Technology. In: Khalil, T.M. (Ed.) *Management of Technology II. The Key to Global Competitiveness.* Proceedings of the Second International Conference on Management of Technology. 161–175.

Chanaron, J.J. and Jolly, D. (1999). Technological Management: Expanding the Perspective of Management of Technology. *Management Decision*. 37/8, 613–620.

Christensen, C.R. and Jansen, A.J. (1987). *Teaching and the Case Method: Text, Cases and Readings*. Boston, MA: Harvard Business School.

Dankbaar, B. (1993). *Research and Technology Management in Enterprises: Issues for Community Policy*. Overall Strategic Review. EUR 15438 EN. Brussels.

Hamel, G. and Prahalad, C.K. (1994). *Competing for the Future*. Boston, MA: Harvard Business School Press.

Linton, J.D. (2004). Perspective: Ranking Business School on the Management of Technology. *The Journal of Product Innovation Management*. 21, 416–430.

Nambisan, S. and Willemon, D. (2003). A Global Study of Graduate Management of Technology Programs. *Technovation*. 23, 949–962.

National Research Council (1987). *Management of Technology: The Hidden Competitive Advantage*. Washington, DC: National Academy Press.

Tidd, J., Bessant, J. and Pavitt, K. (2001). *Managing Innovation. Integrating Technological, Market and Organizational Change*. Chichester: John Wiley & Sons.

Zehner II, W.B. (2000). The Management of Technology (MOT) Degree: A Bridge between Technology and Strategic Management. *Technology Analysis & Strategic Management*. 12(2), 283–291.

Part II

The internal organization of high-technology firms

INTRODUCTION

In order to understand the nature of technology management it is important to know how high-technology firms are organized. Instead of focusing on the market or society, our introduction into the management of technology starts with an internal view of the firm. The internal organization of a high-technology firm encompasses topics such as organizational structure, Human Resource Management (HRM), financial accounting and high-technology marketing. In other words what does it mean to be a market-oriented and customer-focused organization?

High-tech firms have evolved over the years. In the 1960s most products bore standard technology characteristics. In the 1970s competition changed as customers gradually avoided low-quality products. Quality, in addition to price, became an important factor for market success. In the 1980s the competitive struggle changed again as products were put on the markets faster and flexibility became an important source for competitive advantage. Nowadays, it is vital for products to stand out from the competition and manufacturers should strive for product uniqueness. This again puts new requirements on the internal organization of high-technology firms.

Chapter 2 by Bernard Katzy deals with the "design of technological firms" and introduces design rules for such companies. How should they decide on the most relevant design objectives, and how can they achieve the "ends" through selection of the appropriate "means"? To this end the chapter cuts through insights from a variety of academic disciplines, especially entrepreneurship, strategic management theory, organizational theory and operations management. The chapter calls for "technopreneurs" as professionals who are capable of managing the uncertainties associated with designing technological firms.

Chapter 3 by Verburg and Den Hartog aims to provide insight into the role and nature of Human Resource Management for the success of high-technology firms. The importance of employees or "human resources" for organizational performance has attracted much attention over the years. The core of Human Resource Management (HRM) is the recognition of the value of employees for organizational success, which is often defined in terms of creating and sustaining competitive advantage. Selection, development and socialization of employees become core activities of personnel management. Such practices are no longer seen as tasks that are the responsibility of a separate personnel department, but they become part of top management's personnel strategy. Line managers are increasingly expected to play an active role in implementing HR policies and practices. In order to prepare managers of high-technology firms for their task, it is necessary to have a good understanding of human resource issues, such as selective recruitment, reward systems, training and development in advanced technology.

Chapter 4 by Tom Poot deals with cost and financial accounting in high-technology firms and provides insight into the relevant financial issues high-technology firms face. Cost accounting addresses the planning and reporting of financial flows in the firm in order to make decisions on technology and the production structure. On the other hand, financial accounting addresses the reporting of the firm's performance to the outside world: the tax authorities and the shareholders. To meet the requirements of the latter group the management needs to gain knowledge on asset pricing in financial markets because it affects the performance of the firm. Particularly in markets of novel high-technology products, such as UMTS systems in telecommunication markets, investors are very vulnerable to all kinds of news that may affect the share price of the firm positively or negatively. This strongly affects the opportunities of the firm to attract financial flows necessary to invest in new technologies.

The section concludes with a chapter on the foundations for successful high-technology marketing by Mohr, Slater and Sengupta. Companies operating in the high-tech marketplace live on the edge of revolutionary changes that transcend industry boundaries. In order to truly capture the promise that new technological developments have to offer, companies must take careful steps in the commercialization and marketing of their new innovations. Therefore, the purpose of Chapter 5 is to provide the reader with a solid foundation of the critical ingredients of successful high-technology marketing. Topics covered include delineating the scope of marketing activities in the high-tech company, understanding what it means to be market-oriented and customer-focused, and overcoming barriers to successful high-technology marketing. By the end of the chapter, readers should understand that marketing is more than a set of activities; it operates as a philosophy of making decisions in the high-tech organization.

Chapter 2

Design of technological firms

Bernhard R. Katzy

OVERVIEW

In this chapter I take the very specific perspective of the "technopreneur," a scientist or engineer, who designs a new venture to commercialize his or her technology. The objective is to introduce those issues and choices that are typical in the design process of a firm – and which have strong impact on its growth.

The chapter introduces design rules for technological firms; how to decide on the most relevant design objectives, and how to achieve the "ends" through selection of the appropriate "means." To this end the chapter cuts through insights from a variety of academic disciplines, especially entrepreneurship, strategic management theory, organizational theory and operations management. The chapter calls for "technopreneurs" as professionals who are capable of managing the uncertainties associated with designing technological firms.

DESIGN OF TECHNOLOGICAL FIRMS – MOTIVATION FOR A PROFESSION

What is a firm?

The legal and business term definition of *firm* differs from the colloquial English (and most other European languages) use in that *firm* is legally (since the code napoleon) defined as the name – only – under which a salesperson runs his or her business, acts in the market and can be sued or otherwise be held liable. The name has to meet certain requirements. It has to be honest, e.g. does not pretend or is otherwise misleading. It has to be unique in order not to lead to confusion with other firms, and it has to meet integrity standards, especially that no company shall use multiple firms. Simply said, the first creative design choice in designing a technological firm is giving it a name!

In colloquial English the term *firm* includes the entire company, its name as well as legal entity, corporation and organization. We will address the breadth of this colloquial definition, which is in line with the objectives of the field of management of technology, to provide instruments to turn technology into products, subsequently commercialize the products in markets and, ultimately, gain financial return for the effort undertaken. Therefore, firms are addressed as complex systems that bridge multiple domains.

The firm – as a name – can be used by individual persons or corporations. The second important design choice for technological firms, therefore, is the legal entity to which the name is given. Corporations, like individual persons, can be contractual parties in markets and therefore are established and registered by court as autonomous legal entities. Corporate law further offers a choice of basic legal forms, such as Llp., Ltd or trust that a firm can adopt and which makes it partly (in the case of limited liability) or completely (in the case of stock listed firms) independent from those who created and/or own it. Incorporation further allows the roles of those who act in the name of the firm to be defined, including their decision power, responsibilities, share of risk to carry, and share of benefits and reward to receive. The design of the legal structure of a corporation fills a book in its own right. The guiding design principle for a technological firm, however, should always be the growth of its activities, which therefore is the interest of this chapter.

Who designs technological firms?

For further discussion it is helpful to distinguish three distinct roles and refer to those who initiate and create a firm as *entrepreneurs*, or, in order to emphasize the specifics of initiating technical firms, the *technopreneur*. The technopreneur often involves other partners, such as customers, suppliers, key experts, or capitalists into the joint effort as *shareholders* who each make important contributions to the technological firm and own their share of risk, responsibility, resources and – if successful – rewards of the firm. After firms are created they are given into the hand of *managers* who run and operate them.

Allocation and re-allocation of these three roles is an ongoing entrepreneurial design effort for a technological firm. Technopreneur and management together are responsible for designing the business of the technological firm as an economic entity that aims to maximize return on the invested effort and capital. To this end business models have to be designed that define the firm's behaviour in markets, adopt business strategies such as to enter or exit markets, adapt the firm's offerings, compete or cooperate with other firms in the market and so forth.

Firms are associated with *organizations* with one or more members (human beings) who work towards achieving the firm's objectives. Organization is the stable coordinating structure ("machine" in the words of Weber (1947) who is one of the founders of organizational theory) in which members specialize on tasks and efficiently integrate towards the overall product and offering. Designing technical firms therefore entails *organizational design* (OD) as the choice of how tasks are defined, and which rules govern their coordination.

Organizational design tasks are very different from the technical and business design tasks, because they entail social design (Simon, 1981) and leadership. It entails motivating people to join the group, adopt team-behaviour and put effort in the achievement of joint objectives. Scientists and engineers should be aware that they are well trained in technical design tasks that are easily adapted to "hard" business and financial engineering, but not necessarily to "soft" social design.

All dimensions of an organization are interrelated and highly dependent on the specific situation in which a technological firm operates. Take, for example, a science student who develops a technology at university and, with a partner, starts a firm to develop and commercialize products. She, herself, is entrepreneur, shareholder and manager. With initial success personnel are hired, jobs are assigned and she becomes a manager when she supervises employees. Later, when the organization grows she decides to hand over management, and concentrate on product development, but remains owner and shareholder. Or, take the case of an engineer who, in the lab of a large corporation, develops a promising product that is recognized by his management. He writes a business plan, is promoted as head of his newly created department and gets funds to introduce the product. He is technopreneur and manager of the business and designs the organization of the department as it grows, but is not a shareholder.

Aim of the chapter

The objective of this introductory chapter is to discuss basic questions of the design of technological firms. In other words, the focus here is on the *business model*, which is the way to design and maintain a consistent configuration of technology, markets, strategy, organization and people. Later chapters of this book will expand on each of the dimensions. Most theory, including social sciences and management theory is analytical, or about the question "how things are." The title of this chapter, however, indicates that this chapter follows a "theory of design" (Simon, 1981) approach. In other words, it deals with the question "how things ought to be," a more common approach to engineers. Of course it cannot be the aim of such a chapter to prescribe the single or the best design of any technical firm. But the aim is to support rational decision making in the design of technological firms by outlining *design logic* through separating typical *design objectives* as "ends" from the range of existing "means" and analysing the "rules" by which means and ends relate, what is possible or feasible and what not. The chapter finishes with an introduction into the *design process* that can be mastered by professional technopreneurs as a systematic sequence of steps and activities in the search and evaluation of alternative firm designs.

DESIGN LOGIC OF TECHNOLOGICAL FIRMS – MEANS AND ENDS

Technological firms are firms – the economic perspective

Economic theory is simple in that firms are modelled as closed systems that pursue highest possible "rents" or *maximized profit*. Regarding this "end," a technological firm is economically no different than any other firm. This is simple but noteworthy because the professional codex of scientists and engineers is different! Scientists and engineers bind themselves to ethical values of improving human welfare through technology (see, for example, the preface of the bylaws of the international Institute of Electrical and Electronics Engineers – IEEE). Technological firms and their management teams therefore often need to live with a lingering internal conflict of interest between their profit-maximizing business people and the utilitarian, value-driven, technical part of the organization. Similarly, for each engineer or scientist personally, it is an intensive experience to adapt to two worlds of very different values, when moving from science and technology into business management. In short, the dominant economic objective of designing firms is profit maximization in the long run.

Within the management of a firm it is the role of *strategic planning* to plan for success or, in economic terms, achieve above average rents. A traditional means of strategic planning is to position the firm in an industry where low competitive pressure will allow it to be profitable (Porter, 1985). An *industry* is the group of firms (i.e. automotive industry) that produce exchangeable goods and therefore compete in a *market*, the "place" of exchange of similar goods (i.e. New York Stock Exchange).

One important design choice, therefore, is the industry in which the firm does compete. Industries differ in their inherent profitability and not all industries offer equal opportunities for a firm's sustained profitability. Strategic design of the technological firm, therefore, must grow not only from understanding the technology but, equally importantly, out of an in-depth understanding of the rules of competition that determine an *industry's attractiveness*. The standard analytical tool for industry analysis is the five forces analysis (Porter, 1985). In short, an industry is the more attractive the smaller the danger is of (1) entry of new competitors, (2) the threat of substitutes, (3) the bargaining power of buyers, (4) the bargaining power of suppliers, and (5) the rivalry among the existing competitors (see also Chapter 11 on corporate strategy and technology).

The five forces determine industry profitability because they influence the prices, costs and required investment of a firm in that industry. Buyer power influences the prices that firms can charge from their customers. The threat of *substitution* describes the risk of losing investments when another technology or product renders a firm's product obsolete. The bargaining power of suppliers determines the costs of raw materials and other inputs. The intensity of rivalry influences prices as well as the costs of competing in areas such as plant, product development, advertising and sales force. The threat of entry places a limit on prices, and shapes the investment required to deter entrants.

In terms of design logic, Porter's model expresses an *outside-in* rule in that it attributes the competitive position of the firm within the industry as the (only) means to achieve economic rents. Neither the firm itself nor its internal structure is considered in this analysis. In this logic of *structure-follows-strategy* (Chandler, 1962) competitive strategy determines basic, mostly long-term goals and objectives for the firm based on outside force analysis, which, in a second step, need to be implemented inside the firm through allocation of resources such as working force, information, capital, raw materials, buildings, machinery, etc.

It is the task of Organizational Design (OD) to develop and implement an efficient structure to adapt the firm to the needs of the industry. The design rule of organizational *contingency theory* (Galbraith, 1973), therefore, is that the better the *fit* of the firm to the industry need, the higher will be its performance. It is the role of management to maintain a continuous design process to maintain or improve this fit.

The basic organizational design therefore starts with a *task analysis* step, to understand and specify the necessary economic activities in response to the situation in the competitive environment. The beginning of economy as a science is often defined as the observation of Adam Smith (1776) that *specialization* of workers in a pin factory in England increased their productivity manifold. Instead of each worker doing all tasks from cutting the rod to completing the pin, workers in that factory passed the semi-finished needles on from person to person after completion of each manufacturing task. As a result workers were more skilled in doing their individual task and the group produced more pins. Later, Frederick Taylor (1911) made this insight the basis of *scientific management* and introduced analytical methods for studying work and designing tasks. Henry Ford, in the same period, used this knowledge to create the first mass-produced T-model car with an assembly line. Every worker had to carry out one specialized activity in the work rhythm of the line so that every few minutes a T-model left the assembly line.

The second step in the organizational design process is to group and integrate the specialized tasks so that they are *coordinated*. The organizational designer has a number of means at hand to achieve coordination through structuring the organization. In five broad categories these means are (Mintzberg, 1979):

1 *Direct supervision*, i.e. oversee coordination by a manager in person who is responsible for the work of others, who issues instructions and monitors activities.
2 *Standardize work processes*, for example, in the case of the assembly line, by specifying contents of work in detail or even (computer) programming it.
3 *Standardize skills* and develop professional training that is required to perform the specified work. Examples are accountants or truck drivers who are certified as registered accountants or via driving licence, respectively.
4 *Standardize output*, i.e. specify the results of the work and control it, for example, through quality inspection.
5 *Mutual adjustment*, i.e. allow task owners to achieve coordination in teams through informal communication and repeated practice, e.g. like medical doctors in the surgery room.

In other words, coordination is managing dependencies (Malone and Crowston, 1994) between tasks, which the designer can shape by streamlining five *flows* between them. *Material flow* are the parts and other physical goods that are moved between tasks, *information flow* are the documents and forms moved between tasks, *work flow* is the sequence of the tasks, *decision flow* concerns supervisor involvement, and *informal information flow* is the general knowledge exchange inside the organization. Good organizational design will allow for natural flow of processes in the organization and, for example, avoid loops of repeated task execution or abundant material flow.

Already, this short introduction shows that organizational design is a multi-criteria optimization process that allows for many alternative design options. Furthermore, conditions permanently change when the firm's activities expand in new markets, respond to shifting demands, change sources of supply, react to fluctuating economic conditions, new technological developments, or competitors.

The formal outcome of organizational design is the *organigram*, which describes the structure of an organization as units and departments that are arranged in a *hierarchy*. *Chains of command* define how supervision is exercised by superiors and how line managers report back. This visible part of the organization, however, is no more than the tip of the iceberg. Especially in well coordinated technological firms valuable knowledge, capabilities and expertise is deeply embedded in the organization and the complex relationships of its internal members and external partners.

To best further development, therefore, alternative designs of the technological firm need to be evaluated in the course of the design process. The strategic VRIN criteria

	Valuable?	Rare?	Inimitable?	Not substitutable?
Competitive Disadvantage	no			
Competitive Need	yes			
Volatile Opportunity	yes	yes		
Competitive Advantage	yes	yes	yes	
Sustainable Competitive Advantage	yes	yes	yes	yes

Figure 2.1 *Evaluation of firm designs with the VRIN framework* (Barney and Tyler, 1991)

(Barney & Ouchi, 1986) evaluate a firm with respect to the longevity of its competitive advantage. Firm structures need to be valuable (to contribute to the firm's effectiveness or efficiency, otherwise they are a disadvantage) and rare (not widely held, otherwise they are commonly available). When these resources are simultaneously imperfectly inimitable (they cannot easily be replicated by competitors, which is also the idea behind granting patents on ideas) and not substitutable (other alternatives cannot fulfil the same function) *sustained competitive advantage* is achieved (Figure 2.1).

Technological firms are about technology – the trend towards resources that make the difference

The title "technological firm" indicates that this chapter deals with the design of firms that are specific because they embody technology of strategic relevance. The technopreneur therefore can specifically profit from adopting a *resource-based view* (Wernerfelt, 1984) of the firm, which looks inside the organization for sources of sustainable competitive advantage and, therefore, is also referred to as the *inside-out* perspective.

Changing perspectives is a general design technique. Like architects, who regard building designs from different angles to verify spatial proportions, the technopreneur can thus assess validity of the design of the technical firm. In that regard the resource-based view is a more recent school of strategic theory and a complement rather than a contradiction to the outside-in perspective.

The resource-based view acknowledges that the internal structure of a firm can be valuable, is rare or even unique, because it is the result of an extended design effort. *Resources* in the economic sense include "all assets, capabilities, organizational processes, firm attributes, information, knowledge, etc. controlled by a firm that enable the firm to conceive of and implement strategies that improve efficiency and effectiveness" (Barney and Tyler, 1991). It is evident that simple resources, e.g. machines or computer programs, are not strategic resources because they are not rare if competitors can buy them as well. Consequently they cannot be a source of sustained competitive advantage. It further adds to the importance of designing technological firms, that unique resources of strategic interest can only be those that are designed and built within the firm.

Organizational capabilities (Selznick, 1957) or core-competences (Prahalad and Hamel, 1990) are examples of such "sticky" resources that cannot easily be transferred from one firm to another. Capabilities involve complex patterns of coordination between people, and between people and other resources. Perfecting such coordination requires learning through repetition (Grant *et al.*, 1991). It therefore takes the time for experience-based learning from frequent repetition of similar activities to build organizational routines that embed organizational knowledge to solve the specific problems of a context. In other words it is not only the formal structures that are subject to design efforts of technological firms but all patterns of decision-making and problem-solving (Montgomery, 1995).

Differences between firms can be understood from conceptualizing technological firms as *bundles of resources* that are heterogeneously distributed across firms and differently integrated by each firm. Such differences persist over time and can be the sources of sustainable competitive advantage (Wernerfelt, 1984), which cannot be understood from an external industry analysis.

The trend towards technological firms as nodes in networks – the design problem of timing and flexibility

From what we have said so far, it is apparent that technological firms face a constant timing problem. Unique resources, technology and capabilities need be designed inside the firm, which can take ten years or more (Prahalad and Hamel, 1990). Core competences, therefore, can easily become *core rigidities* (Leonard-Barton, 1992) when it gets impossible to adapt the firm fast enough to maintain satisfactory *fit* with changing market needs. Such situations are described as hyper competition (d'Aveni, 1994) or *turbulent competitive environments* (Crowston and Katzy, 2003) where changes in the competitive environment outpace lead times of design and re-design processes within the firm. Under such circumstances design of relationships with other firms, *networks*, gains importance and enlarges the scope of firm design.

The *network* view of the firm (Easton, 1992) is a recent, but fast developing third perspective on the design of the firm. Networks can be designed in many forms through disaggregating existing large firms in more autonomous business units, through subcontracting larger parts of production, such as in the auto or aeronautics industry, through licensing activities, or spin-off creation. Alternatively, networks emerge from cooperation of independent peer partners in joint ventures, through alliance forming or mergers. Regardless of how partner relationships were introduced, the design of firms in networks changes because business functions such as product design and development, manufacturing, marketing and distribution that were typically done by one single firm, are now distributed across several firms within the network.

From the individual firm's perspective, the primary benefit of cooperating in networks is that they can focus and *specialize* on a limited set of activities, core competence or technologies while having access to complementary technologies and resources from the network (Miles and Snow, 1986;

Katzy and Schuh, 1998). Instead of re-designing the firm in the face of market changes, the network functions as a switch-board (Mowshowitz, 2002) to marshal resources to temporary market needs. The network thus provides *flexibility* while preserving relative stability inside the firm.

Networks, however, are not free of cost. In order to leverage networks technological firms need to develop dedicated capabilities to design and maintain network relationships, besides their own technical strategic resources. This includes the capability to operate inter-organizational information systems, which can, for example, radically change a firm's internal business processes and communication pattern. Often such competences are developed on the network level and referred to as broker (Hargadon and Sutton, 1997).

Clusters are networks that develop specific technological competences and are situated in geographic proximity of a region. Most famous examples of clusters are Silicon Valley and Boston in the US and Cambridge and Munich in Europe. Frequent exchanges of people and joint projects within the region creates open accessible know-how (so-called *spill-over effects*) for firms and experts within the region, which in turn fuels innovation. A further relevant design decision therefore is to move the firm into the region of a relevant cluster and profit from the available know-how.

Technological firms and innovation – an evolutionary design perspective

So far the firm has been presented as an institution, with name, legal entity and an organization. The *boundary of the firm* – what is inside and what remains outside – has therefore been a central concept in all three perspectives presented so far. In the outside-in perspective, internal firm design did follow external requirements, while the inside-out perspective suggests searching for favourable external environments based on existing internal structures. The network view puts stress on the relevance of boundary-spanning inter-firm relationships.

A fourth, evolutionary, perspective considers change and developments over time and thus complements the three static perspectives. For example, when strategic resources are no longer valuable in the VRIN framework, e.g. because a technology becomes obsolete, they can become the source of disadvantages and *rigidities* (Leonard-Barton, 1992) that hinder the further competitive development of the technological firm. This process is very likely to happen for technology firms as it lies in their nature to drive innovation and therefore render technologies (including its own prior) obsolete. At the same time research and innovation for new products also contributes to learning and competence building within the firm. More insightful for the design of the firm – than only a snapshot at one point in time – is the specific *development path* of the technological firm. Teece *et al.* (1997) identify *dynamic capabilities* as organizational routines that determine such development paths and hence the competitive performance of the firm in changing markets. Take the example of the printer business of HP which, for many years, was known for such high innovation rates that the strongest competition for each of their models was its own successor. The source of sustained competitive advantage for HP was not the product features of one generation of printers but the high rate of introduction of new printers. Dynamic capabilities can be routines of learning, innovating, introducing new products or production processes.

In this evolutionary perspective firms differ not only by the market or industry in which they are positioned, but also from the phase of product life-cycle in which they are specialized. Miles and Snow (1978) identify four consistent firm designs. The *Pioneer* is strong in explorative experimenting and innovating new product concepts. An example in the computer industry would be Macintosh with its Apple computer and iPod products that pioneered graphical user interface (GUI), or the USB plug with products that only had a limited market reach. The *Analyser* is strong in analysing what small but innovative ideas have mass adoption potential in markets and then scaling up volumes. Microsoft can serve as an example in the computer industry. With the Windows operating system product, for example, Microsoft made the GUI a standard and their Internet Explorer product did the same for Web browser technology. *Marketers* have access to markets and available distribution channels to reach customers, and *Defenders* have special capabilities in introducing highly reliable, high-quality and cost-effective production processes, an example of which is Dell in the computer industry. Dell does not develop new computers but constantly improves global production and logistics processes for reliability and cost leadership.

From an evolutionary perspective it therefore makes sense to pass a product on from one firm to the next, when it grows into the next phase, and move to a new product in the firm's phase of the product life-cycle. Two firms that look like competitors in industry analysis can be complementing partners from an evolutionary perspective because they deliver to the same market.

This chapter has introduced four alternative design logics and related design processes for technological firms. First, an outside-in logic suggests that the firm adapts to the external competitive conditions for best performance. Second, an inside-out logic suggests the firm searches for applications that best suit its internal competences. The network logic suggests not only designing the firm, but also its relationships with other firms. And finally, the evolutionary logic suggests not only designing static structures of the firm but also its development paths over time. These design logics are complementary rather than alternatives in that each of them contributes important insights for the design of a technological firm. Good and consistent designs will need to satisfy all design logics simultaneously.

It would be an extremely complex task to design technological firms from scratch. Therefore, the next chapter will introduce working designs of technology firms, which can be used like blueprints as basic references and be fine-tuned to specific conditions of the firm in the course of the design process.

REFERENCE ARCHITECTURES AS BLUEPRINTS OF THE TECHNOLOGICAL FIRM

Taylor's architecture of specialized (scientific) foremen

Frederick Taylor (1911) formalized the principles of scientific management as a rational fact-finding approach and replacement for what he rejected as "old rule of thumb" and the arbitrary "autocracy" of shop floor managers, who at that time had almost absolute power over workers, but often lacked competence. This is no surprise because management and business education in dedicated schools was only introduced around that time. His studies on the handling of pig iron laid the foundation of scientific method study. He, himself, introduced work analysis with stop-watch timing, what he called "time and motion study," and started systematic classification of work elements. His impact on firm design is the claim that a firm is a system of abstract rules rather than personal relationships between individuals.

Based on his work classification he proposed a system of nine specialized foremen or managers, who are trained experts in the scientific methods of their class. In order for the foremen system to function he set up a framework for organizational design that is still generally applied. (1) task specialization, (2) clear delineation of authority and (3) responsibility for their domain allows foremen to apply their expertise. (4) Separation of planning from operations gives them time and room for scientific management. Foremen implement their planning through (5) incentive schemes for workers, and (6) training of workers to apply general scientific rules so that foremen can constrain to (7) management by exception.

Thompson's model of buffering the technological core from uncertainty

For Thompson (1967), "Uncertainty appears as the fundamental problem for complex organizations, and coping with uncertainty, as the essence of the administrative process." His architecture of a firm is a technical core, which is sealed off and protected from uncertainty by a group of "boundary spanning units." Since complete closure is impossible firms surround their technological core with input and output buffers such as stockpiling purchased materials against uncertain supply quantities. Similarly, warehouses buffer sales and distribution fluctuation from inventory on the output side of production.

The design of boundary-spanning units, their shape and size, can directly be derived from (and thus reflects) the environmental uncertainties that the firm has to deal with. If technological uncertainties are high, research and development departments are large; if market uncertainties are dominant, marketing departments will grow; if skills and training of the work force are a constraint, human resource departments are needed; and if financial resources are scarce, for example in venture capital backed firms, finance departments need to be installed to cope with the uncertainty of funding renewal.

Mintzberg's architecture of coordination of the organization

Henry Mintzberg (1979) extracted a generic organizational architecture. It is widely used and consists of five basic parts as depicted in Figure 2.3. The strategic apex (top management) is charged with coordinating the organization by objectives, to effectively serve its mission and to respect the interests

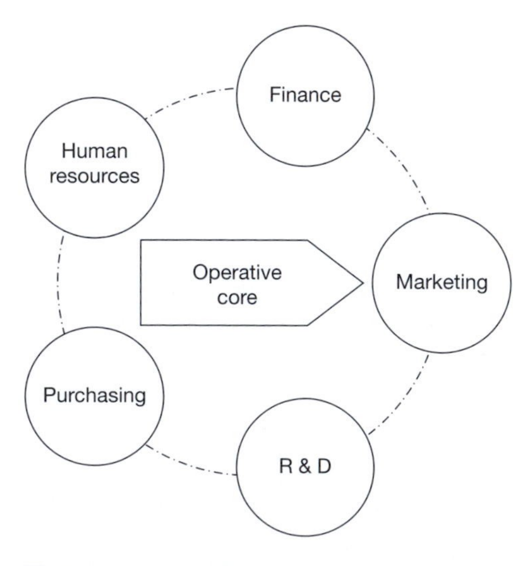

Figure 2.2 *Thompson's (1967) uncertainty-driven organizational design*

of the *shareholders* who own the firm and the other *stakeholders* who have a vested interest in the organization (employees, customers, suppliers, etc.). The middle line connects the strategic apex with the operating core by middle line managers who coordinate through command chains with formal authority. The operating core of the organization are the technical processes and people who perform the work processes directly related to the production of products and services. The continuous design of the organization is the task of the technostructure, with planners, controllers, accountants, organizational designers, or IT specialists who coordinate through designing the general rules and standards of the organization. Support staff summarizes those functions that are not directly involved in the operative core, like maintenance or facility management.

The designer of the firm will adapt this general model to distinct shapes for different types of organizations. Young and small organizations, for example, will have a strategic apex directly connected or even embedded into the operative core and no middle line. Medical practices, cabinets of lawyers and other professional organizations also have a strategic apex embedded in the operative core, but show a distinct support staff when they grow. Old and bureaucratic organizations, finally, detail all functions out and show the structure of Figure 2.3.

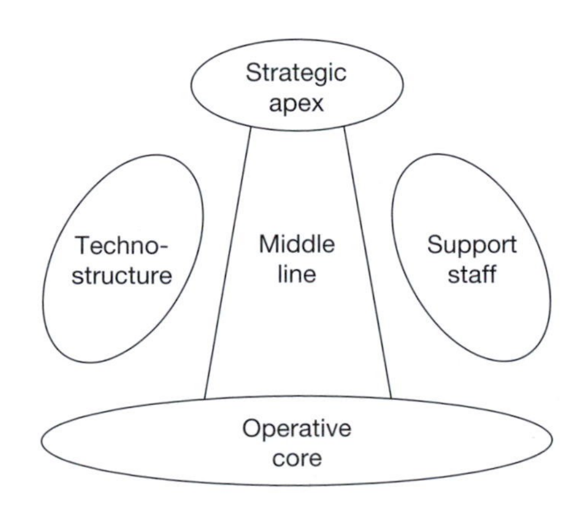

Figure 2.3 *The organizational architecture according to Mintzberg (1979)*

Nadler and Tushman's congruence architecture of problem-solving organizations

Nadler and Tushman (1997) derive a firm architecture from the design rule that *congruence* has to be reached between the problems that the firm has to solve and its organizational components. Here, in essence, we find the *fit* principle back in the shape of a design rule. They distinguish four key components of the organizational system that need to be in congruence: the work (basic activities in the firm), the people (responsible for the working tasks), the formal arrangements (explicit structure, processes, systems and procedures) and the informal relationships (informal guidelines that have powerful influence on the organization).

Matrix organization for synergy

Large conglomerates, in the divisional form, are simply a collection of divisions that are individually designed as autonomous firms with little or no relationships between each other. But technological firms can often profit from technical synergies between divisions that operate in different markets. The basic structure of matrix organization is described in Figure 2.4. Take, for example, the car industry, where each manufacturer operates in a separate market, e.g. for trucks, small or large passenger cars. Many technical components such as engines, brakes, power trains or bodies are similar. Technologies and functions on the one axis

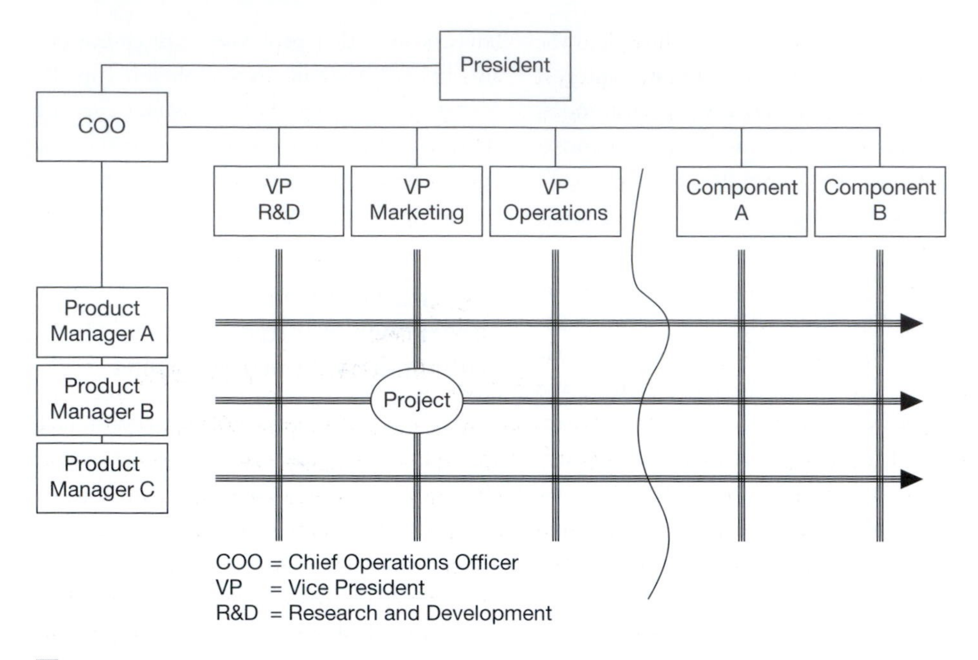

Figure 2.4 *Matrix organization*

and markets on the other then form a matrix, where concrete activities do occur on each intersection as projects.

Matrix organizations show two hierarchies, the traditional or functional line hierarchy on the one axis and a project hierarchy on the other axis. Each member of the organization operates in a double hierarchy, as being a member of a functional unit, but temporarily assigned full-time or part-time to a project as well. The functional manager may be responsible for part of a team member's workload, while a project manager assigns the work associated to the project. The matrix organization is especially effective in creating extensive knowledge exchange between markets and technologies through individuals who move in and out of projects. Matrix organizations also allow for flexibility in capacity planning when projects do not permanently need resources. The challenge of matrix organizations, however, is the much more complex coordination and balancing of the two hierarchies that carry an inherent tension and risk of conflict between the two managers that compete for the same resource.

Project-based or virtual organization for agility

A project-based or virtual organization emerges when individual projects of a technological firm vary so much, that the organization of a project is designed on a case by case basis and no stable second axis of a matrix can be designed. Engineering problems that design firms solve, research projects for research institutes, studies for consulting firms, and court cases for law firms are typical examples where each project varies in competences and capacities needed, so that no stable organization can be held in place for it. Instead, the stable, former functional organization becomes a mere resource platform with project managers as the powerful actors.

Virtual organizations, therefore, are based on *dynamic capabilities* to create temporary cooperations and to realize the value of a short business opportunity that the partners cannot (or can, but only to lesser extent) capture on their own. Three basic elements define this business architecture: (1) the business opportunity and the value that can be created, which is the reason to create (2) a temporary project from (3) the network, or source of resources and competences (Figure 2.5).

This chapter has presented reference architectures of technical firms. It is unlikely that any real case of a firm will exactly match one of the types described here. The design of a technical firm's appropriate business model is contingent on multiple factors such as the market, its size, age, technology or external environment. The aim of presenting concrete types here is to introduce idealized solutions for six common design objectives: (1) specialization, (2) coordination, (3) uncertainty, (4) problem-solving, (5) synergy and (6) agility, that are associated with the presented designs.

Organizational design rarely means choosing between distinct or extreme alternatives. For example, if flexibility becomes a constraint to the development of the technical firm, elements of a virtual organization can be strengthened or adapted. Literature therefore provides numerous gradients in the adoption of organizational means when design constraints change. For example, Van de Ven and Delbecq (1974) found that if task uncertainty increased, coordination by programming and hierarchical means was substituted by horizontal communication in the organization. Lawrence and Lorsch (1967) suggest that dynamic environments tend to lead organizations to adapt less formal controls. And Galbraith (1973) argues that the greater the uncertainty, the greater the amount of new information that needs to be processed, which, in turn, leads to the design of more management control systems. In short, the professional firm designer is

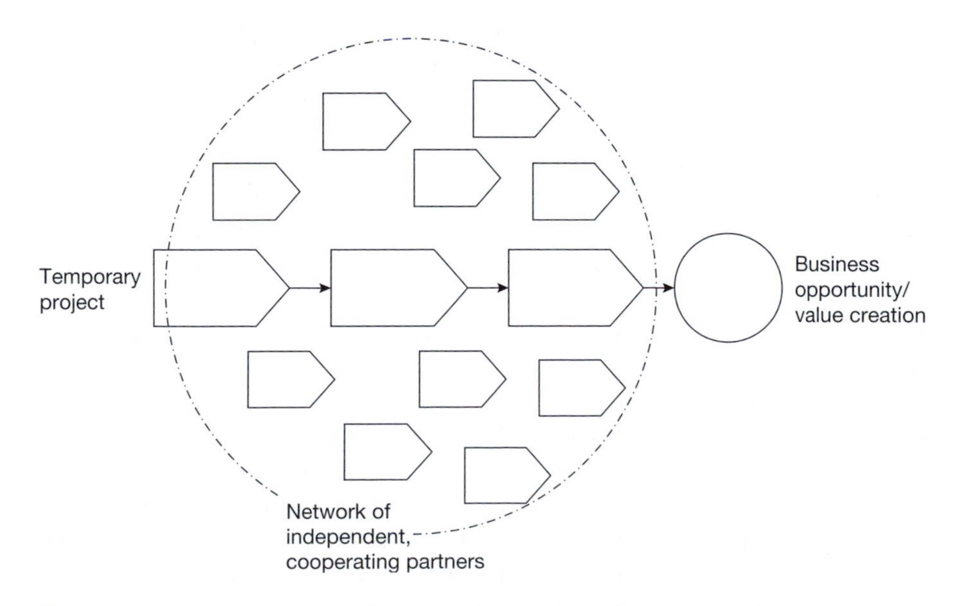

Figure 2.5 *The three elements of the virtual organization* (Katzy and Schuh, 1998)

acquainted with a broad set of such design rules that predict what the impact of applying mastered organizational means will be for achieving design objectives.

THE CAPABILITY OF DESIGNING TECHNICAL FIRMS

This last section shall be dedicated to the design process itself – and how it can be organized and designed. In fact, designing the design process has attracted a lot of attention in recent strategic literature. Under the title *dynamic capabilities* researchers put effort into understanding how competitive advantage can be sustained over time even though markets change (Henderson and Cockburn, 1994; Eisenhardt and Martin, 2000). In the words of Teece *et al.* (1997), the dynamic capabilities framework "analyzes the sources and methods of wealth creation and capture by private enterprise firms operating in environments of rapid technological change". They define dynamic capabilities as a "subset of competences, which allow the firm to create new products and processes, and respond to changing market circumstances."

Dynamic capabilities themselves take different forms, depending on the design activities they are tuned to. Brown and Eisenhardt (1997) found that firms with highly structured processes developed new products efficiently, but that those products often were not well adapted to conditions in fast changing markets. Highly flexible experimental processes are more effective in such high velocity markets (Hargadon and Sutton, 1997) but less efficient for the recurrent design of product variants in more stable markets. From the two types of dynamic capabilities at opposite ends of the market velocity, systematic design processes can be observed as a planned linear design process on the one side and a design process of entrepreneurial growth on the other.

Planning as systematic linear design process

This process pattern is common to a broad variety of design projects from architectural design to organizational design, budget planning or software engineering, where it is known as the waterfall model. This metaphor refers to defined phases of requirements analysis, goal setting, solution design, implementation, and success control, which are sequentially undertaken. Like falling water, in principle, there are no feedback loops to earlier stages. Clearly structured phases allow for specialization and efficient execution of each of the design tasks.

As *long-range planning* this sequential pattern is applied to strategic design of a technological firm: in a first phase the marketing department undertakes market analysis and defines the product program strategy with the product range that the enterprise will address. A critical decision of program strategy is the width of the product range. If it is too narrow, the served market may not be big enough; if it is too wide, focus of the firm may be diluted. Subsequently the *product program* is designed as the group of similar products or services that cover the product range, and sales quantities of each product are forecasted. This marketing plan is passed on to the production/operations department that turns the market requirements into the *production program.* Decisions are made about which parts are to be internally produced and what is to be purchased from suppliers. Production volumes and schedules are calculated so that the need for *production capacity* – machines and personnel – is established. This plan is transferred to the financial department for *investment plan-*

ning of new machines and to the organizational design department for job design, and update of the organigram. Finally, the plans are transferred to the human resource department to hire or train personnel to fit the new organigram, and to the purchasing department for the purchase of machines and equipment.

Such design and planning cycles involve many departments, do take time and are therefore undertaken on an annual basis, normally in autumn. Strictly speaking, long-range planning covers time horizons that are longer than one year, while design and planning activities with a time frame of less than one year are referred to as tactical or operative planning. Market changes that occur in the course of the year, therefore, cannot be addressed through the long-range planning process, but require exception handling projects by management. Furthermore efficient long-range planning follows the same established procedures every year. Nelson and Winter (1982) refer to these procedures as the "genes" of the firm, because the structure of the design process has a strong effect on the planning outcome and, like genes, are hardly changeable. Mintzberg *et al.* (1998) have caricatured the inability of coping with change by comparing the annual planning procedures as a "rain dance" that the management medicine men undertake in autumn of each year.

Design as entrepreneurial growth

Growth of the firm is only limited by the management capabilities that are available to the firm, concluded Edith Penrose (1968) in a book that was not reprinted until 1998. Her argument is that the growth of the firm is governed by the *productive opportunities*, which are all the possibilities that its managers *see* and can take advantage of. In other words, growth of the firm essentially is not the outcome of a sequence of planning steps but of entrepreneurial managers who design the firm to take or leave the productive opportunity. The productive opportunity is restricted to the extent to which management does or does not see it (entrepreneurial versatility), is willing or unwilling (entrepreneurial ambition) and capable (entrepreneurial judgement) to design the firm to react upon the opportunity.

Growth of the technological firm is not only driven by managerial competences, but also drives the need for additional management competences. Greiner (1972) has observed five typical crisis situations that occur when the firm outgrows the managerial competences with which the growth was initiated (Figure 2.6). In other words, re-design of the firm during each revolutionary period determines whether the company will move into the next stage, stagnate in the existing period or even disappear. This framework describes why firm design with certain management styles, organizational structures and coordination mechanisms works successfully at one point in time, but also carries the seed for future re-design need in them.

The first of the five phases of organizational development and growth is growth through creativity. Founders of the organization dominate this stage and the main emphasis is on creating both a product and a market. The founders are usually more focused on technical and entrepreneurial activities. They put all their efforts into making and selling a new product. Management problems occur when workload increases beyond the capacity of the founders and new people have to be hired and directed. Founders find themselves with unexpected management responsibilities.

In the second stage, growth is maintained through direction. In this phase, the new

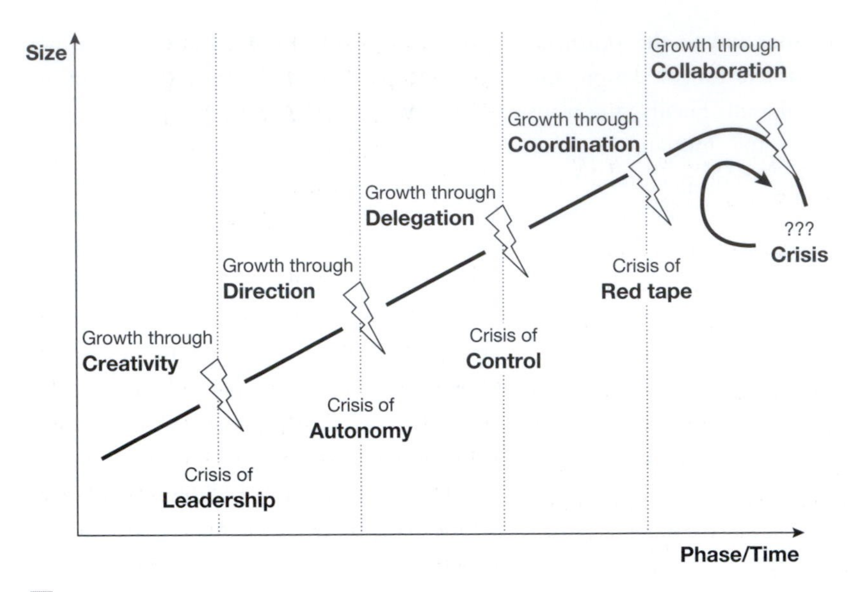

Figure 2.6 *Crises in the growth process* (Greiner, 1972)

manager and key staff focus on giving direction to employees who are functional specialists. Work content is a challenge for employees who demand little autonomy for managerial decision-making. With success in this period, employees become more autonomous and no longer perceive direction as support, but as constraint. This eventually leads to the next revolutionary period – the crisis of autonomy. Employees demand decision-making autonomy and managers become overwhelmed by the details of directing the increasing number of members in the organization.

The third stage of growth therefore requires delegation. In this stage of delegation, the organization usually begins to develop a decentralized inside structure, which fuels motivation and self-initiated activity in all parts of the organization. The next crisis arrives when activities get so numerous that top management loses control over a more and more diversified field of operation and mistakes happen. If the design of coordination and control systems cannot cope with the developments in this stage, the crisis of control may result in a return to central direction and the suspension of growth.

Growth in the fourth stage is driven by coordination and monitoring. This period is characterized by the design of formal systems and standards to improve coordination and efficiency. Focus on control systems, however, renders the organization more rigid and less flexible at the same time and the next crises occur when management gets carried away by formal programs and rigid systems and loses contact with the organization and its people. It is the crisis of red tape.

Further growth in stage five needs renewed collaboration to revitalize the firm. Organizational means, such as team action for problem solving, or cross-functional task teams, are re-introduced. But, again, this phase is bound to reach an internal growth crisis point when organizational members are exhausted by the intensity of teamwork and the heavy pressure for innovative solutions. Eventually this will again lead to growth of stage four with more coordination and monitoring.

Design of corporate renewal with ambidextrous organizations

Some technological firms concurrently host technologies in different stages of development and maturity. Such ambidextrous organizations (Tushman and O'Reilly, 1996) simultaneously operate organizational units for established products with efficient and stable planning processes and small entrepreneurial units for emerging new businesses of the next generation. Technology managers then personally experience the tension between the different planning and design approaches: while established businesses are big, entrepreneurial units are small, while systematic planning processes are highly sophisticated and specialized, entrepreneurial planning is rudimentary and chaotic. Entrepreneurial units act flexibly while established businesses require foresight.

Good knowledge about the particularities of designing technological firms is a way to strike the dynamic organizational balance between new development units and clearly structured efficiency units on the other side. Maintained corporate renewal entails managing the transition between both ends of the ambidextrous organization when entrepreneurial units are repeatedly created, grown and turned into established business to adapt to technological and environmental change (Orlikowski, 1996; Katzy *et al.*, 2003).

VALUE CREATION AS THE CONCURRENT DESIGN OF TECHNOLOGY AND FIRMS

Designing technological firms means bridging the two worlds of science and technology on the one side and business on the other side. Technological enterprises are firms: economic entities that act in markets with management that designs strategies, and organizations of people that put effort in the endeavour. The chapter, therefore, started with economic and managerial design logics of a firm, and reference architectures of how firms as complex systems can consistently be structured.

The two main conclusions of this chapter are, first, that technology is an additional dimension of complexity that distinguishes technological firms. For the design of the firm, technology is more than a technical artefact. It is a complex web of organizational routines, problem-solving expertise and knowledge. This is the source of sustained competitive advantage, if well mastered. Good design of the technological firm therefore is the key to turning technology into a strategic resource.

Technology drives innovation. Innovation, again, is more than the technical invention alone. In order to achieve economic return, firms need concurrently be designed to commercialize the invention. The second conclusion of this chapter, therefore, is that the capability of designing technological firms is a strategic asset in itself. Especially in Europe professionals who are capable of designing technological firms are scarce. This skill is a bright prospect for the future for technopreneurs.

FURTHER READING

Tushman and Anderson (1997) provide a collection of papers that address relevant design choices for technological firms.

Checkland and Scholes (1990) introduce their soft systems methodology as an approach to social design. This book is an interesting exploration of the two distinct design approaches of the technological firm, which is both a technical and social entity.

Nobel Prize winner Herbert Simon (1981) in his book reflects on the "sciences of the artificial" and how a design perspective furthers understanding for the professional.

McGrath and Macmillan (2000) discuss leadership in designing the firm in the face of uncertainty.

Fields (1999) gives practical recommendations and introductions into many of the standard methods and tools of designing a growing technological firm. His book is especially suited for scientists and engineers.

REFERENCES

Barney, J.B. and Ouchi, W.G. (1986). *Organizational Economics.* San Francisco, CA: Jossey-Bass.

Barney, J.B. and Tyler, B. (1991). The Attributes of Top Management Teams and Sustained Competitive Advantage. In M. Lawless and L. Gomez-Mejia (Eds), *Managing the High-technology Firm.* Greenwich, CN: JAI Press.

Brown, S.L. and Eisenhardt, K.M. (1997). The Art of Continuous Change: Linking Complexity Theory and Time-paced Evolution in Relentlessly Shifting Organizations. *Administrative Science Quarterly*, 42, 1–34.

Chandler, A.D. (1962). *Strategy and Structure. Chapters in the History of the Firm.* Cambridge, MA: MIT Press.

Checkland, P. and Scholes, J. (1990). *Soft Systems Methodology in Action.* Chichester: Wiley.

Crowston, K. and Katzy, B. (2003). Competency Rallying for Value Creation: An Action Study of a Virtual Enterprise. Le Tim Working Paper Series, no. 8.

d'Aveni, R.A. (1994). *Hypercompetition: Managing the Dynamics of Strategic Manoeuvring.* New York: The Free Press.

Easton, G. (1992). Industrial Networks: A Review. In B. Axelsson and G. Easton (Eds), *Industrial Networks: A New View of Reality* (pp. 3–27). London: Routledge.

Eisenhardt, K.M. and Martin, J.A. (2000). Dynamic Capabilities: What Are They? *Strategic Management Journal*, 21, 1105–1121.

Fields, R.W. (1999). *The Entrepreneurial Engineer: Starting Your Own High-tech Company.* Norwood, MA: Artech House.

Galbraith, J. (1973). *Designing Complex Organizations.* (vol. 4) Reading, MA: Addison-Wesley.

Grant, R.M., Krishnan, R., Shani, A.B. and Baer, R. (1991). Appropriate Manufacturing Technology: A Strategic Approach. *Sloan Management Review*, 32, 43–54.

Greiner, L.E. (1972). Evolution and Revolution as Organizations Grow. *Harvard Business Review*, 37–46.

Hargadon, A. and Sutton, R.I. (1997). Technology Brokering and Innovation in a Product Development Firm. *Administrative Science Quarterly*, 42, 716–749.

Henderson, R. and Cockburn, I. (1994). Measuring Competence? Exploring Firm Effects in Pharmaceutical Research. *Strategic Management Journal*, 15, 63–84.

Katzy, B.R. and Schuh, G. (1998). The Virtual Enterprise. In A. Molina, J. M. Sanchez and A. Kusiak (Eds), *Handbook of Life Cycle Engineering: Concepts, Methods and Tools* (pp. 59–92). Dordrecht: Kluwer Academic Publishers.

Katzy, B.R., Dissel, M. and Blindow, F. (2003). Dynamic Capabilities for Entrepreneurial Venturing: The Siemens ICE Case. In M.v.Zedwitz, G. Haour, T. Khalil and L. Lefebvre (Eds), *Management of Technology: Growth through Business, Innovation and Entrepreneurship* (pp. 3–22). Oxford: Pergamon Press.

Lawrence, P. and Lorsch, J. (1967). *Organization and Environment*. Boston, MA: Harvard University Press.

Leonard-Barton, D. (1992). Core Capabilities and Core Rigidities: A Paradox in Managing New Product Development. *Strategic Management Journal*, 13, 111–126.

McGrath, R.G. and Macmillan, I.C. (2000). *The Entrepreneurial Mindset – Strategies for Continuously Creating Opportunity in an Age of Uncertainty*. Boston, MA: Harvard Business School Press.

Malone, T.W. and Crowston, K. (1994). The Interdisciplinary Study of Coordination. *ACM Computing Surveys*, 26, 87–111.

Miles, R.E. and Snow, C.C. (1978). *Organizational Strategy, Structure and Process*. New York: McGraw-Hill.

Miles, R.E. and Snow, C.C. (1986). Organizations: New Concepts for New Forms. *California Management Review*, 28, 62–73.

Mintzberg, H. (1979). *The Structuring of Organizations – A Synthesis of the Research*. Englewood Cliffs, NJ: Prentice-Hill.

Mintzberg, H., Ahlstrand, B. and Lampel, J. (1998). *Strategy Safari: A Guide Tour through the Wilds of Strategic Management*. New York: The Free Press.

Montgomery, C.A. (1995). Of Diamonds and Rust: A New Look at Resources. In C.A. Montgomery (Ed.), *Resource-based and Evolutionary Theories of the Firm* (pp. 251–268). Boston, MA: Kluwer Academic Publishers.

Mowshowitz, A. (2002). *Virtual Organization: Toward a Theory of Societal Transformation Stimulated by Information Technology*. Westport, CT: Quorum Books.

Nadler, D.A. and Tushman, M.L. (1997). A Congruence Model for Organization Problem Solving. In M. L. Tushman and P. Anderson (Eds), *Managing Strategic Innovation and Change* (pp. 159–171). New York: Oxford University Press.

Nelson, R.R. and Winter, S.G. (1982). *An Evolutionary Theory of Economic Change*. Cambridge, MA: The Belknap Press of Harvard University Press.

Orlikowski, W.J. (1996). Improvising Organizational Transformation Over Time: A Situated Change Perspective. *Information Systems Research*, 7, 63–92.

Penrose, E. (1968). *The Growth of the Firm*. Oxford: Basil Blackwell.

Porter, M.E. (1985). *Competitive Advantage – Creating and Sustaining Superior Performance*. New York: The Free Press.

Prahalad, C.K. and Hamel, G. (1990). The Core Competence of the Corporation. *Harvard Business Review*, 68, 79–91.

Selznick, P. (1957). *Leadership and Administration*. New York: Harper & Row.

Simon, H.A. (1981). *The Sciences of the Artificial*. (2 edn) Cambridge, MA: MIT Press.

Smith, A. (1776). *The Wealth of Nations*. Harmondsworth: Penguin.

Taylor, F.W. (1911). *The Principles of Scientific Management*. New York: Norton & Company.

Teece, D.J., Pisano, G.P. and Shuen, A. (1997). Dynamic Capabilities and Strategic Management. *Strategic Management Journal*, 18, 509–533.

Thompson, J.D. (1967). *Organizations in Action*. New York: McGraw Hill.

Tushman, M.L. and Anderson, P. (1997). *Managing Strategic Innovation and Change – A Collection of Readings*. 656. Oxford, Oxford University Press. Ref Type: Serial (Book, Monograph).

Tushman, M.L. and O'Reilly, I.A. (1996). Ambidextrous Organizations: Managing Evolutionary and Revolutionary Change. *California Management Review*, 38, 8–29.

Van de Ven, A.H. and Delbecq, A.L. (1974). A Task Contingent Model of Work-Unit Structure. *Administrative Science Quarterly*, 19, 183–197.

Weber, M. (1947). *The Theory of Social and Economic Organization*. New York: Oxford University Press.

Wernerfelt, B. (1984). A Resource-based View of the Firm. *Strategic Management Journal*, 5, 171–180.

Chapter 3

Human Resource Management for advanced technology

Robert. M. Verburg and Deanne N. Den Hartog

OVERVIEW

This chapter is about Human Resource Management and focuses on managing employees in the context of advanced technology firms. The core of Human Resource Management (HRM) is the recognition of the value of employees for organizational success, which is often defined in terms of creating and sustaining competitive advantage. Selection, development, and socialization of employees become core activities of personnel management. Such practices are no longer seen as tasks that are the responsibility of a separate personnel department, but they become part of top management's personnel strategy. Line managers are increasingly expected to play an active role in implementing HR policies and practices. In this chapter, we will discuss several key points, such as the nature of HRM and its practices, the link between HRM and firm performance, HRM dilemmas of advanced technology firms, and future trends.

The success of organizations is to a large extent dependent on the effort and performance of the people who work for them. Having a knowledgeable, productive, and flexible workforce can form a source of competitive advantage for firms. Top managers increasingly realize that their people can make a difference by creating value for the organization and that therefore treating people as valuable resources can quite literally "pay off." As such, managing the employment relationship is a crucial challenge for organizations and the design of strategies for the optimal management of people has become an issue on the agenda of many boardrooms.

Regardless of the market a firm operates in, and whether it is a relatively small high-tech start-up employing just a few people or a much more established enterprise, all firms need to manage their workforce. Starting up a new business venture that involves hiring and managing a group of employees would, for instance, entail planning how many and what type of employees are needed to effectively perform the work, and designing interesting jobs for them. It would entail recruiting, selecting, and socializing these employees to become productive members of the new organization. Once the workforce is in place, the organization needs to continually ensure employees' optimal functioning, for instance, through setting and discussing performance goals, evaluating, motivating, training, and rewarding employees. Systems that aim to ensure optimal functioning in the future, for instance, through developing employees' capacities, skills, and careers, are also needed. Jointly, such people management tasks that aim to create and sustain a capable, creative, and high performing workforce are labeled "Human Resource Management."

Clearly, managing people in the workplace is a demanding task for any type of organization. However, doing so in today's turbulent business climate is especially challenging for advanced technology firms. The developing information technology, globalization, and other economic factors are changing the nature of work and organizations as we know them in a pervasive and long-lasting manner (e.g. Malone, 2004). Organizations are becoming increasingly flexible. Increasingly, organizations are comprised of temporary systems whose elements (people as well as technology) are assembled and disassembled according to the shifting needs of specific projects or tasks (e.g. Keegan and Den Hartog, 2004). These flexible forms put the traditional hierarchical structures under pressure and make managing, planning, and coordinating the efforts of employees more complex. Due to the technological possibilities, work is no longer necessarily performed in the office building and people working together no longer need to be together physically. ICT applications such as groupware have made it easier to work from other locations and with others without being co-located (Verburg *et al.*, 2005). This means that people from different countries and cultures can work together in virtual teams. Direct supervision of such intellectual tasks performed at various locations at different times will be very difficult (Den Hartog and Koopman, 2001). Such a virtual and diverse work environment also creates new challenges for the management of employment relationships.

Along with the form in which it is organized, the content of work itself and the corresponding characteristics of the workforce are changing. Much twenty-first-century work in the Western world will be intellectual rather than physical (House, 1995). This changing nature of work also offers challenges for people management, for example, how does one ensure that the workforce can continually develop and keep up with the fast pace of change? The workforce has also become more diverse and organizations need to deal with this increased diversity (e.g. in terms of ethnic background). All these developments in the way we organize and perform work make people management a more difficult yet also more crucial task than ever before. This chapter describes Human Resource Management and focuses on managing employees in the context of advanced technology firms. In this chapter, we will discuss several key points:

- What is HRM?;
- HRM practices;
- HRM and firm performance;
- A model of ideal typical HRM systems;
- HRM dilemmas of advanced technology firms and future trends.

WHAT IS HRM?

The term Human Resource Management (HRM) emerged more than three decades ago in the context of the American labor market and has increasingly replaced the term personnel management. Although HRM can be seen as a form of personnel management, the two terms are often contrasted. Personnel management is then seen as a bureaucratic and procedurally oriented function of the organization, focusing on the execution of employment rules and regulations, and administrative tasks. Control, compliance, and standardization are the responsibilities of a centralized personnel department. In contrast, the HRM approach stresses commitment, flexibility, individualized arrangements, and employees' willingness to go beyond contract. The responsibility for HRM tasks lies with line managers, who may be supported by the personnel department (Storey, 1992). From the start, the HRM approach stressed the strategic value of employees for organizational success, for example, in terms of creating and sustaining competitive advantage. This made HRM a crucial and strategic boardroom issue relevant to all managers, rather than a necessary but menial administrative function. In line with this, HRM can be defined as "a strategic approach to managing employment relations which emphasizes that leveraging people's capabilities is critical to achieving sustainable competitive advantage, this being achieved through a distinctive set of integrated employment policies, programmes and practices" (Bratton and Gold, 2003, p. 3). By definition, the nature of the employment relationship is of central importance to HRM. The exchange relationship between individuals and their employing organizations goes beyond an economic exchange or legal contract, which is discussed in Box 3.1.

Speaking about *human resources* instead of *personnel* clearly reflects the increased (financially driven) appreciation for employees and the increased strategic importance placed on employees nowadays. Some even use the term *human capital* when talking about the skills, abilities, and traits that people bring to the workplace (and that could yield profits if properly invested in). However, these terms are sometimes challenged as they are so strongly financially driven. For example, as Paauwe (2004) puts it, human resources are more than just malleable "resources." They are active individuals with different experiences, preferences, and values that are only partially governed by the institutions they work for. As such, there is more to HRM than increasing shareholders' value. HRM is also about shaping the exchange relationship between employee and organization while ensuring a fair and equitable deal for both parties involved (achieving fairness at the individual level), and about helping organizations abide by the demands set forth in collective bargaining agreements and labor legislation (achieving legitimacy at a more collective level). Thus, HRM aims to achieve added value in financial terms, but also moral value in terms of fairness and legitimacy (Paauwe, 2004). These concerns for fairness and legitimacy also, for instance, include employee health and safety as important points to have on the HRM agenda.

In the context of organizations in the advanced technology area, HRM usually also needs to meet two further demands that are crucial to organizational functioning, namely ensuring a high level of flexibility and innovativeness of the workforce. Advanced technology firms often operate in markets that are far from stable. Rapid technological developments, and fast changing demands of the market, will have an impact on the internal organization of the firm. In order to survive, such firms need to continually adapt to these changes and this can only be done through HRM practices that support flexibility of skills, knowledge, and commitment of their employees.

BOX 3.1 THE EXCHANGE RELATIONSHIP BETWEEN INDIVIDUALS AND ORGANIZATIONS

Different ways exist to describe the exchange relationship between individuals and organizations. One can, for example, see this exchange as an economic relationship (the exchange of work for pay), a legal one (focusing on the formal employment contract), a psychological one (focusing on implicit obligations and promises) or a social relationship (a relationship between people).

The legal employment contract is explicit and describes agreements on issues such as salary, working hours, and other mutual obligations between the two parties. Besides these explicit obligations, implicit ones also tend to exist. Parties in relationships expect certain things from each other. The term psychological contract is used to refer to such more implicit mutual expectations between organizations and their employees. An individual employee's psychological contract reflects their personal beliefs regarding terms and conditions of the reciprocal exchange agreement between them and their organization. These beliefs may differ from others in the same organization, as individuals have different experiences, needs, and preferences. Psychological contracts are implicit (not "written down"), subjectively understood, and dynamic (Rousseau, 1989). They develop through interactions, for example, between employees and representatives of the organization (e.g. supervisors). In essence, the psychological contract describes what employees feel the organization offers them and has promised them in exchange for their loyalty and hard work. Employees may, for example, see it as the organization's obligation to offer them a promotion in exchange for good performance when that was promised to them earlier or when they see colleagues receiving this in a similar situation. Not meeting these perceived obligations can be seen as a violation of the psychological contract, which can lead to feelings of betrayal and can have serious negative consequences (e.g. Robinson, 1996). It is therefore important for organizations to be aware of the existence of such expectations and carefully manage these. HRM can play an important part in doing so, for example, by outlining for prospective employees what this organization will have to offer them.

HRM PRACTICES

HRM involves the use of many different policies and practices that shape the employment relationship. The HRM approach assumes a central role for top and line management in this process. As the definition implies, HRM is "a strategic approach to managing employment relations" that is of key importance to line management rather than the task of a separate organizational department or a group of specialized personnel officers. Although many organizations do employ HR professionals, the HRM approach implies that line managers are primarily responsible for managing the employment relationship and that top management treats HRM as an intricate part of their business strategies aimed at achieving sustainable competitive advantage. HR professionals play an important advisory role to support the management of the core business at different levels in the organization, and help to develop and administer HR policies and practices. However, the actual practitioners of HRM, the ones using these practices as an integral part of their job are often line managers. They recruit new employees, do their performance appraisal, and are responsible for the development of their staff. Thus, nowadays, having some understanding of HRM is not only relevant to HR managers, but also for any (line) manager who is responsible for one or more employees.

So, what are the "tools of the trade?" Which policies and practices do firms use to manage their employees? Providing a comprehensive overview of all specific HRM practices is impossible, as a vast number of specific practices are used and new ones are constantly developed to meet changing demands of the market and workforce. Here, we provide a basic overview linked to three main phases of employment management: entry, performance, and exit of employees. In the further reading section, readers interested in more detailed information on the different and more specific practices will find relevant sources.

Entry

The entry phase encompasses HRM practices aimed at the recruitment and selection of new employees into the firm as well as HR planning, which is concerned with predicting the number of employees needed to effectively perform work. In some firms the demand is relatively predictable; however, in others, planning and job design can be a challenge, especially for the more flexible firm whose demand for employees may vary from project to project.

Although recruitment and selection are often mentioned together, they are very different activities and the emphasis placed on either may vary over time. When the labor market is tight, suitable employees are harder to find and increased emphasis may be placed on recruitment. Yet, when job candidates are readily available, selection is emphasized. Firms have different ways available of recruiting their prospective employees. They could, for example, advertise positions in general media, specialized trade journals, on their company website or a general recruitment website. They may also directly approach candidates, offer internships, or even outsource the search for candidates to specialized firms, such as executive search firms or employment and temp agencies. The way recruitment is organized will differ depending on the nature of the organization and, for example, their philosophy regarding developing staff versus buying in new people. The most effective way to recruit also depends on the type of job that needs to be filled, for instance, on the level of specialization and hierarchical level of the position. For top management positions organizations often outsource recruitment to executive search firms that have the expertise to design realistic job profiles and have the required resources to locate good candidates. When recruiting, it is important to know which elements of the prospective job tend to appeal to the targeted group, and where and how this group can be reached most effectively. For example, are they likely to read newspapers or trade journals and, if so, which ones?

Once candidates are found, selection starts. Of all personnel management practices, selection has the longest tradition. The first intelligence scale was already introduced by French psychologists in Paris in 1905 and the US Army introduced its first tests (Army Alpha and Beta) to select new recruits in 1917. Currently there are numerous tests and assessment procedures available and selection is widely used. Sources of information about candidates in selection include interviews, different types of tests (e.g. IQ or personality tests, work sample tests, and assessment centers), and information from letters, résumé's or recruitment forms, and references. Using such information to predict future success of job candidates forms the core of selection. Important in selection is to reduce the number of errors in hiring decisions. Two types of errors need to be minimized: (1) hiring unsuitable candidates (e.g. people who fail on the job even though they met the selection norms), and (2) not hiring candidates that would have been successful on the job (e.g. because the norm was too strict). As Figure 3.1 indicates, these two errors are related. The stricter the norms in selection (e.g. the higher the required score on a test), the fewer unsuitable candidates will be hired. On the other hand, the chance one rejects candidates that would have successfully performed the job will also increase.

Both errors can be costly. Ensuring one does not hire unsuitable candidates is crucial when the cost of such mistakes is high (e.g. when employees fail or quickly leave positions requiring expensive training programs, such as airline pilots). This is also what employers tend to focus on. However, not hiring people who would have been a success may also be costly, since in a tight labor market organizations might not be able to afford turning down people who would have been successful. It can also be very de-motivating to candidates to be wrongfully turned down.

In selection, using methods that are reliable (i.e. consistently lead to the same result) and valid (i.e. predict future success on the job and measure what they are intended to measure and not other things) is important (e.g. Cooper and Robertson, 1995). As stated, the key to selection is to predict success: does this person make a good engineer? A good sales manager? For many jobs, intelligence and certain personality traits are key selection criteria that are tested using standardized selection tests. The personality traits that are important vary somewhat for different types of jobs (e.g. Judge *et al.*, 1999). For example, the broad trait of conscientiousness (i.e. being organized, responsible, cautious as well as performance- and excellence-oriented) seems to predict success in many jobs, especially those high on autonomy, and

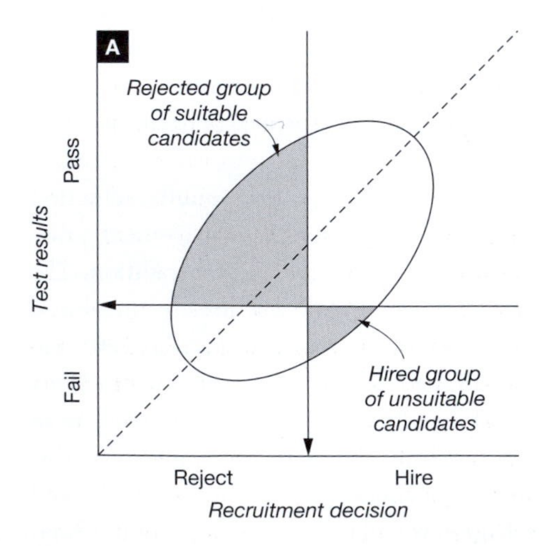

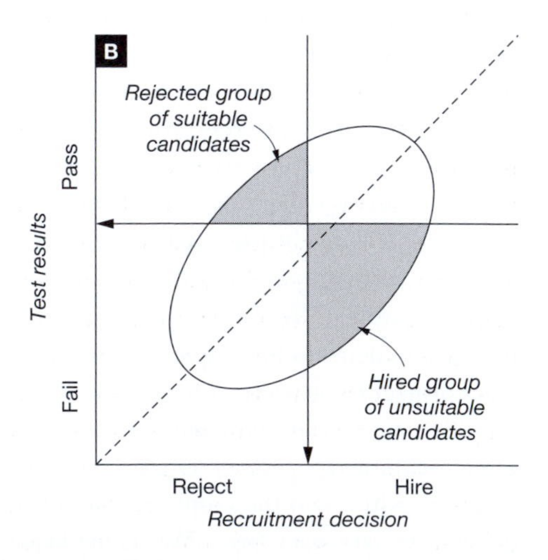

Figure 3.1 *The effect of using stricter selection norms, compare A and B*

a trait such as extraversion (i.e. being energetic, active, talkative, and sociable) tends to predict success in positions implying a social role, such as many sales-oriented positions.

Performance

A key set of practices focuses on managing employee performance, and ensuring performance remains at desired levels over time. The process of measuring and subsequently actively managing organizational and employee performance in order to improve organizational effectiveness is often labeled Performance Management. Baron and Armstrong (1998) emphasize the strategic and integrated nature of Performance Management, which in their view focuses on "increasing the effectiveness of organizations by improving the performance of the people who work in them and by developing the capabilities of teams and individual contributors" (pp. 38–39). Performance Management is a continuous process focusing on the future rather than the past.

Performance appraisal is one of the key elements of Performance Management. The process of Performance Management involves managing employee efforts based on *measured* performance outcomes. Thus, determining what constitutes good performance and how the different aspects of high performance can be measured is important (e.g. Den Hartog *et al.*, 2004). The specific criteria on which performance is rated will, of course, be different for different jobs. For many jobs nowadays output is hard to measure, which complicates performance appraisal. Organizations with an emphasis on long-term-oriented knowledge work and strongly interdependent teams working together on different tasks may find it is difficult to measure individual employees' performance.

Linked to appraisal are the reward and career development systems that are in place. Rewards as well as promotion opportunities can be tied to employee appraisal. Reward systems may have several different functions in organizations. Retaining competent employees by paying them well and stimulating their performance (e.g. through performance related pay), are central functions of rewards. However, rewards can also help in attracting new employees by offering them a more attractive reward package than they have elsewhere. In addition, rewards may be used to compensate employees for learning new work-related behaviors and skills or for having difficult circumstances on the job (e.g. paying more for working the night shift or doing hazardous work). Sometimes financial incentives may also be used to solve conflicts at work (e.g. Baron and Kreps, 1999). Pay-for-performance is often used as an instrument for motivating employees. Although this can be very successful, there are some things to keep in mind. For example, where output is difficult to measure, subjective judgments of superiors may form the sole basis for rewards and employees who are liked by their boss or those with better "political skills" may more easily influence these judgments in their favor. This may lead to jealousy and anger among others. Pay-for-performance may also lead employees to overemphasize personal, short-term targets and competition at the expense of long-term, joint targets and cooperation. Therefore, pay-for-performance systems need to be implemented carefully. They are most suited where performance goals can be set and their attainment can be clearly and fairly appraised. Performance targets need to be clear and specific as well as challenging and attainable to be most motivating (e.g. Locke and Latham, 2002). Also, the link between employees' performance and the rewards they will receive in return needs to be unambiguous, and employees need to have the idea that they can influence their performance. Finally, these systems will only work to motivate if rewards are sufficiently desirable – where only very little extra reward is offered it will not form much of an incentive for employees (Richardson, 2001).

Employee development in a broad sense is a crucial element of HRM, especially in rapidly changing contexts. Increasing expertise within the firm is important for the organization. Also, most employees value opportunities for some form of development at work, increasing the attractiveness of employers able to offer such opportunities. Training and development of employees have several (related) aims (Mabey and Salaman, 1995). The most general is to ensure required levels of knowledge and skills within the organization and keep these up to date. Training can also help introduce or speed up the implementation of change in the organization. An emphasis on training and development will signal the importance of continuous learning and can help create an innovative and learning-oriented climate. Finally, having a good employee development system (both in terms of learning opportunities and potential career and promotion opportunities) may help attract the desired new employees to the firm. Training and development programs usually form part of a larger system of career development in the organization. Especially developing so-called "high potentials," who are expected to end up in the key (top management) positions in the future often receives much attention in special talent development or management development programs.

Exit

A generally less favorable but very important phase of employment management is the exit of employees. Although employee development generally proves to be more (cost) effective than a relentless hiring and firing of employees (e.g. Pfeffer, 1994), companies need to be aware that a certain employee turnover enables the influx of new employees. Especially, advanced technology firms may often benefit from a regular influx of new employees with up-to-date skills, state of the art knowledge, and new ideas and perspectives. However, high turn-over rates are generally not favorable in organizations as in many cases important skills and knowledge are lost for the company when experienced people leave to pursue their career outside the firm. Often the process of hiring, socializing, and training new employees is more costly than organizations realize. Thus, the decision whether to develop employees within the firm to meet new demands or to hire new employees from outside can be a difficult and costly challenge.

HRM AND FIRM PERFORMANCE

Many HRM scholars agree that HRM can help fulfill business objectives as it can lead to positive outcomes such as high performance or even sustained competitive advantage. However, exactly *how* HRM yields these positive effects on performance is less clear. Often, the assumption is that HRM practices affect relevant employee skills and abilities, as well as employees' motivation and the effort they expend in work-related activities. This, in turn, is expected to result in achieving superior performance and sustained competitive advantage. As an example of the way this relationship tends to be modeled, Figure

3.2 presents a model developed by David Guest (1997). His model shows a sequential link running from a firm's HRM strategy (i.e. what the firm intends to achieve), via HR practices (what they actually do) and different types of outcomes, finally resulting in a potential impact on the bottom line.

In Guest's (1997) model, HRM practices (e.g. selection and training) ensue from HRM strategies, such as differentiation, focus, or cost-reduction. A different strategy would lead to a different set of implemented HR practices. A strategy aimed at innovation, for example, may lead to the implementation of practices that aim to develop new skills or hire new staff to bring in such skills. Next, HRM practices are assumed to result in HRM outcomes, in particular employee commitment, employees' quality, and flexibility of the workforce. Such HRM outcomes then result in employee behavior in terms of willingness to expend effort and high motivation to work on behalf of the organization as well as employees' willingness and ability to cooperate with others in the work place. The behavioral outcomes influence performance outcomes such as productivity, product and service quality, innovation, employee absence due to illness, and employee turnover. Finally, the last step in the causal chain is the financial outcomes, such as profits and return on investments (ROI) that follow from such performance.

It is important to keep in mind that different (groups of) employees may perceive or value the practices offered by the organization differently. What one employee finds desirable, the other may not like as much (e.g. high flexibility may appeal to some but not to others). Thus, HRM practices can be seen as "messages" or "signals" of the organizations' intentions towards its employees and are interpreted as such by individual employees (e.g. Den Hartog *et al.*, 2004). However, employees do not necessarily perceive such "signals" similarly and the impact of HR practices on employees' commitment and performance depends on employees' *perception* and *evaluation* of these practices (e.g. Guest, 1999). Attention to this is important to help achieve desired effects. Variation may exist in employees' perceptions of HRM practices or benefits offered by the organization even when, in objective terms, what is offered to different employees is very similar. Individual differences in such perceptions may, for instance, be based on employees' previous experience, comparisons to others within or outside the firm, employees' life stage, or their type of employment contract.

Besides the question how the process by which HRM affects performance unfolds, another question regarding HRM and performance is whether it is possible to find a specific set of practices that leads to high performance regardless of the context in which they are implemented or whether HRM can only affect performance if it is specifically adapted to various elements of the organizational and institutional environment. These

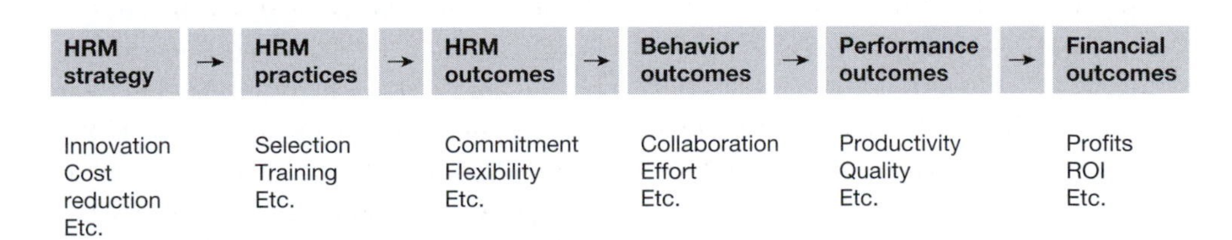

Figure 3.2 *The theoretical link between HRM and performance according to Guest (1997)*

two possibilities reflect the two dominant views regarding the relationship between HRM and performance that we will address below: the "best practice" and the "best fit" approaches (Boxall and Purcell, 2003).

"Best" or high performance work practices

As said, there are numerous practices and it is a challenge for firms to identify and implement the best possible practices for dealing with the three main phases of employment management. This is not an easy challenge. Some authors have proposed a number of "best practices" (often labeled high performance work practices). The basic premise of the best practices approach is that certain HRM practices will universally lead to a better firm or employee performance, no matter in what context these practices are implemented. Based on studies, literature reviews, and personal observations, the American scholar Jeffrey Pfeffer (1998), for example, lists seven dimensions of the HRM systems of companies producing "profits through people." He holds that companies tend to outperform competitors through explicit attention to one or more of the following:

1. offering employment security;
2. selective hiring of new personnel;
3. decentralization of decision making and self-managed teams;
4. offering relatively high compensation contingent on organizational performance;
5. offering extensive training;
6. reduced status distinctions;
7. extensive information sharing with regard to company performance.

Other research on high performance work practices focuses on similar practices, although there is some variation in exactly which (sets of) practices are proposed to lead to high performance. For example, Delaney and Huselid (1996) mention employee participation and empowerment, job redesign including team-based systems, extensive employee training, and performance-contingent incentive compensation as practices that are likely to improve organizational performance.

Coherent combinations of such "best practices" that lead to high firm performance are often labelled "high performance work systems" or "high involvement work systems" (e.g. Huselid, 1995). High performance or high involvement work systems are thought to increase employees' organizational commitment and motivation which, in turn, enhances performance. High performance work systems tend to stress practices in the area of employee development, autonomy, and participation, as well as having a motivating reward system that ensures that employees' hard work "pays off," both in terms of financial compensation and career opportunities. Strict selection, work designed so that employees have discretion and opportunity to use their skills in collaboration with other workers, and an incentive structure that enhances motivation and commitment may also be seen as part of such systems (e.g. Batt, 2002; Den Hartog and Verburg, 2004).

The legal or institutional context forms a constraint for the implementation and effectiveness of some such high performance work practices. Although the use of a given practice may vary widely in one country (thereby giving an employer offering this practice a competitive edge compared to one who does not), the same practice may be required by law or regulated through collective bargaining in others. Distinguishing oneself from other employers on such regulated practices may prove harder as there is less room to maneuver (see also Chapter 9 with regard to labor relations). For example, in the Netherlands a system of works councils ensures a limited form of employee participation (every firm of over fifty employees is obliged by law to have a works council) and representatives of employees have a number of legal rights towards management. Strict laws on working conditions regulate maximum numbers of working hours, safety, etc. Trade unions participate in collective bargaining at the sector level and collective agreements lead to relatively high job security (e.g. Boselie *et al.*, 2001). Thus, like many other European countries, the Netherlands is an example of a nation that has a relatively strongly institutionalized context. Legal requirements and collective bargaining systems are different in other nations. An important realization is that legislation and regulations may, in some areas of the world, prescribe part of the content of some practices (e.g. employment security, minimum wages), which implies that to distinguish themselves from other firms, employers will need to creatively develop the content of these practices (while remaining within the legal boundaries) and may also need to focus partly on other practices to gain a competitive edge.

Best practices or best fit?

As stated above, some authors in the HRM field argue for (systems of) "best" practices. However, others argue that such universal best practices do not exist, and that the effectiveness of HRM practices depends on the context in which they are embedded (e.g. Delery and Doty, 1996). The challenge is therefore to achieve a certain "fit" or alignment between the HRM practices and their context. Different types of fit are described (see e.g. Wood, 1999). Different types of fit that are distinguished include:

- *Internal (horizontal) fit*: Alignment of all HR practices within the HRM system. In this view, a coherent system of HR practices is needed and the different HR practices and their "signals" to employees should not contradict each other.
- *Strategic (vertical) fit*: The link between the HRM system and organizational strategy. In this view, aligning HR practices with the overall business strategy helps focus efforts and stimulates people to fulfill organizational goals. For example, if firm strategy involves a strong emphasis on innovation, the HR practices in place should help employees be innovative, learn, and remain open to change.
- *Environmental fit*: How the HRM system is adapted to the external environment. According to this perspective, the HRM system should adapt to the changes in the environment and to the rules and expectations of the institutional context that affect the organization. For example, conforming to legislation.
- *Organizational fit*: The fit between the HRM system and other relevant systems in the organization. Relevant systems might, for instance, include the technological system, organizational structure, and the production and control system.

All forms of fit may be relevant in developing HRM systems to maximize performance.

IDEAL TYPICAL BUNDLES OF HRM PRACTICES

Despite the convincing arguments for the benefits of fit and alignment, it is often hard to understand what such fit entails in practice. What does it mean for an organization? How does one accomplish fit? To further develop the idea of fit and show why it may be of use, we present a model of four ideal types of HRM systems that are likely to be effective within different contexts. The model described in this chapter (depicted in Figure 3.3) proposes four types of HRM systems. The two dimensions underlying our model of ideal types of HRM are: (1) the overarching goal of the HRM policy (does the company have a "commitment philosophy" or a "compliance strategy") and, (2) the locus of responsibility for development and employability (the company or the individual).

The first dimension of the model describes whether the overall objective of the employment strategy is to reach employee commitment or to maintain control and compliance. This difference between aiming for control or commitment was introduced by Walton (1985) and it has recurred in the study of HRM since. Walton contrasts so-called traditional work systems that are characterized by close supervision, narrowly defined jobs, and no career development, with high commitment work systems that highlight more broadly defined jobs, concern for learning and growth, and a larger role of the team in performance evaluations.

The second dimension of our model contrasts whether the company or the individual employee is seen as responsible for development. In the past, organizations were typically tall hierarchies. It was quite common for people to spend their entire career in one or two organizations. Careers progressed in linear stages, monitored and planned by the organization. Success was defined by the organization and measured through promotions and increases in wages (Sullivan, 1999). Later, many corporations started to downsize and restructure in order to become more flexible in response to environmental factors such as increasing competition and more rapid technological changes. For employees in many lines of work this lead to increased job insecurity and the need to be adaptive. Ideas of "cradle-to-grave" employment were no longer valid (e.g. Hall and Mirvis, 1995). Organizations have started to stimulate

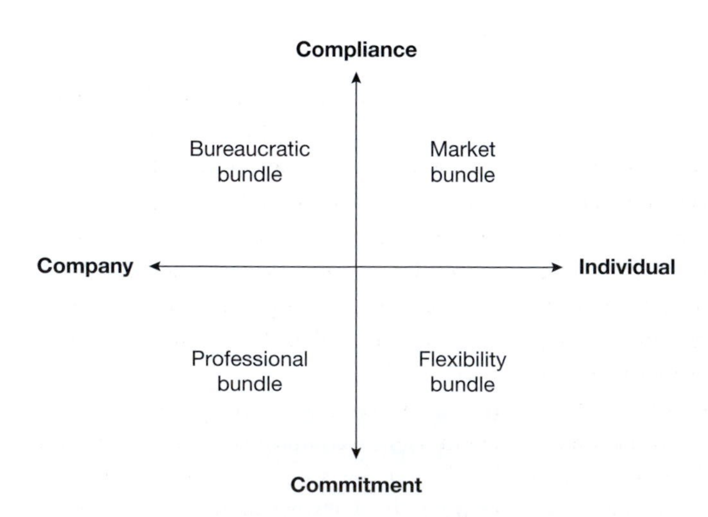

Figure 3.3 *Typology of bundles of Human Resource Management*

employees to be less dependent on them for job security and personal or career development. Typical career patterns are changing. In the "traditional" career, workers had job security and planned advancement and offered their performance and loyalty in exchange. Increasingly, however, workers will exchange their performance for the opportunities they are offered for continuous learning and marketability rather than for job security. Employees become responsible for their own career management rather than giving the responsibility for the management of their career into the hands of the corporations they work for. Employees will therefore increasingly stress developing transferable rather than firm-specific skills. Learning is mostly done on the job rather than through formal programs and success is measured by the feeling of doing psychologically meaningful work and no longer only through status, pay, and promotion (Sullivan, 1999). This is an important part of what the employability HRM dimension deals with. It describes the locus of responsibility for employee development. In some organizations, strong corporate responsibility is still assumed for the (career) development of employees, whereas in others, individual employees are now themselves primarily responsible for their own (career) development.

This dimension also more generally describes the level of flexibility. As stated, for many organizations, environmental turbulence has increased and flexibility is a key concern. Flexibility is related to the pervasiveness of rules and regulations. Companies that are characterized by many rules and regulations stress corporate responsibility for outcomes. In such organizations, rules and procedures form the basis for coordination of efforts (standardization of behavior). Such organization-centered structures tend to be bureaucratic and low on flexibility. In contrast, organizations with less emphasis on rules and regulations tend to put more responsibility for outcomes in the hands of individual employees. In control terms, organizations relying heavily on rules or direct supervision emphasize behavior and/or output control (Snell, 1992).

Behavior control assumes managerial knowledge of cause–effect relationships, making it possible to prescribe and judge appropriate behavior. If desired outcomes or standards of performance are clear and measurable, organizations may also use output control, where rewards are based on reaching predetermined performance targets. In turbulent environments, desired behavior and performance standards may be ambiguous or changeable, which makes relying on predetermined rules or performance targets to coordinate employees' efforts more difficult.

Four types of HR systems

Combining the two dimensions leads to four ideal typical forms of HRM, which are labeled the *bureaucratic* bundle, the *market* bundle, the *professional* bundle and the *flexibility* bundle (see Figure 3.3). Note the similarity with the work on the structuring of organizations by Mintzberg (1979). Both the *professional* and the *flexibility* bundle assume a commitment philosophy. They differ in who is seen to be responsible for employee development and careers. In the flexibility bundle, individual employees are, to a large extent, responsible for their own development and few rules govern the process, whereas in the professional bundle, development is seen as a task of the organization and is more carefully regulated. The *bureaucratic* and *market* bundles stress compliance. The bureaucratic bundle can be seen as "classical personnel management" traditionally found in large bureaucratic organizations. Extensive sets of rules and regulations are common. The market bundle has fewer rules in place, direct supervision is common. Individuals are responsible for their own development. We will describe the four types in more detail below.

The *bureaucratic* configuration consists of practices that act as behavior control mechanisms, such as rules and regulations for coordinating large groups of lower-skilled employees. Rules govern the production process and help maintain efficiency. Employees' tasks are narrowly defined and required skills are limited. As tasks and skill sets are narrowly defined employees usually do not have many opportunities for further development. Large personnel departments are responsible for developing and executing formal procedures that prescribe how to manage the workforce. Examples of this model are most likely to be found in traditional bureaucratic organizations (e.g. in the heavy industrial sector). Recruitment and selection are formal, yet simple procedures. Recruitment tends to be relatively easy, as the core production process primarily consists of standardized and relatively simple tasks. The personnel department is leading in the selection of employees and the use of more expensive methods of recruitment and selection (e.g. tests, outsourcing) is rare. Rather, standardized application forms and interviews are used. The standardized production process calls for a system of compulsory, formalized training emphasizing technical skills. Employees are not primarily responsible for taking the initiative to update their knowledge and skills. Besides updating of technical skills, overall opportunities for promotion and development are limited and the organization spends little money on broader training and development of employees. Extensive management development programs aimed to increase functional flexibility of managers are rare, as the stable market does not call for such flexibility. A strongly formalized reward system is standard. Group appraisals and rewards are rare as the production process is broken down into narrowly defined separate tasks. Performance measures are clear and performance evaluation involves the use of standardized evaluation forms. The overall HRM policy is designed in order to maximize compliance and control over the production process rather than to enhance commitment.

The *professional* configuration also has many rules and regulations, but far less emphasis is found on behavior control mechanisms. A system of output control is more common as core production consists of complex tasks. HRM policy aims to enhance employee commitment. A large, central personnel department is responsible for setting out policy, although execution tends to be decentralized. The company usually employs highly educated professionals and aims to accommodate and develop them. This type of personnel management may, for instance, be found in public service organizations with a strong focus on quality within a stable market environment (e.g. hospitals, research institutions). Recruitment and selection aim to attract an exclusive group of highly skilled professionals and are mostly decentralized. The central personnel department assists units trying to hire newcomers (providing information, forms, etc.). Selection is primarily done through interviews. Tests and assessment centers may be used. Although the highly specialized professionals feel personally responsible for keeping their knowledge and skills up to date, the organization also provides support for training and development activities. The number of compulsory training courses is limited. The organization offers sufficient promotion opportunities. Promotion and development are often embedded in formal systems of career development, such as management development programs. Performance appraisals are done by supervisors, aided by systems and performance criteria developed by the personnel department. Team performance may be part of the appraisal process. Employees tend to be rewarded through a fixed salary tied to a formal job classification system, bonus systems are usually limited. The company may have developed a formalized HRM policy and mission statement. As the core activities of the organization call for high levels of specialized knowledge, the emphasis in the employment policy is on employee development. Commitment is also an important driving force for the policy as this specialized group of employees is difficult and expensive to replace.

In the *market* bundle, the HRM policy is hardly developed and few formal procedures with regard to HRM practices are in place. It is often the owner/founder of the business who is responsible for setting out the employment policy. High levels of direct supervision are found, combined with a limited number of explicit rules and regulations to maintain flexibility. The organization may offer its employees nothing more than a salary for their efforts. Core production calls for a lower skilled workforce. Employees usually do not perceive clear opportunities for further development within the firm. Examples of companies with a market model may include small businesses or start-ups in the area of construction, restaurants, or industrial services.

Informal procedures with regard to recruitment and selection are typical. The lower-skilled (and often temporary) employees tend to be relatively easy to recruit and selection is done through an informal interview with "the boss." Application forms or formal tests are not used. There are mostly no explicit funds for employee development. Compulsory training in order to obtain some necessary skills may be provided, but such needs are determined and dealt with on a case-to-case basis. Only few opportunities for promotion tend to be available due to the small scale of the business. Rewards are based on informal procedures and no formal system of job specifications or classifications exists. The informal nature of reward practices may include ad hoc bonuses. Performance appraisal is based on the manager's perception of employees' performance and is often based on whether a set of pre-determined goals or targets was met. Overall policy aims at remaining (numerically) flexible and not at enhancing commitment.

Finally, the *flexibility* configuration emphasizes commitment of individual employees. Few formal rules and procedures are in place. The company is in a continuous process of alignment with customer needs in a turbulent market. Tasks are broadly defined and functional flexibility of employees' is a prerequisite for success. Rather than developing specialized skills and knowledge that are of use in stable and well-defined tasks, employees will need to be broad professionals that can perform different tasks when needed. Although organizations do not offer much job security in this turbulent environment, they do aim for commitment in order to keep their well-trained and flexible employees (as long as they are needed). Creating their own job security is seen as the responsibility of individual employees, who need to keep themselves "employable" for future jobs within or outside of the present company. The company offers opportunities for development, but employees themselves need to take the responsibility and initiative to develop in any given area. Rather than having a large personnel department, the responsibility for the execution of many HR tasks lies in the hands of line managers, who are supported by a few internal or even external HR consultants. Examples of firms that may use the flexibility model can be found in the area of advanced technology, international trade, and professional services. Recruitment and selection of well-educated and able professionals is central to this bundle of HRM practices. Psychometric tests and assessment centers are often used in selection, and activities in this area are also often outsourced to a specialized recruitment and selection firm. There are few formal procedures. Line managers rather than the personnel department are responsible for hiring their own new staff. As jobs are very broadly and flexibly defined, there is usually no formal system of job classifications or task descriptions. Rewards tend to combine a base salary and substantial pay-for-performance. Both individual and group bonuses may be used. Company or departmental results play an important role in performance appraisals. There are no standard forms and there is no formal system of sanctions. There are no compulsory training courses. Employees are personally responsible for keeping up to date in their field, though often the company does fund development activities at the individual's request. There is no formal management development system, but ample opportunities for promotion and development remain (mostly at the initiative of the employee). The overall aim of the policy is commitment of employees, who are highly educated and difficult and expensive to replace.

The HRM systems of many advanced technology firm are likely to be in line with the flexibility model although the practices of larger organizations in this area, such as large laboratories, may more closely resemble the professional bundle. The many start-ups in advanced technology, however, may also reflect the HRM of the market bundle before they start to develop more extensive HRM systems. In the next section we will describe some of the particular HRM challenges advanced technology firms have to deal with.

HRM CHALLENGES FOR ADVANCED TECHNOLOGY FIRMS

IXEurope plc provides datacenter services in four countries – UK, Germany, France, and Switzerland – through a network of datacenters. IXEurope delivers specialized datacenter services to systems integrators and IT consultants, telecom companies, internet service providers, and large corporations such as Hewlett-Packard, Nikon, and AOL. IXEurope is a private company founded in 1998. The company received a first round of venture capital in July 1999 and has continued to grow rapidly ever since. In September 2000, IXEurope acquired a second round of financing and began a rapid roll-out across Europe, including two acquisitions in Germany and Switzerland. Today, IXEurope's portfolio consists of seven highly sophisticated datacenters, located in London, Paris, Frankfurt, Düsseldorf, Morfelden, Zürich, and Geneva. Running IT systems is a challenge for any business nowadays, and network access, staff costs, and developments can all take the focus off the core business. Helping organizations deal with this challenge forms IXEurope's core business. IXEurope specializes in delivering industry-leading services enabling their clients to focus on their customers rather than their back office. IXEurope currently employs a staff of 95 specialists. It is vital for IXEurope's solutions to enable complete alignment with the business models of their various clients. Flexibility and scalability are therefore important drivers for their business.

Organizations such as IXEurope work within the turbulent and demanding context of advanced technology. In this context HRM needs to help the organization meet the demands that are crucial to organizational functioning and success, namely ensuring a high level of flexibility and innovativeness of the workforce. HRM needs to be able to deal with rapid growth and change, and ensuring flexibility of employees is a prerequisite for success. Rather than developing specialized skills and knowledge that are of use in stable and well-defined tasks, employees will often need to be broad professionals who can perform different tasks when called upon. The recruitment and selection of professionals meeting these criteria forms a big challenge for such firms. On the one hand, the rapid growth rate of the business calls for fast procedures enabling the company to employ the people who are needed to handle the increasing work load as soon as possible. On the other hand, the unique profile of the preferred broad, knowledgeable, and flexible professionals implies more elaborate and more time-consuming recruitment and selection practices. Only such extensive methods are likely to support the decision to hire suitable people. Rather than putting out job advertisements in newspapers, a more focused and detailed search through specialized channels might be a more beneficial recruitment strategy. In case of selection, tests or assessment centers will probably be more reliable than interviews alone, when one needs to establish whether the flexibility and attitude of prospective employees matches the demands of the job.

Keeping their staff up to date in the rapidly changing field forms another key challenge for advanced technology firms. In a market with rapid technological developments, there is a clear and limited "expiration date" of employees' technological skills and knowledge. The ability to ensure and update necessary skill and knowledge levels is, therefore, an important prerequisite for the success of advanced technology firms. In other words, these contexts lead to a significant link between HRM and Knowledge Management.

HRM and Knowledge Management

Knowledge Management can be defined as "the effective learning processes associated with exploration, exploitation and sharing of knowledge (tacit and explicit) that use appropriate technology and cultural environments to enhance an organization's intellectual capital and performance"(Jashapara, 2004, p. 12). Knowledge Management is an interdisciplinary field that combines insights from disciplines, such as information management, management science, psychology, and economics. Knowledge Management is a key factor in the performance of advanced technology firms involved with innovation processes (see Chapter 14). There is a clear link between Knowledge Management and HRM although they are not the same. In this chapter we will explore the link between the two. Further information on Knowledge Management can be found in Chapter 14 of this book "Managing knowledge processes."

In relating Knowledge Management to HRM, it is interesting to distinguish between explicit and tacit knowledge in firms. In contrast to *explicit* knowledge that has been codified in objects, words, and numbers, *tacit* knowledge consists of mental models, skills, and behaviors of employees. Such tacit knowledge is more difficult to transfer as it resides to a larger extent in the heads of people and to a lesser extent is explicitly written down. Clearly, dealing with tacit knowledge and finding ways to make such knowledge available throughout the organization is an HR challenge. Typical HRM practices, such as the recruitment and selection of new employees not only imply the introduction of new people but also the import of new knowledge into the organization. Other HRM practices, such as training and human resource development are also clearly linked to knowledge development.

Research on the link between HRM and Knowledge Management in 300 Malaysian companies indicates that highly knowledge-intensive companies require a different managerial approach than less knowledge-intensive organizations. Knowledge-intensive companies focus their training on the development of people who are capable of turning internal and external knowledge into organizational knowledge. Training is therefore aimed at stimulating creativity, problem solving skills, and quality initiatives. Performance appraisal and rewards systems also play an important role for Knowledge Management as these are used to support transformation in employee's behavior towards knowledge sharing, application, and creation (Yahya and Goh, 2002). Traditional production processes are much easier to monitor and measure than inter-personal processes such as the exchange and development of knowledge among employees. Dealing with the HRM and Knowledge Management link is thus clearly a challenge for advanced technology firms.

E-HRM

The internet, as the almost universal medium for interaction across boundaries, has created an infrastructure that enables organizations to launch all kinds of online HRM activities. Many companies post job advertisements and application forms on their public company website. Prospective employees can browse through job profiles and may even be able to apply online. However, recruitment is just one example. The full range of HRM practices and policies using web-based technology is central to the concept of Electronic Human Resource Management, or E-HRM.

Some authors confuse E-HRM with the growing number of available HR information systems. Such systems are used to support HRM business processes in the same way as any other enterprise resource planning software and are aimed at back office activities of personnel departments. However, E-HRM is explicitly aimed at delivering HR-related services to employees through electronic media. Ruël *et al.* (2004) argue that the rise of E-HRM stems from the transition of Western economies from being industry-oriented towards being knowledge-oriented. They further show that in Europe and the US, more than 90 percent of all companies use their website for recruiting people. Other popular web-based HRM practices involve training and development, and intranets are also used for information dissemination with regard to personnel issues. General E-HRM strategies are still rare as selection, performance management, career issues, rewarding, and coaching do not yet tend to be supported through the web in most firms (Ruël *et al.*, 2004). An increase in E-HRM activities in the coming years is likely, as more and more companies will explore or enhance the possibilities of using web-channels for elements of managing the employment relationship. Especially for companies in the advanced technology area, exploring the possibilities for extending their web-based activities may be of interest, for example, for firms working with virtual teams or with tele-workers that already make extensive use of the web for other areas of business.

Future trends

Above we argued that in the turbulent context of advanced technology firms, HRM needs to be able to deal with rapid growth and change, and to ensure employees' flexibility in order for firms to be able to survive. The knowledge intensity of many advanced technology firms calls for the support of knowledge processes, and HRM practices, such as selective recruitment, rewards, training, and development play a leading role here. Finally, HRM benefits from the possibilities of using online applications for offering its services to employees. Online recruitment and information sharing are only a beginning as many companies explore the possibilities of E-HRM in offering training, career development tools, and in new ways of performing performance appraisal.

Another important trend is that external collaborations between firms are becoming increasingly important, as the traditional "stand-alone" model for innovation is under pressure. Advanced technology firms can no longer afford to rely entirely on their own ideas and competences to advance their business, nor can they restrict their innovations to a single path to market (Chesbrough, 2003). Companies need to strategically leverage internal and external sources of ideas and take them to market. In this sense, HRM may not only be about the internal organization of the firm but may play a major role in the support and stimulation of knowledge networks between companies.

FURTHER READING

Bratton, J. and Gold, J. (2003). *Human Resource Management: Theory and Practice*, 3rd edition, Houndmills: Palgrave Macmillan. This book provides a broad introductory overview of the theory and practices of HRM.

Paauwe, J. (2004). *HRM and Performance: Achieving Long-Term Viability*, Oxford: Oxford University Press. This book is an up-to-date text on the link between HRM and outcomes.

Gratton, L. (2000). *Living Strategy: Putting People at the Heart of Corporate Purpose*, London: Pearson Education. This classic text provides a number of corporate examples on the link between people and business performance.

REFERENCES

Baron, A. and Armstrong, M. (1998). Out of the box. *People Management*, 23, 38–41.

Baron, J.N. and Kreps, D.M. (1999). *Strategic Human Resources: Frameworks for General Managers*. New York: Wiley.

Batt, R. (2002). Managing customer services: human resource practices, quit rates and sales growth. *Academy of Management Journal*, 45 (3), 587–597.

Boselie, P., Paauwe, J., and Jansen, P. (2001). Human resource management and performance: lessons from the Netherlands. *The International Journal of Human Resource Management*, 12, 1107–1125.

Boxall, P. and Purcell, J. (2003). *Strategy and Human Resource Management.* Houndmills: Palgrave Macmillan.

Bratton, J. and Gold, J. (2003). *Human Resource Management: Theory and Practice,* 3rd edition. Houndmills: Palgrave Macmillan.

Chesbrough, H. (2003). *Open Innovation: The New Imperative for Creating and Profiting from Technology.* Boston, MA: Harvard Business School Press.

Cooper, D. and Robertson, I.T. (1995). *The Psychology of Personnel Selection.* London: Routledge.

Delaney, J.T. and Huselid, M.A. (1996). The impact of human resource management practices on perceptions of organizational performance. *Academy of Management Journal,* 39 (4), 949–969.

Delery, J.E. and Doty, D.H. (1996). Modes of theorizing in strategic human resource management: tests of universalistic, contingency, and configurational performance predictions. *Academy of Management Journal,* 4 (39), 802–835.

Den Hartog, D.N. and Koopman, P.L. (2001). Leadership in organizations. In: N. Anderson, D.S. Ones, H.K. Sinangil, and C. Viswesvaran (eds), *Handbook of Industrial, Work and Organizational Psychology,* volume 2. London: Sage.

Den Hartog, D.N. and Verburg, R.M. (2004). High performance work systems, organisational culture and firm effectiveness. *Human Resource Management Journal,* 14 (1), 55–78.

Den Hartog, D.N., Boselie, P., and Paauwe, J. (2004). Performance management: a model and research agenda. *Applied Psychology: An International Review,* 53 (4), 556–569.

Guest, D.E. (1997). Human resource management and performance: a review and research agenda. *The International Journal of Human Resource Management,* 8 (3), 263–276.

Guest, D.E. (1999). Human resource management: the workers' verdict. *Human Resource Management Journal,* 9, 5–25.

Hall, D.T. and Mirvis, P.H. (1995). Careers as lifelong learning. In: A. Howard (ed.), *The Changing Nature of Work.* San Francisco, CA: Jossey-Bass, pp. 323–361.

House, R.J. (1995). Leadership in the twenty-first century: a speculative inquiry. In: A.Howard (ed.), *The Changing Nature of Work.* San Francisco, CA: Jossey-Bass, pp. 411–450.

Huselid, M.A. (1995). The impact of Human Resource Management practices on turnover, productivity, and corporate financial performance. *Academy of Management Journal* (38) 3, 635–672.

Jashapara, A. (2004). *Knowledge Management: An Integrated Approach.* Harlow: Prentice Hall.

Judge, T.A., Higgins, C.A., Thoresen, C.J., and Barrick, M.R. (1999). The big five personality traits, general; mental ability, and career success across the life span. *Personnel Psychology,* 52, 621–652.

Keegan, A.E. and Den Hartog, D.N. (2004). Transformational leadership in a project-based environment: a comparative study of the leadership styles of project managers and line managers. *International Journal of Project Management,* 22, 609–617.

Locke, E.A. and Latham, G.P. (2002). Building a practically useful theory of goal setting and task motivation. *American Psychologist,* 57, 705–717.

Mabey, C. and Salaman, G. (1995). *Strategic Human Resource Management.* Oxford: Blackwell.

Malone, T.W. (2004). *The Future of Work: How the New Order of Business Will Shape Your Organization, Your Management Style, and Your Life.* Boston, MA: Harvard Business School Press.

Mintzberg, H. (1979). *The Structuring of Organizations: A Synthesis of the Research.* Englewood Cliffs, NJ: Prentice Hall.

Paauwe, J. (2004). *HRM and Performance: Achieving Long-Term Viability.* Oxford: Oxford University Press.

Pfeffer, J. (1994). *Competitive Advantage Through People.* Boston, MA: Harvard Business School Press.

Pfeffer, J. (1998). *The Human Equation: Building Profits by Putting People First.* Boston, MA: Harvard Business School Press.

Richardson, R. (2001). *Performance Related Pay: Another Management Fad?* Inaugural addresses. Research in Management, nr. EIA-2001–01-ORG. ERIM, Erasmus University Rotterdam.

Robinson, S.L. (1996). Trust and breach of the psychological contract. *Administrative Science Quarterly,* 41, 574–599.

Rousseau, D.M. (1989). Psychological and implied contracts in organizations. *Employee Responsibilities and Rights Journal,* 2, 121–139.

Ruël, H., Bondarouk, T., and Looise, J.K. (2004). *E-HRM: Innovation or Irritation.* Utrecht: Lemma.

Snell, S.A. (1992). Control theory in strategic human resource management: the mediating effect of administrative information. *Academy of Management Journal,* 35, 292–327.

Storey, J. (1992). *Developments in the Management of Human Resources.* Oxford: Blackwell.

Sullivan, S.E. (1999). The changing nature of careers: a review and research agenda. *Journal of Management,* 25 (3), 457–484.

Verburg, R.M., Den Hartog, D.N., and Koopman, P.L. (2004). *Configurations of Human Resource Management Practices: A Theoretical Model and Empirical Test.* Paper presented at Academy of Management Meetings 2004, New Orleans, LA.

Verburg, R.M., Andriessen, J.H.E., and De Rooij, J.P.G. (2005). Analyzing the quality of virtual teams. In: M. Koshrow-Pour (ed.), *Encyclopedia of Information Science and Technology.* Hershey, PA: Idea-group publishing, 117–122.

Walton, R. (1985). Toward a strategy of eliciting employee commitment based on policies of mutuality. In: R. Walton and P. Lawrence (eds), *Human Resource Management, Trends and Challenges.* Boston, MA: Harvard Business School Press, 35–65.

Wood, S. (1999). Human resource management and performance. *International Journal of Management Reviews,* 4 (1), 367–413.

Yahya, S. and Goh, W.K. (2002). Managing human resources toward achieving knowledge management. *Journal of Knowledge Management,* 6 (5), 457–468.

Chapter 4

Cost and financial accounting in high-technology firms

Tom Poot

OVERVIEW

All business managers, including those involved in high-tech firms must make important decisions about financial issues, such as monitoring company performance, allocating budget funds and appraising capital investments. Financial goals and constraints play a major role in the day-to-day decision-making process of a high-tech firm. This chapter provides a brief introduction to management accounting and financial accounting. Because of the broad scope of accounting, we will introduce a few of the most important concepts of management and financial accounting, as a first step in understanding the importance of accounting for managerial decision-making in high-technology firms.

AN INTRODUCTION TO FINANCIAL MANAGEMENT OF HIGH-TECH COMPANIES

Why should an engineer bother about accounting?

Most technological development takes place within private enterprises. From the perspective of the enterprise, the development of technology is not a goal on its own, but a means to transform inputs into goods and services to be sold on the market and to earn profits. This is aside from other goals, such as business continuity or technological leadership. From an economic point of view, technology is one of the many resources that companies use to generate products or services.

One of the major features that inputs and other resources have in common is that they are scarce. Companies are often short on money (and other valuable assets) to buy all the inputs that they require. To put it differently, such companies must make choices on how to spend their limited resources, since the inputs to efficiently produce goods and services will inevitably exceed the resources available. The process of converting input into output (products or services) is called the production process.

To manage and monitor the production process a system of accounting is required. For instance, we need to know how much input is needed to generate a certain production volume. Typical questions that accountancy deals with are:

- If we want to generate products or services efficiently (to produce the largest output with the least input) how do we know if our production process is efficient?
- Are there any tools to measure efficiency?
- If the demand for a particular product increases, how can we increase the production?
- What additional costs are involved, how much input is needed, and do we still make a profit? After all, it is obvious that the conversion process varies depending on the technology used, the type of company, and the market conditions.

This is where accounting, business economics and financial management meet the world of technology. Engineers are confronted with all kinds of constraints in the process of solving technological problems. A technical solution has to meet not only technological but also economic and financial requirements. In this chapter we will concentrate on the financial requirements.

Management accounting versus financial accounting, two sides of the coin

To get a clear understanding of the tasks, methods and objectives of accounting, it is essential to distinguish between management accounting and financial accounting.

Management accounting focuses on the costs of production and on the operating aspects of the production process such as decision-making, control, performance evaluation and capital investment appraisal. Financial accounting, on the other hand, focuses on the overall financial position and results of the enterprise, such as its assets, liabilities, cash flows, profits and losses. Financial accounting copes with the legal obligation to inform the shareholders about the financial status of the company, their return on investment, and at informing financiers, suppliers and other stakeholders about the liquidity and solvency of the company.

Accounting and high-tech companies

In principle there are no differences in financial accounting among the various types of enterprise. High-tech companies have to meet the same accounting standards as other companies. Also, the same principles of management accounting and management control apply to every company. There is no distinction between high-tech enterprises and others. However, there are differences to consider, such as uncertainties about demand and financial constraints. This is the case especially when the high-tech enterprise is a start-up without a sound (financial or commercial) reputation, with limited financial resources, and with high expenditures for the development of new products or services. These differences and the consequences for accounting will be dealt with in the final section of this chapter.

MANAGEMENT ACCOUNTING AND MANAGEMENT CONTROL

Costs as management information

The basic approach used in management accounting is to identify the relevant costs of all activities related to the production of goods or services. This raises two questions: (1) what are costs, and (2) what is the relationship between costs and production activities? Costs can be defined as the monetary value of the inputs necessary to produce goods or services to be sold on the market. Note that we are talking about monetary value and not about expenses. In accounting there is a key distinction between costs and expenses. We will elaborate on this later on, when we discuss the accounting principles underlying financial accounting.

Cost calculation is important for three reasons:

1 as a tool to calculate profits or losses;
2 as a means to get information to support the management decision process and to manage the conversion process;
3 as a means to provide the data for cost determination, price setting and calculation of margins.

In practice, every company has to deal with changes in the prices of its inputs and outputs. It is important to know how changing input and output values affect the cost system. Higher input prices affect the costs of raw materials and can imply higher output prices if everything else remains equal. Most of the time, managers are keen to find compensating measures. For instance, if a company succeeds in increasing its input efficiency by lowering the amount of labor needed, the total costs per unit can be kept at the same level. Another scenario is to expand total production and thereby achieve economies of scale. This may also offset rising input costs. In fact, cost calculations are made to analyze current cost structures and to adapt to changing supply and demand and to new market opportunities. Before delving into this matter more deeply, we will first address the various categories of costs that enable a systematic approach to cost calculation.

There are many cost categories to consider. The transformation process differs from one company to another and likewise the way companies classify costs. To be useful as a tool for management decision-making, we have to categorize the various costs in a systematic way. In this chapter we will focus on the four basic categories that every firm uses:

- direct and indirect costs;
- variable and fixed costs;
- opportunity costs;
- sunk costs.

Direct and indirect costs

The first cost category focuses on the accountability of costs, distinguishing between direct and indirect costs. Aside from their "definition," costs can be described as the use of resources with a monetary value for a particular purpose. If a particular product involves the use of a single piece of machinery, the costs of using that machine (the operating costs) are categorized as the direct costs of that product. In a situation where a piece of equipment is used to produce more than one final product, direct costs do not provide an adequate description of the total cost structure. Instead of direct costs we then have to deal with indirect costs. The costs are not directly related to the manufacture of a single product when a firm produces more than one type of product.

Examples of indirect activities (activities not related to a single product) are the costs of R&D or of customer service. In the latter situation an allocation system has to be used to calculate the costs of a product or service. Suppose a company produces two products: a product that is developed with rather basic technology, and a second product that is very sophisticated and based on state-of-the-art technology. The first product generates 80 percent of total turnover, while the sophisticated product generates the remaining 20 percent. However, the engineering costs of the first product are only 10 percent of total engineering. It would be sensible in that case to allocate 90 percent of the indirect engineering costs of the R&D department to the sophisticated product, instead of allocating the indirect costs based on the percentage of sales. In this case the allocation of engineering costs is based on the amount of engineering labor hours needed to develop the products. In the customer service department, however, a more sensible allocation of costs would be based on the share of sales in total turnover.

In general, direct costs are mainly made up of the raw materials used to produce a particular product. Indirect costs consist of labor (unless part of the workforce is dedicated to the production of a particular product); building costs, finance costs and costs that are linked to more than one product. This implies that single product enterprises do not have indirect costs. Note that such firms are rare.

Variable and fixed costs

Variable costs are directly proportional to the volume of production whereas fixed costs are not. Examples of fixed costs are the rent of an office building, investments in machinery, and employees with a permanent contract. Variable costs include the costs of raw materials, electricity and temporary employees. When production increases, fixed costs remain unchanged whereas variable costs increase as the production level increases. Later on, when discussing the cost function, we will elaborate on the relationship between the level of production and the variability (or lack of) of costs. For now, we can conclude that, when cost driver activity level changes, the total fixed costs do not change; they do change per unit of activity. Total variable costs, on the other hand, will change, but the cost per unit (cost driver) will not change.

When talking about fixed costs it is important to realize that these remain the same as long as the production capacity is not exceeded. Fixed costs remain fixed only within a relevant production range. The boundaries of the relevant range do not only depend on the production capacity and the associated cost system but are also limited by time. For instance, a yearly change in tax rates or labor costs will change fixed costs and thus the relevant range.

Opportunity costs and sunk costs

The final two cost categories, opportunity costs and sunk costs, are especially relevant to capital investment appraisal. Opportunity costs are a special category of costs that have no relevance from a strict accounting perspective. From an economic point of view, however, opportunity costs are very important for the selection of business opportunities and investment projects.

Opportunity costs can be described as the costs incurred by spending the money on alternatives. Suppose a company has the opportunity to either invest in a piece of equipment or in a new production line. The costs of the investment consist of the purchase and installation of the machinery and the costs of personnel to operate and maintain the machinery. The benefits of the investment project are the sales of a new product. The investment project is feasible if the profits from sales exceed the initial investment. But even if the revenues exceed the investment, the question remains whether those benefits are substantial enough to justify the initial investment.

Would a different investment opportunity have a higher yield? Instead of investing money in a new piece of machinery with uncertain revenues, a company could place its money on a bank deposit and receive interest in return. The interest from a bank deposit is known in advance and involves less uncertainty. The interest earned on a bank deposit is called the opportunity cost of the investment project. From a business point of view, investment projects should generate revenues at least equal to or exceeding the interest from the bank to be feasible. For this reason opportunity costs are an important cost category in case of evaluating business opportunities or investments.

Sunk costs are costs that were incurred in the past and cannot be recovered. Such costs are therefore irrelevant for decision-making purposes. Expenses on marketing and on R&D are usually labeled "sunk costs." As these two types of expenses are a major concern for high-tech enterprises, it is useful to examine why costs incurred in the past are irrelevant for decision-making today. Suppose the R&D department is working on a new device to resolve a technical problem. Unless the technical problem is resolved, the device is not economically feasible. Management has to decide on an additional budget to continue the R&D effort to solve the technical difficulty, regardless of how much has already been spent on the R&D project. Of course the total amount of money spent on the project is important as a guideline for deciding on future projects.

Cost accounting systems

Costs are very much related to the process of converting raw materials or semi-finished goods (inputs) into final products or services (outputs) for the market. The conversion process consists of all activities that are related directly or indirectly to the production of goods or services. The classification of costs related to the conversion system is called the cost accounting system. The cost accounting system is the basic tool of management to extract financial information for decision-making and financial reporting purposes. As an example, Figure 4.1 shows the transformation process and some of the associated costs of a manufacturing company. The company produces various types of electronic switching devices, which are sold to electro-engineering companies. The devices are very similar as far as the cost system is concerned. In the example this means that the transformation process and the related cost system can be presented as the production of a homogeneous product. In the production of electronic switching devices, the company uses raw materials such as thin metal sheets of different alloys to produce resistors. The company does not manufacture all components itself. The company buys circuit boards and other components from other suppliers. These are depicted as indirect resources because the components are used in every product that the company manufactures.

Beside a direct path from inputs and conversion or transformation to outputs, most firms also have auxiliary departments or activities to support the main transformation process. These may include R&D, the design of products, services and processes, marketing, distribution and customer service. As an introduction to cost systems, we present the conversion process in Figure 4.1 as a single-stage process. A single-stage manufacturing process can be found in firms producing one type of product in different varieties. In this case all activities are more or less related to part of a particular product or service. Besides the resources with a direct link to

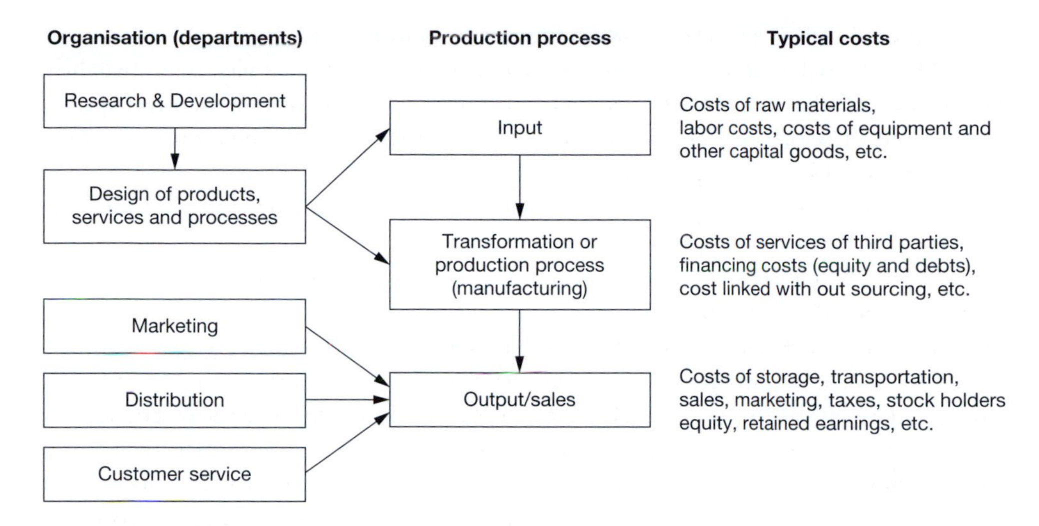

Figure 4.1 *The transformation or production process, a traditional cost system*

the transformation process, there are also indirect resources, such as factory overheads. Examples are factory buildings, auxiliary machinery use, administration and marketing. Labor costs can also be indirect, such as the costs of forklift drivers, plant guards, and customer service employees who provide internal services to all products produced.

In calculating product cost, those indirect costs have to be allocated as well, but this can be an arduous and costly task. More important than the activities or departments themselves are the related costs and their cost drivers. A cost driver is any output (finished or semi-finished products) that leads to costs or the use of costly resources. It is important to note that we are talking about causes of costs, meaning that we want to identify the relationship between an activity and the costs involved. Typical cost drivers are engineering hours, labor hours, kilowatt-hours of electricity, number of advertisements placed by the marketing department, and new product proposals issued by the R&D department. Cost drivers are a means of allocating different direct and indirect activities to products.

The way indirect costs and cost drivers show up in the cost system is closely related to the organization of the enterprise and the objectives of management. The above example shows that direct and indirect costs are the primary cost categories in the transformation process. It also shows that classifying costs as direct or indirect may be quite cumbersome when the transformation process becomes more complicated.

Activity-based costing (ABC)

The need to respond more quickly to the diverse needs of customers and the growing importance of product differentiation to meet those needs has forced many firms to alter their manufacturing processes. With the introduction of new types of manufacturing, such as Flexible Manufacturing Systems (FMS), Computer-Integrated Manufacturing (CIM) or Just In Time (JIT), the need for a different approach to management accounting has become apparent. In order to manage flexible manufacturing processes, managers need different types of information. For instance, information on the costs of expanding the product portfolio while maintaining the same level of efficiency as in traditional mass-production, single-item enterprises. The advent of FMS and multi-product firms has had a profound impact on the cost system and the associated accounting system. With the introduction of FMS, indirect costs are becoming more important, but it also becomes more difficult to calculate the exact cost of single items. In general there is a shift from traditional cost calculation to activity-based costing (ABC).

To illustrate the impact of multi-stage transformation on accounting, Figure 4.2 presents a simplified multi-stage transformation process that is rather common to large multi-product enterprises. This figure shows the transformation process of a fictitious company that produces electronic switches, current regulators and uninterruptible power supplies (UPSs), to illustrate the relationship between the organization of production, the production process and the accounting system. Figure 4.2 makes clear that the mix of activities that are directly and indirectly related to the various products constitute various cost drivers, each at a different level.

Figure 4.2 shows that there are various indirect resources, which may be connected to more than one activity. The cost drivers associated with indirect resources and activities are unique. In Figure 4.2 the cost drivers are depicted as arrows. Each step in the transformation process has cost drivers connected with it, and the transformation process is identified by the combination of

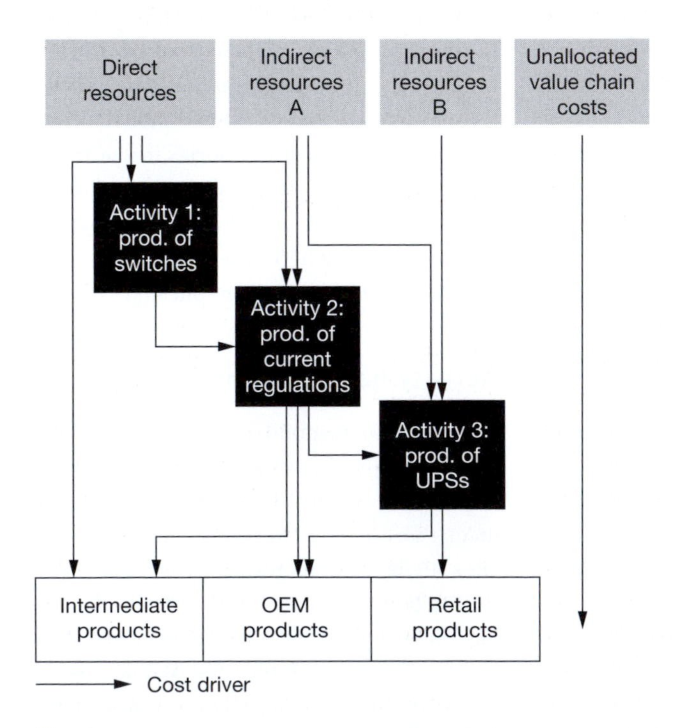

Figure 4.2 *Activity-based cost system (ABC)*
Adapted from Horngren *et al.* (2002)

associated cost drivers. In our example, the production of current regulators can be identified by the cost drivers associated with the indirect resources A, the costs associated with components (such as resistors, circuit boards bought from suppliers, the internal production of switches used as a component in the current regulators), and direct resources devoted to the production of current regulators, such as the labor needed to assemble the regulators. By adding all the cost drivers we can estimate the total costs of a particular activity and the associated products. The activity-based cost system provides the tools to analyze the transformation process and shows how it responds to changes in inputs and outputs. It also provides the information needed to prepare the financial statements that will be discussed later on.

Figure 4.2 also shows unallocated costs. These are costs for which we cannot identify a relationship to a cost objective. Examples of unallocated costs are expenses on R&D, process design, and legal, accounting or information services. Unallocated costs are associated with indirect or auxiliary activities that fall outside of the principal transformation process depicted in Figure 4.1. The major difference between Figure 4.1 and Figure 4.2 is that in the former we talk about departments and the transformation process, whereas in the latter we talk about economic activities, direct and indirect resources and cost drivers. Figure 4.2 portrays the accounting perspective of companies and their production processes. Depending on the organization of a company and its conversion process, some companies will speak of unallocated costs, while others may use the terms direct or indirect costs. Thus, the distinction between unallocated costs and direct and indirect costs is company-specific and to some extent arbitrary.

Technological developments in manufacturing have a profound influence on accounting practices. Especially, the introductions of Computer-Integrated Manufacturing (CIM) and Flexible Manufacturing Systems (FMS) have changed the way the costs of a manufacturing process are generated. Direct labor costs become less important because a large variety of products are manufactured with the same equipment. It is not the labor costs of operating a machine that are important but the amount of time the FMS needs to complete a product. It is not the direct labor costs, but, rather, the effective operating time of the FMS and Computer Numerically Controlled (CNC) machines that are the important cost driver. On the other hand, machine hours start to drive more overhead activities, and indirect costs rise. The accounting system has to be adjusted to the growing importance of indirect costs, and the overhead allocation base needs to reflect the actual operating time devoted to the production of a particular item (Bruggeman and Slagmulder, 1995; Tayles and Drury, 1994).

Cost–volume relationship

As said above, the classification of costs is a very important tool for decision-making. Identification of the relevant cost drivers provides the necessary insights into what costs are critical to the overall production costs. The direct and indirect costs provide management with information about the cost structure of an enterprise. What is lacking, however, is a dynamic perspective on the cost structure in relation to changes in the volume of production. A cost structure split into direct and indirect costs does not take into account the fact that the cost structure does not only depend on the organizational structure of the firm (and on the associated cost drivers) but also on the level of production.

Suppose a machine has a maximum capacity of 1,600 units per annum. In order to increase production to 2,000 units, a second machine is needed, increasing direct manufacturing costs by a factor of two. Total production, however, increases by merely 25 percent. Therefore, due to capacity constraints, the relationship between the total production costs is not necessarily proportional to the volume of production. An important aid for analyzing the relationship between costs and production volume is the cost function. The cost function and the analysis of cost behavior rely heavily on understanding fixed, variable, average and marginal costs. These cost categories enhance the accounting system and provide management information about the volume of production and the associated cost structure.

Fixed, variable, average and marginal costs

We will now focus on the stylized relationship between cost structure and volume of production (see Figure 4.3).

Figure 4.3 shows how average fixed costs (AFC), average variable costs (AVC), average total costs (ATC = AFC + AVC) and marginal costs (MC) relate to the relevant range of production (units of output). Variable costs vary in direct proportion to the volume of production. Fixed costs do not vary within a given range, but the fixed costs per unit (AFC) decrease as shown in Figure 4.3. Average variable costs, however, follow a unique pattern.

At low output levels, average total costs decrease as the level of production rises until a certain turnaround point. This point

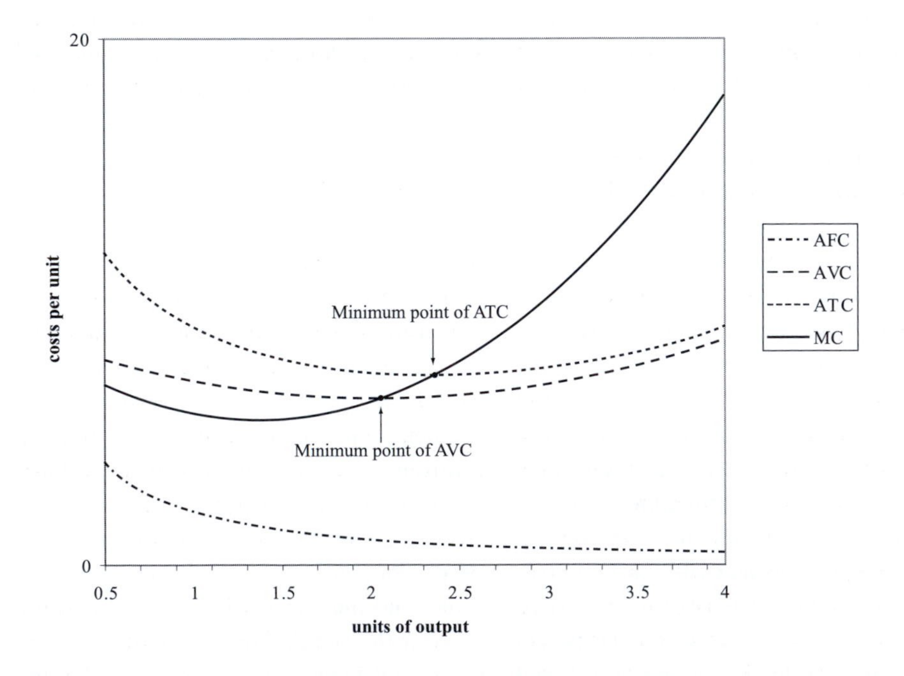

Figure 4.3 *The cost function of average fixed (AFC), average variable (AVC), average total (ATC) and marginal costs (MC)*

reflects so-called development reverses that cause average costs to rise with increasing production. At low levels of production the full potential of the production process is not reached, and economies of scale will improve the production process and will lower the variable costs per unit of output. The marginal costs are below the average total costs. This will increase efficiency. At some point, however, an optimum is reached. In this case the production process is at its most efficient level, producing output with a minimum amount of resources and costs per unit of output (ATC = MC). Raising production levels beyond this point will result in a diseconomy of scale, meaning that a larger production will be less efficient per unit costs. Average total costs per unit of production will thus rise, ultimately resulting in a loss. The pattern of average variable costs (AVC) can be explained by looking at the utilization of a machine in relation to the cost to operate. For most equipment, maintenance costs are rather constant over a certain range of production. When a piece of equipment is used far below its capacity (units per hour), the costs of operation (labor costs) are high in relation to the output produced. This ratio improves when the potential capacity is utilized to a fuller extent. When the equipment is used almost to its maximum capacity, the cost of maintenance tends to rise. The costs will rise even more sharply when the machine is operated at its maximum capacity. In this case the labor costs of operating the machine are still proportional to the units produced but the costs of maintenance rise to disproportionate levels. Thus, the average operating costs including labor and maintenance costs are relatively large when a piece of equipment is used below its capacity. At some level of production, the costs of maintenance tend to rise sharply, so the average costs will increase as well (see Figure 4.3).

Conclusion

Direct and indirect costs are used to calculate the cost of products or services, both for budgeting purposes and for management decisions regarding the organizational efficiency of an enterprise. Fixed and variable costs are used to relate costs to the volume of production. The cost function in which fixed and variable costs are combined provides management with information about the optimal volume of production. The cost curves that result from the analysis of fixed and variable costs show production costs at each level of output using the available technology.

FINANCIAL ACCOUNTING

Introduction

The primary objective of financial accounting is to communicate the financial status of a company to parties outside of the company, such as stockholders, suppliers, banks and other stakeholders. Clear and concise communication calls for a language that can be understood by outsiders. Therefore, financial statements must be in line with established accounting principles, so everyone can rely on their accuracy and trustworthiness. In order to provide a clear and accurate picture, the financial statements must be published according to US GAAP (Generally Accepted Accounting Principles) or IFRS (International Financial Reporting Standards). American companies and companies outside the US that are listed on the New York Stock Exchange, NASDAQ or another US stock exchange must comply with US GAAP and the regulations of the Securities and Exchange Commission (SEC). In Europe every country has its own legislation but the various regulations do comply more or less with IFRS. There are indications, however, that the IFRS rules and European accounting practices are gradually converging to US GAAP (Shields, 1998). The growing number of European companies that are listed on an American stock exchange reinforces this trend. For efficiency reasons these companies generally use the same accounting system to prepare financial statements that comply both with US GAAP and IFRS.

The three principal elements of a financial statement are the balance sheet, the income statement and the cash flow statement. Together with the explanatory notes these three elements together constitute the annual report. The explanatory notes describe the company's operating results during the past year and discuss business policy, new developments and future expectations.

To prepare the financial statements, the transactions of a firm have to be recorded. These bookkeeping records provide the essential figures that the financial statements are based on. Without a proper and accurate recording of transactions, financial statements will never meet US GAAP or IFRS standards. This implies that the bookkeeping of transactions and the allocation of indirect costs to finished products or services must comply with those accountancy rules.

As an introduction to the basics of financial accounting, the three statements will be discussed individually. The relationship that exists between the three elements will then be described, showing that the balance sheet, the income statement and the cash flow statement together provide the information needed to make an informed judgment about the financial health (or lack thereof) of a company.

Balance sheet

The first questions that we want to address are: what does a balance sheet look like, why is it important, and what information does it reveal and what not? We will clarify the concept of a balance sheet by discussing its main components. According to US GAAP and IFRS regulations, the financial statements have to be presented in a fixed order although the details vary from one company to another. In addition, the financial statements must include explanatory notes that provide the information that is necessary to make an informed judgment (see Table 4.1).

The balance sheet lists the assets that a company owns and the related liabilities and equity at a particular point in time. This may be the situation as per December 31, but not necessarily. In Table 4.1 the balance sheet contains information about the financial situation on December 31 of the years 2000 and 2001. It provides the reader with information about the changes that have taken place in an entire accounting period. The importance of having information on a full accounting period will be discussed later, in the section on the interpretation of the financial statements.

By definition, total assets must equal total liabilities plus equity, meaning that the two sides are in balance for every year listed. That is why it is called a balance sheet. This is called the balance sheet equation.

Assets (A) = Liabilities (L) + Stockholder's Equity (SE)

SE equals the original ownership claim plus the increase in ownership because of profitable operations (retained earnings), or:

Assets (A) = L + Paid-in Capital + Retained Earnings

The assets, shown on the left side of the balance sheet, provide information on what items a company has spent its money on. The

Table 4.1 *Abbreviated balance sheet*

Balance sheet per December 31					
Assets	2001	2000	Liabilities and Equity	2001	2000
Cash and securities	14.0	21.0	Accounts payable	84.0	42.0
Short-term investments	0.0	91.0	Notes payable	154.0	84.0
Accounts receivable	525.0	441.0	Accruals	196.0	182.0
Inventories	861.0	581.0	**Total current liabilities**	**434.0**	**308.0**
Total current assets	**1,400.0**	**1,134.0**	Long-term bonds	1,055.6	812.0
Net plant and equipment	1,400.0	1,218.0	**Total debt**	**1,489.6**	**1,120.0**
			Stockholder's equity	238.0	238.0
			Retained earnings	1,072.4	994.0
			Total common equity	**1,310.4**	**1,232.0**
Total assets	**2,800.0**	**2,352.0**	**Total liabilities and equity**	**2,800.0**	**2,352.0**

assets are listed in the order of their liquidity or the time it takes to sell assets and receive liquidities (cash) in return. The liabilities and equity tell us the sources of funds or how the assets are financed: externally (liabilities) or with the company's own financial resources (equity). The latter consist of equity from shareholders and profits from previous periods. There are two major groups of liabilities: debts to be paid on short notice (mostly within a year) and long-term debts. The liabilities are listed in the order in which they must be paid. In accounting, current means within one year. This applies to assets as well to liabilities.

We will now have a look at the individual lines of the balance sheet. On the assets side of the balance sheet we first find cash and securities. These consist of money and short-term loans that are secured by marketable assets. It will be clear that cash or bank balances are the only real liquidity a firm can have. Short-term loans can be sold to other parties in return for cash, although with a certain loss equal to the risk of uncollectibility. This means that the liquidity or money value of assets other than cash is not known in advance and has to be estimated (as if they were sold on the market). If a short-term loan or its marketable assets are regularly sold, its market price will be a reasonable indicator of its liquidity. Uncertainties about the liquidity of loans or their marketable assets tend to rise as the number of market transactions decreases. Although plant and equipment are listed as assets that can be sold, it will be obvious that their liquidity is very low. Only few companies, presumably operating in the same market as competitors, are potential buyers, but these have a strategic incentive to underbid and thus lower the value of the assets. The valuation of assets is beyond the scope of this chapter, but it is important to realize that the cash value of assets is more or less arbitrary. Various accounting rules under US

GAAP and IFRS refer to the valuation of assets.

Another asset category is accounts receivable. Accounts receivable arise whenever goods are sold on credit. Inventories consist of raw materials, work in progress and finished product that are not yet sold. Net plant and equipment consist of land and buildings, machinery, installations, fleets of cars and the like. Note that whenever goods are sold on credit (accounts receivable), two accounts are created: an asset item entitled accounts receivable appears on the books of the selling company, and a liability called accounts payable appears on the books of the purchaser.

On the liabilities and equity side of the balance sheet we find items such as accounts payable. Accounts payable are amounts owed to suppliers (to be paid within 30 days), while notes payable are cash owed to others (to be paid within 90 days).

Accruals consist of accrued wages and taxes (provisions for payments in the near future). Long-term debt includes bonds, bank loans and marketable bonds. Stockholder's equity is the nominal value of the company's own shares listed on the stock market or owned by other parties such as private firms, venture capitalists, pension funds or other public institutions. The value of the shares on the balance sheet is equal to the cash paid in return for the shares when originally issued. This value may differ from the price as listed on the stock market. The last line under equities is retained earnings, which is profit earned that will be used for future operating purposes.

Income statement

Unlike the balance sheet, which is a snapshot at a specific moment in time, the income statement records the flow of resources over

Table 4.2 *Abbreviated income statement per December 31*

Income statement	*2001*	*2000*
Net sales	4,200.0	3,990.0
Cost of sales (incl. general and administrative expenses, excl. depreciation)	3,662.7	3,495.8
Depreciation	140.0	126.0
Total operating costs	3,802.7	3,621.8
Earnings before interest and taxes (EBIT)	397.3	368.2
Less interest	123.2	84.0
Earnings before taxes (EBT)	274.1	284.2
Taxes (40%)	109.6	113.7
Net income available to preferred dividends	**164.5**	**170.5**
Preferred dividends	5.6	5.6
Net income available to common stockholders	**158.9**	**164.9**
Common dividends	80.5	74.2
Addition to retained earnings	78.4	90.7

time and provides information about earnings and expenses. In the example presented in Table 4.2, the income statement reports on operations during the calendar year. However, an income statement can be prepared over any period of time, for instance quarterly, thus providing management information more frequently and giving the opportunity to respond faster to changes in the financial situation of a company.

The first line in the income statement is net sales. In this abbreviated income statement only the most important expense categories are listed. The cost of sales, or the expenses needed to realize the products that are manufactured, is the largest expense category. In this example the cost of sales also includes selling expenses such as marketing or distribution. Depreciation expense is listed on a separate line as it is a non-cash adjustment to total earnings. Depreciation involves no money transfer. Nonetheless it is an expense and has to be deducted from net sales in order to arrive at earnings before interest and taxes (EBIT).

The issue of non-cash expenses such as depreciation raises the question how income must be measured. The determination of income is based on the principle of accrual accounting. Accrual accounting recognizes the impact of transactions on the financial statements in the period when revenues are generated and expenses are incurred instead of when the company receives or pays cash. Put differently, revenues and expenses are recorded in the period when goods or services are sold, irrespective of the timing of the related cash flow. The accrual principle is the framework for matching accomplishments (revenues) with efforts (expenses).

In the case of fixed assets it is not the initial investments (procurement) that are important, but the contribution over time to production or revenues. The preferred approach is to spread the costs over the expected useful time of the assets in a linear fashion, although other methods of depreciation do exist. Depreciation represents the monetary value of the contribution of fixed assets to the sales in a particular accounting period. In this example the total operating costs consist of the cost of sales and depreciation.

Subtracting the total operating costs from net sales yields earnings before interest and taxes (EBIT). EBIT is a popular indicator among financial analysts of the profitability of the operating activities of enterprises and is used in business and share valuations. Because important expense categories are left out, this indicator is today regarded as questionable. Nonetheless, it is still widely used.

Important expense types are interest and taxes, which are listed on the next lines. Deducting these from EBIT yields the net income available to the owners of the company. The income statement shows that not all stockholders are equal as a distinction is made between preferred and common stockholders. The preferred stockholders have special rights such as guaranteed dividends or special voting rights to influence board decisions. The common stockholders, on the other hand, do not have special rights. Sometimes they are not even entitled to dividends, depending on the dividend policy of the company.

On the final line we see that a substantial part of the income available to common stockholders is added to retained earnings. The retained earnings provide the company with the cash needed to purchase new inputs and to pay debts and labor costs. The balance sheet shows us that although total turnover or net sales increased from 2000 to 2001, the addition to retained earnings actually decreased. The board of directors decided to raise the dividend, apparently to increase the share price. In general, a higher dividend tends to elevate the price of the share. The price that investors are willing to pay to buy shares depends, among other things, on the future value of dividends or profits. The higher the current and expected dividends, the more attractive the shares will be.

Cash flow statement

The financial statements also include a cash flow statement. Some textbooks treat sources and uses statements separately from the cash flow statement. Both statements provide an overview of how a company obtains its cash (sources) and how it spends its cash (uses) during an accounting period. We will only discuss the cash flow statement here. This statement presents information on the relationship of net income to the changes in cash balances. The cash flow statement reports on past cash flows as an aid to evaluating the generation and use of cash, determining a company's ability to pay interest and dividends, and to repay debts when they are due, or to predicting future cash flows and operating results.

At this point one might ask why we need a cash flow statement on top of the income statement. A major problem with the income statement is that it includes accruals that do not constitute cash flows, such as depreciation. Only cash flows that relate to the sale of goods are included. Other cash flows, such as investments or financial transactions, are excluded from the income statement, even though these do have consequences for the company's cash position. At best the income statement provides only a partial view on the cash flows. The cash flow statement, on the other hand, lists all cash flows: not only those from operating activities, the primary concern of the income statement, but also those from investment and financing activities. As a matter of fact, the cash flow statement is structured along operating, investment and financial activities. According to US GAAP or IFRS, all cash activities should be listed within these categories.

The fundamental approach to the cash flow statement is simple: list the activities that have led to an increase of cash during the accounting period, those that have led to a decrease of cash, and, finally, put the cash flows in the proper category (operating, investment and financing activities). The cash flow statement starts with cash flows from operating activities. As the list indicates, cash flows from operating activities are generally the effects of transactions that impact the income statement, such as sales and wages. Depreciation is a non-cash expense that was deducted when calculating net income. Therefore it must be added back to show the correct cash flow from operations. Working capital is made up of current assets minus current liabilities. Adjustments to working capital affect cash flows because an increase in a current asset decreases cash, and an increase in a current liability increases cash.

The cash flow in the statement below reveals an important message. Although the net cash flow (net income plus non-cash adjustments such as depreciation) is positive (€164.5 + €140.0 = €304.5), the net increase in working capital is very large, resulting in a negative cash flow from operating activities (minus €3.5). If this negative trend persists, this company will eventually go bankrupt even though it may well report a net profit in the income statement at the same time. Profits as reported on the income statement can be "massaged" by such tactics as depreciating assets too slowly, not recognizing bad debts promptly, and the like.

The second section deals with the cash flows generated by investment activities. In this example we only have investments in long-term fixed assets. In many cases, a company is also involved in acquisitions, takeovers or divestments. The resulting cash flows would be listed in this section.

Financing activities include resources obtained from creditors and owners and providing owners with returns on their investments in the form of dividends. The financing activities consist of cash inflows such as borrowings from banks and issues of new shares, and cash outflows such as repayments of short-term and long-term borrowings, repurchase of equity shares and payment of dividends.

The listing and classification of activities is not without problems. Quoting from Horngren *et al.* (2002, p. 639):

> Perhaps the most troublesome classifications are the receipts and payments of interest and dividends. After all, these items are associated with investment and

Table 4.3 *Abbreviated cash flow statement per December 31*

Cash flows from operating activities	*2001*
Net income	164.5
Adjustments to reconcile net income to net cash provided by operating activities	
Depreciation and amortization	140.0
Deferred income taxes	0.0
Due to changes in working capital	
Adjustments in accounts receivable	–84.0
Adjustments in inventories	–280.0
Adjustments in accounts payables	42.0
Adjustments in accruals	14.0
Net cash provided by operating activities	**–3.5**
Cash flows from investing activities	
Additions to property, plant and equipment	–322.0
Acquisition or disposition of business	0.0
Net cash used by investing activities	**–322.0**
Cash flows from financing activities	
Net change in short-term borrowings	70.0
Net change in long-term borrowings	91.0
Adjustments in stockholders equity	243.6
Cash dividends paid	–86.1
Net cash provided by financing activities	**318.5**
Net adjustments in cash	**–7.0**
Cash at beginning of accounting period	21.0
Cash at end of accounting period	14.0

financing activities. After much debate, the Financial Accounting Standards Board (FASB) decided to include these items with cash flows from operating activities. Why? Mainly because they affect the computation of income. In contrast, payments of cash dividends are financing activities because they do not affect income.

This quote implies that financial statements are prepared in accordance with accounting rules that are more or less arbitrary. Many rules do exist in order to ensure consistency and to avoid disputes. For reasons of understanding and clarification, every financial statement has to comply with the mandatory rules.

Preparing and interpreting financial statements

As stated before, the basic objective of financial statements is to inform management, stockholders, debtors and other stakeholders about the financial position of the company. The financial statements have to comply with the accounting standards in order to provide an accurate and reliable picture of the financial position of the company. In this section we will discuss how the balance sheet, the income statement and the cash flow statement are related and what these tell us about the financial position of a firm.

Recall that accrual accounting is the framework of the balance sheet and the income statement. The income statement measures the performance (sales minus associated expenses) during a specific period (weekly, monthly, quarterly, or any other time interval). The balance sheet takes a snapshot at a certain moment, usually at the end of the accounting period. Therefore, adjustments are required. To measure income under the accrual principle, accountants have to make adjustments to deal with implied expenses such as unpaid wages, prepaid rent and interest earned. These expenses are not associated with day-to-day activities but must be taken into account in order to properly measure the efforts and revenues. Following Horngren *et al.* (2002), we identify four types of principal adjustments:

1 expiration of unexpired costs;
2 recognition (earning) of unearned revenues;
3 accrual of unrecorded expenses;
4 accrual of unrecorded revenues.

Assets are resources waiting to be used in the production process and are expensed when the products manufactured with the aid of those assets are sold. Unexpired costs are assets that are expensed in the future and that expire when sold. Other examples are equipment and various prepaid expenses such as prepaid insurance and prepaid property.

An example of the second type of adjustment is a customer's payment in advance for products to be delivered later. In principle, the cash received has to be recognized as a liability, because the company still has to deliver the goods. This means that on the balance sheet the money from the advance payment will be recorded as an asset (cash) and as a liability (the value of the goods to be delivered). Note that adjustment type 2 mirrors adjustment type 1. Wage expense for wages earned by employees but not yet paid is an example of an accrual of unrecorded expenses (3), while interest earned but not yet received is an example of an accrual of unrecorded revenues (4).

The relationship between balance sheet and income statement

As a matter of convenience we start with the balance sheet of the previous accounting period. This provides us with information about assets and liabilities and equity. In order to prepare the balance sheet of the current period we have to know what happened in between. To prepare the financial statements, accountants record the transactions (revenues and expenses) of the company. The income statement lists all revenues and expenses. Using the four principal adjustments listed above, revenues (sales) are matched with the expenses needed to realize those sales.

The last line of the income statement shows the entry "retained earnings" of the accounting period. Table 4.2 shows a positive net income of €78.4. These retained earnings are added to the equity of the company. Both the balance sheet and the income statement have this entry in common. This is therefore the link between the balance sheet of the previous accounting period and the current one. To prepare a new balance sheet, the retained earnings have to be adjusted accordingly. The bold arrow in Figure 4.4 shows this primary link between the income statement and the balance sheet. Thus, on the balance sheet total equity increases from €994.0 in 2000 to €1,072.4 in 2001 (Table 4.1).

However, this presents a problem with the balance sheet equation, which requires that total assets equal total liabilities and equity. If the equity is adjusted according to the income statement, other lines have to be adjusted as well. There are many possibilities for this. Net income can be used to buy assets, or to repay debts to lower total liabilities, or both. The income statement does not provide the information necessary to adjust the other lines of the balance sheet and thus to get assets and liabilities plus equity in balance. For that information we need the cash flow statement. Figure 4.4 shows the principles of the adjusting process of all financial statements in a glance.

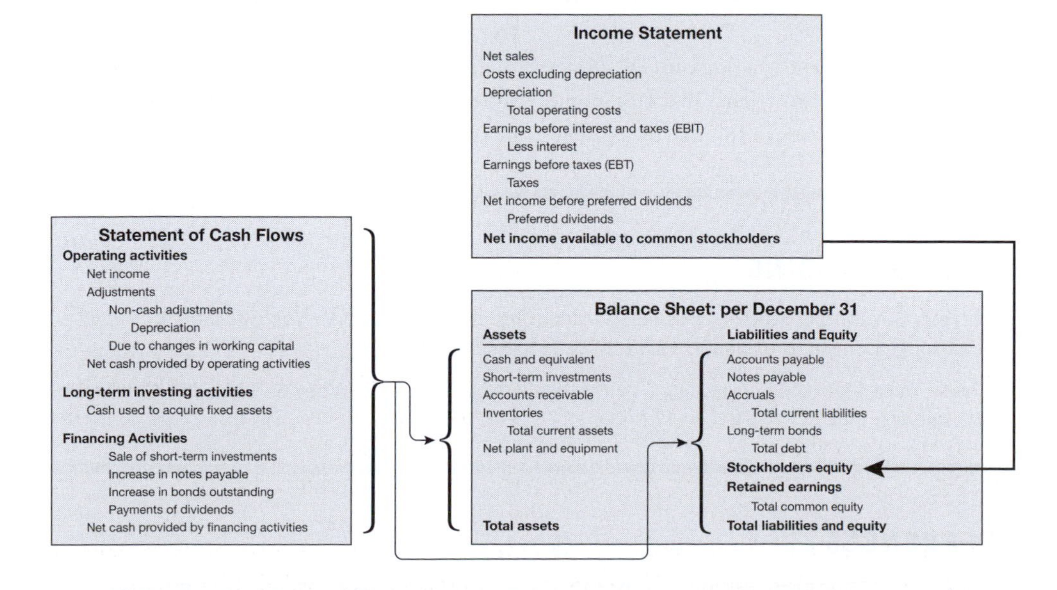

Figure 4.4 *The accounting system*

To illustrate the role of the cash flow statement, we will look at the entry "Net plant and equipment." From the balance sheet we know that in 2000 the firm had €1,218 in net plant and equipment. How should we adjust this item in the 2001 balance sheet? The cash flow statement tells us that due to investment activities, total property, plant and equipment has increased by €322. At the same time, the operating activities category shows us that depreciation expense of €140 was recorded. The net increase in net plant and equipment is thus €182 (€322 minus €140). The balance sheet is therefore adjusted accordingly. From 2000 to 2001 the value of net plant and equipment increased from €1,218 to €1,400.

ACCOUNTING IN THE HIGH-TECH INDUSTRY

Uncertainties about technology and market prospects, the limited ability to attract money, cash shortages, and managing growth are the major challenges that a start-up firm has to cope with. The question is: has financial management anything to say about this? Yes, it has!

If a high-tech firm is a start-up, it has to deal with financial constraints that larger and more established firms do not have. The development of high-tech products involves many uncertainties: will all technological problems be resolved, and if so when and at what cost? Well-established firms with longtime experience may be better aware of the technological and commercial risks involved in order to anticipate.

Very innovative products and services, i.e. products and services that are unlike existing products and services, have to establish a position on the market. Established firms with a longstanding reputation for good products or services may have an advantage over the market-innovative products of start-up firms. Start-up firms are occasionally confronted with high levels of investments and at the same time with cash flows that are not adequate to finance ongoing research and development. On the balance sheet we see very few assets and perhaps a high debt ratio (higher liabilities compared to assets).

Banks and other financiers are reluctant to finance start-up companies without a track record of R&D, sales, cash flow and profits. In other words, start-ups face tremendous financial challenges in bringing their ideas to the market and knowledge of the financial rules of the game may help them to survive.

FURTHER READING

Catherine Gowthorpe (2003), *Business Accounting and Finance for Non-Specialists*, Thomson Learning, London, 640 pages, ISBN 1–86152–872–8, £30.

Michael Jones (ed.) (2002), *Accounting for Non-Specialists*, John Wiley & Sons Ltd, Chichester, 595 pages, plus XIX, ISBN 0–471–49572–7, €45 (£29.95).

REFERENCES

Brigham, E.F. and Ehrhardt, M.C. (2002). *Financial Management, Theory and Practice*, 10th edition. South-Western: Thomson Learning.

Bruggeman, W. and Slagmulder, R. (1995). "The impact of technological change on management accounting," *Management Accounting Research* (6), 241–252.

Horngren, C.T., Sundem, G.L. and Stratton, W.O. (2002). *Introduction to Management Accounting*, 12th edition. New Jersey: Prentice Hall International.

Shields, M.D. (1998). "Management accounting practises in Europe: A perspective from the States," *Management Accounting Research* (9), 501–513.

Tayles, M. and Drury, C. (1994). "New manufacturing technologies and management accounting systems: Some evidence of the perceptions of UK management accounting practitioners," *International Journal of Production Economics* (36), 1–17.

Chapter 5

Foundations for successful high-technology marketing

Jakki J. Mohr, Stanley F. Slater, and Sanjit Sengupta

OVERVIEW

Companies operating in the high-tech marketplace live on the edge of revolutionary changes that transcend industry boundaries. In order to truly capture the promise that new technological developments have to offer, companies must take careful steps in the commercialization and marketing of their innovations. Therefore, the purpose of this chapter is to provide the reader with a solid foundation of the critical ingredients of successful high-technology marketing. Topics covered include delineating the scope of marketing activities in the high-tech company, understanding what it means to be market-oriented and customer-focused, and overcoming barriers to successful high-technology marketing. By the end of the chapter, readers should understand that marketing is more than a set of activities; it operates as a philosophy of making decisions in the high-tech organization. Boxes embedded in the chapter and suggested resources are meant to provide supporting detail to the key concepts.

Consider the case of a new company that has developed a new software program to automate the labor-intensive process of creating meaningful images from digital satellite data. Known as feature automation software, the sophisticated computer algorithm on which it is based provides incredible savings in time and labor costs for customers who use or manipulate GIS (geographic information systems) data across a broad range of market applications, including military, forestry (fire fighting and vegetation mapping), research labs, agriculture (vegetation management), weather forecasting, and emergency response, to name a few. Key questions facing this new high-tech start-up are:

- What industry application(s) should it target with its initial commercialization efforts?
- How sophisticated should the feature set be for the initial release of the software?
- Should the new company consider partnerships to assist with its initial commercialization efforts?

- What will be the likely response of established industry players to this potentially disruptive technology?
- What resources can the new start-up bring to bear in the marketing effort? As importantly, who will bring needed expertise to inform the marketing effort?

This particular company initially decided to target forestry applications, as the software algorithm had been developed by a computer scientist on a university campus where the Forestry Department was known internationally for its work in collecting and manipulating GIS data for various forestry-related applications. The founder of the company (the computer scientist who developed the algorithm) also wore the hat of "chief marketing officer."

In talking to potential customers, the initial reaction of customers was "this is too good to be true; the software couldn't possibly do what you are telling me that it can." Even with simulations and demonstrations, customers thought there must be some hitch. The possibility of saving 99 percent of the time and effort in a labor-intensive process via software automation seemed implausible to customers whose mind-set was based on the established, labor-intensive way of manipulating the digital data into useful images. Because customers were disbelieving, the owner became somewhat disheartened, realizing that bringing a new high-tech product to market was going to be more difficult than previously thought.

This opening vignette captures the complexity of the situations facing many new high-tech start-ups. Although new technologies often offer compelling benefits over incumbent technologies – benefits that innovators believe should be obvious to customers – customers are "balky" (Dhebar, 1996). Because "balky" consumers have doubt and uncertainty about whether the new technology will truly function as promised, they hesitate to embrace the new technology as quickly as developers expect. Moreover, it can be extremely difficult to unseat established competitors whose revenue streams are based on the legacy technology or practices. Firms who face the threat of obsolescence with new innovations will work fiercely to protect their turf. Because of this, the prospect of potential partnering arrangements with industry incumbents poses something of a Catch-22. On the one hand, the new company can benefit from a partnership with an established company who has a known presence in the market. On the other hand, new companies can also face the very real risk that their partners will not push the new technology quite as quickly or aggressively as they might otherwise, due to revenue streams derived from the legacy technology.

Additional complications arise from the reality that many high-tech companies, be they large or small, tend to be more technology savvy than marketing savvy. What this means is that the requisite skills, resources, and expertise to effectively market the new technology are lacking or underdeveloped and, as a result, the company is not fully equipped to handle the difficulties it will face.

The combination of these factors creates an extremely difficult situation that looks something like double-jeopardy: the complicated environment in which high-tech marketing occurs implies that the need for marketing prowess is greater than in less complicated environments. At the same time that high-tech companies need this more sophisticated marketing prowess, these very companies either: (1) lack the talent and expertise, or (2) in the cases where experienced marketing personnel are hired, do not give all the support and resources marketing personnel need to be effective in their role. The engineering brilliance that created the new innovation in the first place takes on a higher status in the organization relative to the needed marketing skills. Either implicitly or explicitly, the preference for engineering-related knowledge and skills becomes a type of core rigidity, a barrier to the cultivation of marketing talents and expertise (Leonard-Barton, 1992).

Research on the role of marketing in technology-driven environments is very clear: technological superiority alone does not guarantee successful commercialization of high-technology products. Rather, it is the combination of marketing acumen and technological superiority that leads to successful commercialization of high-tech products (e.g., Dutta *et al.*, 1999; Gatignon and Xuereb, 1997). Moreover, if marketing is viewed as an afterthought in the technology commercialization process (e.g. develop the technology first, think about which customers to market/sell it to after development is done), the odds of successful commercialization are only 1 in 60 (Booz Allen and Hamilton, 1982). However, when marketing considerations are viewed as part-and-parcel of the development cycle, the odds improve to 1 in 7. (Note that these are still rather low odds.)

In light of this very complicated environment in which high-tech marketing occurs, the purpose of this chapter is to overview the issues and considerations with which high-tech firms must wrestle in order to create a supportive marketing environment that enhances the odds of successful commercialization. In particular, the specific objectives of this chapter are to address:

1 the scope of marketing activities in high-tech firms;
2 the characteristics of a market-oriented firm in high-tech industries, and the benefits of and barriers to becoming market-oriented;
3 critical barriers to marketing success in high-tech firms and how to overcome those barriers.

A caveat: the specific tools and strategies that a high-tech marketer can bring to bear in successfully penetrating the marketplace (such as segmenting the marketplace; selecting a specific target market; refining the value proposition; developing an advertising campaign; or establishing a pricing structure, for example) require more detail than a single chapter can give. So, interested readers are referred to the suggested resources at the end of this chapter to gather additional detail on these particular aspects of high-tech marketing.

SCOPE OF MARKETING ACTIVITIES IN HIGH-TECH FIRMS

At its heart, *marketing* is viewed as an organizational function and a set of processes for creating, communicating, and delivering value to customers, and for managing customer relationships in ways that benefit the organization and its stakeholders. Marketing issues are salient at three levels in any company, as shown in Exhibit 5.1.

Strategic. The first is the strategic level. At the strategic level, the high-tech company must consider issues such as:

- In which market will we compete?
- Which segments will we serve?
- What will our competitive position in the marketplace be (relative to the established technologies and ways of doing things)?

EXHIBIT 5.1 SCOPE OF MARKETING ACTIVITIES

1 Strategic: Proactive consideration of where the best opportunities in the market lie, and how to best develop and position the company's products to enhance the odds of success; decisions that guide the thrust of the company's efforts in the marketplace.
- In which market will we compete?
- Which segments will we serve?
- What will our competitive position in the marketplace be (relative to the established technologies and ways of doing things)?

2 Functional: Focus on marketing as a function (including the marketing mix, or the 4 Ps), as well as the product development function specifically; issues include the level of collaborative cross-functional interaction between various departments in the company, including:
- Interaction between the technical development teams and personnel charged with marketing responsibilities (R&D/Marketing interaction).
- Coordination between operations, customer service personnel, manufacturing/production, and marketing and product development.
- Delivery by personnel in all functional areas of a satisfying customer experience in all customer interactions ("moments of truth").
- Use of market-based information to guide decisions across all functional units.

3 Tactical: Development and implementation of marketing tools; executed consistently with strategic and functional decisions.
- Development of marketing brochures, collateral materials, website, etc.
- Decisions regarding which trade-shows to attend, where to place advertisements, and so forth.

These questions must be proactively considered in charting the firm's strategic direction. In addition, because the needs of customers in different market segments will vary, and because the company's relative position (vis-à-vis competitors) will also vary by segment, the answers to these questions provide guidance to the product development team.

In some larger high-tech companies, there may exist a strategic planning group or a formal marketing department that has primary responsibility for answering these questions. In other companies, the consideration of these questions may be addressed by the top management team (such as the company's founders) or by a product development group. Regardless of who in the firm provides insight into these questions, it must be done. However, despite the need for strategic direction in allocating the firm's resources across customer market segments and across product development efforts, all too many high-tech companies do not proactively wrestle with these questions. Rather, they may find that there is no clear consensus within the company about where to focus efforts and, as a result, the company's forays into the marketplace are diffused across multiple market segments and product development projects. Ultimately companies may see this diffused approach as a way to hedge their bets in the marketplace (so to speak); however, it typically is a recipe for disaster. Because the company never comes to truly understand in a rich fashion the specific customers' needs in any one segment, and because its efforts are not focused, this diffused approach means that the company is less likely to succeed in any area at all than when it has proactively defined a strategic direction.

So, to effectively operate at the strategic level, the company first must ensure that responsibility for strategic guidance is formally vested with some group in the organization (be it a marketing department, product management group, strategic planning group, or top management team). Moreover, the company must commit to developing a competence in market segmentation, targeting, and positioning (e.g., through hiring qualified personnel, relying on a consultant, reading appropriate resources, attending trainings, or even seeking the assistance of a capable university student as an intern or for a university-related project). Finally, the company must be willing to implement the decisions arising from the strategic planning process with focus, discipline, and requisite resources. The strategic marketing challenges in a global market are exacerbated by discernible differences in consumer needs and behavior across major regions such as Asia, Europe, and North America. For example, US consumers lag consumers in certain Asian and European countries in the adoption of specific technologies (Yang *et al.*, 2004).

Functional. The second level of the scope of marketing activities encompasses the functional area of marketing, as well as the product development function specifically. As stated by Peter Drucker (1954), there are only two functions in any organization: marketing and innovation, both of which create a relationship with the customer. In that sense, there are only two types of people in any organization: those who serve the customer, and those who serve those who serve customers (Bruner *et al.*, 1998). This focus on the functional level highlights three critical areas: what the marketing function encompasses (the 4 Ps, discussed next); who performs the marketing activities in the organization; the effectiveness of cross-functional collaboration within the organization, and the degree to which the various functional units rely on market-based information to guide their decisions.

The functional area of marketing includes four arenas (known as the 4 Ps of marketing, or the marketing mix): Product, Price, Promotion, and Place (see Box 5.1 on The 4 Ps of Marketing). Note that the product

BOX 5.1 THE 4 PS OF THE MARKETING MIX

Marketing includes four sub-areas:

1 *Product decisions*: Product decisions address issues related to the new product development process (innovation management); licensing strategies with potential partners; intellectual property rights; services provided to augment the revenue stream from base-products; product name/brand decisions; development of complementary products by partners; creation of industry standards; packaging; and so forth. The critical need is to develop a stream of products with the right set of features to satisfy customer needs in a compelling yet simple fashion.

2 *Price decisions*: Price decisions establish price points for the set of a company's products, and must address issues related to the cost to produce/manufacture the goods; margins along the distribution channel; competitors' prices (pricing relative to a specific firm's market position); customer value; total cost of ownership for the customer; prices for product bundles; and profitability.

3 *Promotion decisions*: This area of the marketing mix includes advertising (both media and messaging decisions), sales promotion (price deals, trade incentives, etc.), personal selling (recruiting, training, compensating sales people), and public relations/publicity (garnering favorable trade press, attending trade shows, engaging in cause-related marketing, etc.). Specific issues can include developing a strong brand name, decisions about the timing and focus of new product pre-announcements, co-branding decisions with potential business partners (including cooperative advertising with channel members), leveraging the Internet and other new media to gain awareness, developing collateral materials, and so forth.

4 *Place decisions*: Place decisions focus on getting the right product to the right customers at the right time, and are commonly known as distribution channels and supply chain management. Good channels strategy is focused on effectively meeting end-user customer needs in a cost efficient fashion. However, successful channels can be difficult to attain when channel partners often have different objectives, margins create conflicts between channel members, and new channels (such as the Internet) can cannibalize revenues from existing channels. Best practices supply chain management is demand-driven, harmonizing upstream logistics and manufacturing with end-user requirements

Example of the 4 Ps: LG Electronics markets Liquid Crystal Display (LCD) flat panel TV sets in many different countries. The following provide examples of the types of functional decisions it has to make in each country market for each of the 4 Ps:

1 *Product.* What diagonal screen sizes should it offer? 17 inch? 23 inch? 30 inch? Larger? How will the product be designed to fit aesthetically into the consumer's home?

2 *Price.* What should be the price for each LCD model given the costs of product development, the expectations of consumers, and the desired positioning relative to competitors?

3 *Promotion.* How should the value proposition be communicated to the target market, through what combination of offline and online media, trade shows, and press releases?

4 *Place.* Which retail chains should be allowed to carry the product to reinforce the brand and its positioning? Should there be a website and how should it complement offline channels?

The critical issue in managing across the marketing mix is ensuring consistency in all decisions that support the product's position in the marketplace. For example, a product positioned as high-quality must have a price that conveys that image, with high-end distribution channel members that provide appropriate levels of support and service, with advertising message and media focused on the premium image.

development arena is considered to be a subset of the marketing function.

Most high-tech firms are founded and originate because they have some superior technology that has been developed that will "revolutionize" the industry. Typically, a product development group – comprised of scientists, engineers, or programmers, for example – is charged with research-and-development activities. Some companies also vest the product development group with other marketing activities (such as collecting market research, conducting market segmentation, targeting and positioning activities, etc.).

Other high-tech companies may have a formal marketing department. However, even with a formal marketing function, the input of the marketing personnel is sometimes neither solicited nor (if solicited) respected/valued by personnel in other functional areas. Some of the disparaging comments can be found in the form of jokes, which can capture common stereotypes (Workman, 1993). For example:

> What is marketing? What you do when your products aren't selling themselves.

The implication of this "joke" is that marketing is not something one should have to bother about unless the product is not good enough to "sell itself." However, this belief operates on the assumption that marketing equates to selling, but as we are learning in this chapter, marketing operates at a much more strategic, fundamental level in the organization. Regardless of who is responsible for it, a key aspect of marketing is to carry the voice of the customer into the decisions, so that a customer-orientation permeates the company's efforts, from initial product development, through production, and eventual commercialization.

Another consideration at the functional level is the degree to which various units in the organization collaborate effectively. For example:

- How effectively do the technical development teams (software developers, information technology, engineers, and scientists) interact with personnel charged with marketing responsibilities?
- How effectively do operations, customer service personnel, manufacturing/production, etc. interact with marketing and product development?

One useful way to assess the level of functional collaboration between technical personnel and marketing personnel (or, in the case of a very small company, to assess the level of integration of a marketing philosophy into the day-to-day thinking of the personnel) can be seen in Table 5.1. Scoring on the questions in the table can give an indication of the level of functional collaboration in a company. Achieving R&D/marketing integration is a worthy goal, and strategies to accomplish this can be found in Mohr *et al.* (2005).

A high-tech company that is able to effectively coordinate its efforts across functional units with the goal of delivering superior customer value as the ultimate objective is more likely to be successful than firms that try to optimize decisions within each functional area in a more piecemeal or independent fashion. When functional areas have independent goals and objectives, it can lead to incompatibility across functional units, conflict, and ultimately, the erosion of a customer-based focus.

Attention to collaborative cross-functional interaction between all units in the organization (operations, customer service,

Table 5.1 *Questions for marketing/R&D interaction*

Rank your company on the following attributes (1 = Strongly Disagree; 5 = Strongly Agree)

- I understand the role of marketing in the product development process.
- Marketing is an integral part of the product development process.
- Marketing is more than an afterthought to product innovation.
- Marketing and engineering have a common language for talking about customer needs.
- There is mutual respect for the knowledge my marketing/engineering counterpart brings to the table.
- Knowledge of customers and markets permeates our corporate ethos.

Note: Higher scores reflect a more collaborative cross-functional environment

accounting, sales, marketing, engineering/ product development, technical support) is especially important in high-tech companies. Recall that high-tech environments are characterized by a set of complicating factors: customer doubt about how to use the technology and whether it will function as promised; backlash from established competitors whose legacy technologies may become obsolete, and so forth. In such environments, every single touch-point that a customer has with a company becomes a "moment of truth" (Bruner *et al.*, 1998); what this means is that the company either responds effectively to the customer's needs in that specific instance and, therefore, strengthens the relationship with the customer, or responds ineffectively to the customer request, which undermines the relationship with the customer. Regardless of which functional area is handling the customer interaction at any point in time, company personnel in all departments must be trained to understand that each individual has the capability to either support the company's efforts in cementing customer relationships (through responsive customer handling) or damage those relationships.

A final concern in many high-tech companies with respect to this functional level is *the degree to which the various functional units in the organization use market-based information to inform and guide their decisions*. The importance of using market-based information, and how to infuse the company with this competence, is covered in the next section on market orientation. The issues here pertain to the ability of the firm to gather, use, and disseminate market-related information to guide all functions in the organization.

Tactical. Finally, at the tactical level, the actual implementation of specific marketing tools is accomplished, such as the development of marketing brochures, collateral materials, and a website, decisions regarding which trade-shows to attend, where to place advertisements, and so forth. Many high-tech companies equate "marketing" with only these tactical considerations, and they relegate marketing input to reactive development of communications devices. Hence, the plea, "We need to hire a marketing person," essentially means hiring someone to operate at the tactical level. Companies who operate only at this tactical level likely have not made some of the harder strategic decisions to guide the company's efforts, and in that sense, are less likely to be successful in the marketplace (regardless of how effective their advertising or trade-show strategies, for example, may be).

In summary, companies who view marketing input as a strategic consideration recognize that success in high-tech markets comes from proactive consideration of where the best opportunities in the market lie, and how to best develop and position the company's products to enhance the odds of success. They work to facilitate collaborative cross-functional interaction between not only marketing and development teams, but all functional areas. And, the tactical considerations are executed in a manner consistent with the strategic foundation of the company. At its heart, companies who operate in this manner are said to be market-oriented.

MARKET ORIENTATION

Market-oriented companies are characterized by a customer-centric culture that places customer information at the forefront of decision making. When issues need to be resolved, a customer perspective is the beacon for resolution. Moreover, market-oriented companies are savvy with respect to competitors in the marketplace. They have a good sense of who the competitors are, what competitive strategies are used, and how their own company's products are positioned vis-à-vis the competitors' positioning (see Box 5.2). In addition, market-oriented companies continuously monitor their business environment for trends, so that emerging opportunities and threats are spotted early and addressed proactively.

BOX 5.2 A NOTE ABOUT COMPETITION IN HIGH-TECH MARKETS

Managers in high-tech firms suffer from three types of myopia about sources of competition.

1. *We have no competitors.* All too often, managers in high-tech markets are heard to say: "our technology is so new we have no competitors." However, this type of thinking reflects a rather myopic view of the marketplace in that customer needs are typically being solved already in some fashion, either with an older-generation technology, or in some cases, by doing nothing. Indeed, entrenched customer habits are sometimes harder to dislodge with new technology than if a competitor did exist.
2. *The new technology being commercialized by new competitors will not pose a large threat.* Another common phrase heard from high-tech managers in established companies is: "the new technology [developed by a new start-up] won't amount to a hill of beans" (or some similar disparaging phrase). However, the high-tech field is littered with the corpses of incumbent firms who underestimated new entrants in the market. During the hey-day of the Internet, this idea was captured with the saying, "You've been Amazoned," meaning that a new competitor came in and stole an incumbent's business with a new technology or business model. The "innovator's dilemma" has come to mean the difficulty market leaders have in developing disruptive innovations, due to their investments in current-generation technologies that are serving existing customer markets reasonably well (Christensen, 1997). In order to avoid this incumbent's curse, high-tech firms must create an organizational culture

that stimulates a willingness to engage in creative destruction, or the willingness to introduce next-generation technology despite the potential cannibalization of a firm's existing revenue source. A key predictor of such a culture is managers' fears of obsolescence, which is strongly indicative of their willingness to cannibalize existing revenue streams by introducing radically new innovations (Chandy and Tellis, 1998, 2000; Chandy *et al.*, 2003). Bifocal vision refers to the simultaneous ability to serve current customers with current products, and to develop next-generation technologies that will serve the customers of tomorrow.

3 *That competitor is in a different industry, and its strategies don't/won't affect my business.* Managers in some high-tech firms suffer from a third type of blindness when it comes to understanding the competitive environment, viewing their industry from a specific product/technology lens rather than from a customer viewpoint. Instead, a broadened view – known as "product form" competition (versus "brand competition") – acknowledges that customer needs can be solved using different underlying technology platforms. For example, in the Internet industry, broadband providers supplying a DSL connection may view their competitors as only other DSL providers. However, from the customer's perspective, cable modem, new forms of wireless and satellite, and even dated dial-up access all meet the user's needs, each with a different technology solution (and in some cases, a different price/performance ratio). Each represents a different product class. In high-tech fields, convergence often means that new competition will be found in different product classes as technological developments add more capabilities to existing products (i.e., now the Internet is competing with telephone companies in providing calling plans via Voice Over Internet Protocols, commonly known as VOIP).

What is the key lesson from these three types of competitive myopia? It is vital that managers think broadly when thinking about competition in high-tech markets.

Empirical findings show that in highly uncertain markets, a customer orientation has a positive influence on the successful commercialization of a new product (Slater and Narver, 1994). Given the importance of this customer or market orientation, a key question is: what are the characteristics of a firm that is "market-oriented?" As shown in Figure 5.1, a firm that is *market-oriented* emphasizes the gathering, dissemination, and utilization of market information as the basis for decision making (Kohli and Jaworski, 1990; Slater and Narver, 1994).

Gathering of information. Market-based information can be gathered in a variety of ways, including from existing, secondary data (as in the case of industry studies available from government sources or third-party research providers/industry analysts in an industry); trade shows and industry publications; competitive benchmarking studies (either purchased or generated internally by company personnel), or through collecting primary market research. Figure 5.2 shows a variety of tools that can be used to collect primary research in high-tech markets

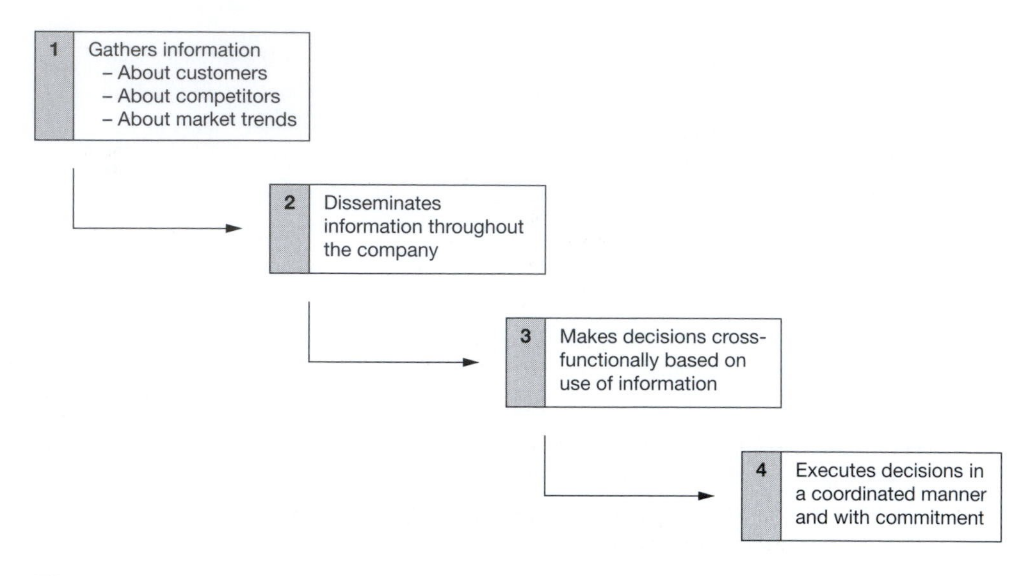

Figure 5.1 *Market-oriented firm characteristics*
Mohr *et al.* (2005), reprinted with permission from Pearson Education

(Leonard-Barton *et al.*, 1995). In order for a high-tech company to be successful in its efforts, it must base its decisions on solid information. A lack of solid information (i.e., relying on too little information, or potentially flawed or biased speculations about the industry) is tantamount to disaster.

Information dissemination. Once the information is collected, it is vital that it be widely disseminated to all the personnel in the organization. Again, consistent with the idea that there are only two types of people in any company (those who serve customers, and those who serve those who serve customers), all company personnel must have access to customer information and market-based data in order to be consistent in the underlying approach to decision making. In addition, such dissemination enables all company personnel to be equally well-equipped to deliver meaningful "moments of truth" (e.g., Bruner *et al.*, 1998) in any potential customer interaction.

Information that challenges existing assumptions about the market, that invites discussion and even dissent, can be a good thing; it can force to the surface underlying areas of ambiguity that additional information can potentially address. However, it is a fact in high-tech markets that information will be incomplete. In such a case, a market-oriented firm will actively encourage debates and the generation of scenarios to tease out underlying issues that decisions must address.

Utilization of information. Although one might think that information that is both gathered and disseminated will be used, such is not always the case. For example, during the design of the Deskjet printer at Hewlett-Packard, marketers gathered information on early prototypes from research in shopping malls to determine user response. They shared data with the engineers and product developers on the twenty-one changes users identified as important for the product to be successful. However, even after gathering and sharing the data, they were not used by the engineers. The

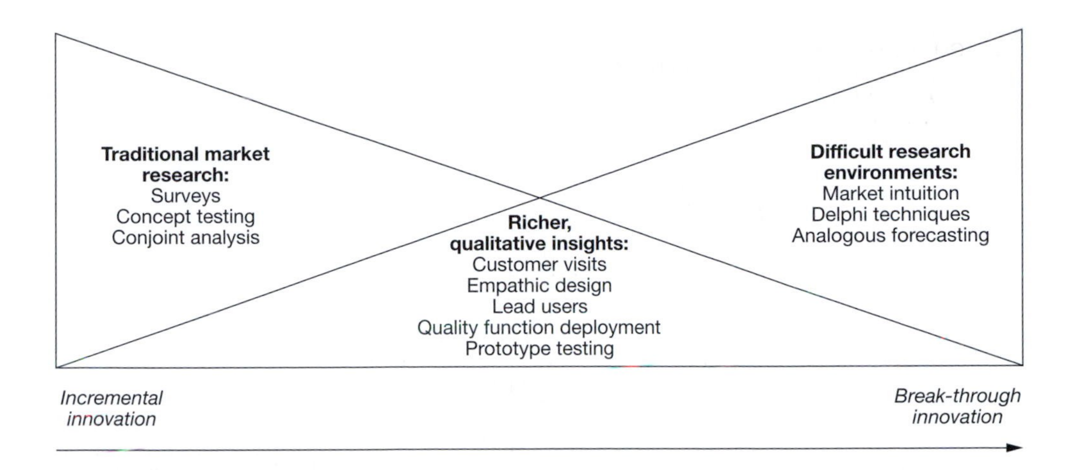

Figure 5.2 *Tools for collecting market-based information on customers*
Adapted from Leonard-Barton *et al.* (1995)

engineers believed that only five of the changes were important (Leonard-Barton, 1992). Again, for a variety of reasons, the information that marketing personnel bring to the design process may go unheeded, which can have negative consequences. So, one cannot assume that simply because the information is there, it will be effectively utilized as a basis for decision making. A market-oriented firm uses the information that is available.

Execute decisions with commitment. Finally, a market-oriented firm executes its decisions with commitment. In Hewlett-Packard's case, the engineers were encouraged to participate in additional market research themselves, and after hearing the same information from the actual customers themselves, the designers incorporated the other sixteen requested changes (Leonard Barton, 1992). Involving a wide array of company personnel in collecting, disseminating, and utilizing the information invites commitment to the information used for decision making and to the decision itself.

Barriers to and facilitating conditions for being market-oriented

Although being market-oriented may seem like an obvious trait that high-tech companies should cultivate, significant barriers to being market-oriented exist. The good news is that there is also a set of facilitating conditions that can assist with the development of a market-oriented culture in the company. These barriers and facilitating conditions are shown in Figure 5.3.

Barriers. First, people in organizations often hoard information to protect their turf, mistakenly believing that access to and control of information provide power and status within the organization. Second, companies may have entrenched habits and routines (commonly known as "straitjackets") that lead to the disregard of information about users; this can be particularly prevalent in companies that value engineering acumen over customer insights. Third, many companies suffer from "marketing myopia," which is the tendency to focus too specifically on solving

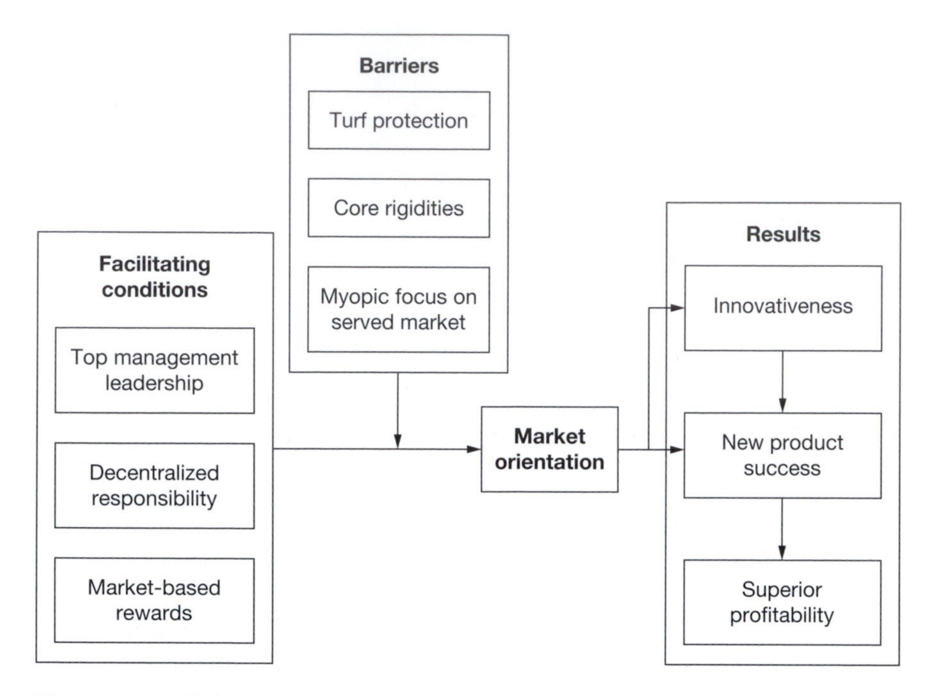

Figure 5.3 *Barriers to and facilitating conditions of market orientation*
Mohr *et al.* (2005), reprinted with permission from Pearson Education

current customers' needs with a current technology. This myopic focus obscures the possibilities that customer needs may change over time and may be solved in radically new ways, and that new customer segments may emerge in the marketplace that should also be addressed.

Facilitating conditions. Three key variables positively affect the development of the market orientation competence (e.g. Jaworski and Kohli, 1993). First, *top management* must be unequivocally and visibly committed to customers and to market-based information. Michael Dell, CEO of Dell Computers, exhibits this kind of unequivocal commitment: "We have a relentless focus on our customers. There are no superfluous activities here . . . Our employees feel a sense of ownership when they work directly with customers" (Mears, 2003).

Second, market-oriented behaviors are more likely to occur when a firm is more *decentralized* in its organizational structure. Fluid job responsibilities, extensive lateral communication, and cooperative relationships between units facilitate a company's efforts to become market-oriented.

Third, of all the variables affecting a firm's market orientation, the *compensation system* of a company has the greatest impact. Employees must be rewarded for generating and sharing market intelligence, for achieving high levels of customer satisfaction and loyalty, and for working in a coordinated fashion across functional areas.

Adopting a customer perspective

To truly have a customer-orientation means that a company adopts a customer perspective in guiding its decisions. However, all too often, the enthusiasm a high-tech company has for its latest innovations is not matched by the customer. One of the key reasons for this "disconnect" is that the company has not genuinely considered the variety of factors that influence the customer's adoption decision for the new high-tech product (Moore, 2002; Rogers, 2003). The factors that affect the *adoption and diffusion of new technology* are overviewed in Box 5.3.

Taken as a whole, the prior section highlighted many of the considerations that high-tech firms face in becoming market- and customer-oriented. The next section of this chapter addresses additional barriers to marketing success in high-tech companies, and how these barriers can be overcome.

BOX 5.3 FACTORS AFFECTING CUSTOMER ADOPTION

1 *Relative advantage*: This factor takes a comprehensive look at all possible benefits and costs the customer will incur in making the decision to adopt – from the customer's perspective. High-tech companies often erroneously over-weight the benefits while under-weighting the costs. *Benefits* can include functional (attributes/features), operational (reliability/durability), financial (for example, favorable credit terms), and/or personal (selecting a "safe" vendor) considerations. *Costs* include not only the monetary outlay, but also non-monetary costs, such as risks of failure, training costs, downtime in equipment transfer, and perceived riskiness.
2 *Complexity*: This factor examines, from the customer's perspective, how difficult it will be to learn to use the new technology in order to derive the intended benefits. Again, high-tech companies that are proficient with the technology tend to underestimate complexity from the customer's perspective. In addition to training on new technology products, additional complications arise from the fact that many high-tech products require multiple components in a solution in order for the customer to receive the benefits. For example, a business customer who wants to add an e-commerce capability to its existing distribution channel not only needs to develop a website; it also must consider who will host the website (unless the company wants to also purchase its own server); who will handle order fulfillment and customer service inquiries from the website (tasks that were likely handled by distribution channel partners in the past), potential search engine positioning to ensure the site is visited, and myriad other issues. So, a company that offers web design services may misunderstand "complexity" from the customer's perspective simply because it does not view its product offering in the context of the end-to-end solution (or "whole product") that the customer must have in place in order to receive the benefits of the isolated product.
3 *Compatibility*: This factor examines whether or not the new technology is compatible with the customer's established ways of doing things (i.e. existing habits or business routines), as well as compatibility with legacy technologies. To be sure, it is difficult to change entrenched customer habits, and new technologies that ask customers to do things in new ways find a more difficult road to market acceptance than those that are

compatible. Moreover, as mentioned previously, one of the costs that customers calculate (either implicitly or explicitly) is the cost of obsolescence of existing products. So, incompatibility can mean that a customer has to "write off" investments made in previous versions of technology that will no longer be functional in the case of a new technology that is incompatible with older products.

4 *Trialability*: One way to lower a customer's perceived doubts about whether the benefits of the new technology will be realized is to offer a trial version of the product on a limited time basis. Certainly, this strategy will not be viable for many high-tech products, but a company that thinks creatively about how to enhance trialability finds it favorable to do so. For example, Apple offered a 30-day "test drive" of a Macintosh computer when it was introduced in the marketplace. Some software companies are able to offer a version of the software for free for a limited time period (via licensing arrangements).

5 *Ability of the firm to communicate product benefits to potential adopters*: This factor captures the ease and clarity with which the benefits of owning and using the new product can be communicated to prospective customers. High-tech firms often make the mistake of speaking in highly technical language, emphasizing features of the products rather than the benefits to the customer. In other words, what customers need to hear is how product features translate into tangible benefits, in a language that is user-friendly.

6 *Observability*: To the extent that a customer is able to clearly see the benefits the new technology generates, adoption rates are enhanced. In addition, adoption rates are enhanced when customers in a marketplace can see the benefits other customers have received from adoption. Steps high-tech firms take to verifiably document performance or productivity enhancements from adopting new technology facilitate observability of benefits (Anderson and Narus, 2004), and hence, customer adoption rates.

Systematic consideration of these factors will facilitate the rate of adoption and diffusion of innovation in the marketplace.

OVERCOMING OTHER BARRIERS TO HIGH-TECH MARKETING SUCCESS

As noted at the outset, success in high-tech markets is elusive for many reasons: the high-tech environment is characterized by a high degree of uncertainty; lead times for development can be long in high-tech industries, often resulting in missed windows of opportunity; expensive R&D efforts – often viewed as a sunk cost – may complicate the pricing decision for a high-tech company, resulting in prices that do not accurately capture the customer's perception of value in adopting the new technology. The list of barriers to effective high-tech marketing is potentially quite daunting. In addition to complications noted previously in this chapter, as well as the barriers noted in Figure 5.4, this section focuses on some of the most common and most important barriers.

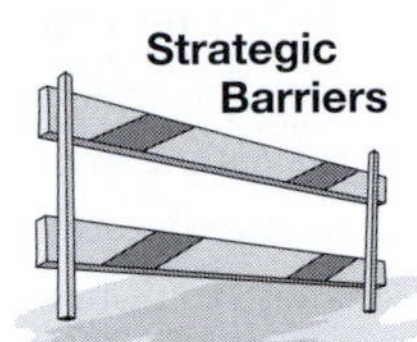

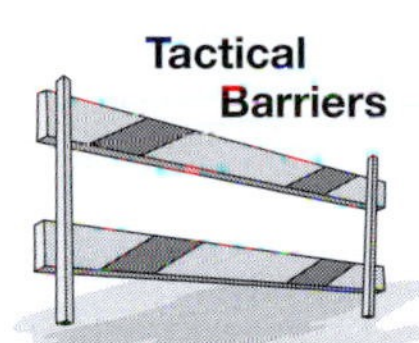

Figure 5.4 *Barriers and complications to effective high-technology marketing*

Overcoming barrier 1: lack of marketing expertise

Marketing is not only more difficult in high-tech environments than in more traditional marketing contexts; marketing sophistication is also more important in high-tech environments in order to adequately manage these complications. Yet, all too often, marketing expertise is lacking in high-tech companies. Although this barrier was noted at the outset, because of its critical importance, it is discussed further here. Many (if not most) high-tech companies do not have a well-developed marketing competence that operates at all the levels mentioned at the outset of this chapter.

Some common responses by a high-tech company to overcome the lack of marketing expertise are to:

1 move technical personnel to a marketing role, even if they have no training or experience in marketing;

2 expect a member of the senior management team to assume the marketing responsibility (again, even if they have no training or experience);
3 hire a marketing expert.

Unfortunately, in some cases, companies delay the hiring of a marketing person because there are insufficient funds to do so (this conundrum can be long-lived in that without time and effort devoted to marketing, it is hard to generate revenues). Even in the most positive scenario, when a company hires a marketing expert, top managers second-guess his or her decisions or provide insufficient resources to do the job properly.

A company does not need a large marketing budget or even a formal marketing department to have marketing sophistication. Many high-tech companies charge product teams with marketing responsibilities, which can be effective if the personnel given this responsibility have:

- adequate training to perform the marketing responsibilities;
- sufficient respect internally to effect the necessary decisions;
- a market-oriented mindset that captures the customer perspective;
- resources that allow them to do their job.

Overcoming barrier 2: crafting – and delivering – an effective value proposition

A second barrier to effective marketing in high-tech companies is the lack of a clear, compelling *value proposition* for the company's technology, as articulated from the customer's perspective.

Many high-tech companies are founded on the basis of a great, new technology, with little consideration as to exactly which customers' needs will be served (i.e. a "technology push" situation). In its forays into the marketplace, the company can articulate quite clearly the specific features the new technology offers relative to the incumbent/legacy technology; however, to potential customers, the specific benefits these sophisticated features will offer is often not clear. Further exacerbating this technology-centric focus is a disregard for the potential risks and drawbacks of adopting the new technology that a customer will face (recall Box 5.3 on factors affecting customer adoption decisions).

This difference between the enthusiasm of company personnel for the new technology and the customer's lukewarm reaction is a typical sign of a company that is not market-oriented. At the extreme, companies might say, "My customers are stupid; they just don't get it." However, it is not the customers who are stupid; rather the company has not done the thorough, credible job of incorporating a customer-orientation into its efforts. By the time the company is actively marketing its product in the marketplace, it should already be aware of what the key benefits of the new technology are – from the customer's viewpoint – and, importantly, what the barriers to adoption will be, and some plan to mitigate them.

Another version of this barrier is a value proposition that is so vague as to be meaningless to the customer, such as, "our company offers the highest quality product at an attractive price"; or, "our company is committed to the highest service standards in the industry." Value propositions (or positioning statements) that are meaningful to the customer must take into account the customer's needs and provide specific insights about meeting those needs (Anderson and Narus, 2004). For example, Hewlett-Packard's website states as its value proposition: "We will reduce our customers' technology acquisition costs and will enable them to achieve measurable improvements in their technology operating costs, allowing them to reap improved business results through their information-technology infrastructure." Support for this value proposition can be found in the numerous customer testimonials on the Hewlett-Packard website.

Overcoming barrier 3: lack of funding for marketing/no budget

Certainly, marketing efforts can carry a steep price tag, particularly when a company is considering strategies such as an advertising campaign, trade-show exhibits, marketing research studies, and so forth. Many companies find themselves in a chicken-or-egg situation: they cannot fund marketing efforts without revenues, yet they cannot capture revenues without marketing. Although many companies have been creative in coming up with visible, effective marketing strategies without spending excessively, one practical way to start implementing marketing strategies on a limited budget is to carefully carve out pilot programs that allow a company to test some of their ideas in the marketplace, to refine them prior to committing larger amounts of funding, and to grow revenues incrementally. The key is to carefully identify where a company's efforts will have the greatest odds of success, to implement its efforts in a focused fashion, and to ensure that the pilot has sufficient support to provide a realistic assessment of the outcome. For example, if one wants to test an advertising campaign in a trade journal, make sure that the creative strategy behind the campaign is well-developed, and that the media plan carries sufficient weight to ensure that the ads are actually seen (i.e., not tiny ads that run only once).

Companies must be careful to view marketing expenditures as an investment and not merely as an expense. The expense view is often a manifestation of an underlying cultural disdain for marketing activities, and hence is a symptom of a larger, more problematic issue. Companies who cite a lack of funding for marketing are often the same companies who view marketing at only the tactical level. It is important that marketing is considered to be a philosophy for guiding the business, and that the customer perspective serves as a basis for the decisions that are made across the organization.

As a company builds its financial support for marketing activities, it will develop its capabilities for more sophisticated marketing strategies. For example, the development of a branding campaign to build brand equity has been shown to protect a firm from the price compression inherent in many technology markets, allowing it to charge a premium on its products and creating a stronger connection with customers (Keller, 2003). Yet, this type of marketing campaign can be costly and requires the understanding that marketing is a long-term investment into the company's presence in the market. As a company implements pilot projects, such as those described by Mohr *et al.* (2005) in their section on advertising campaigns on a limited budget, it will develop a comfort level, allowing it to move forward on the marketing sophistication trajectory.

Overcoming barrier 4: viewing the customer purchase decision in isolation

Regardless of how sophisticated or revolutionary a new technology is, customers do not buy products so much as they buy solutions to their problems. This view of consumer purchase decisions invites the high-tech company to understand the product it is selling through a more holistic lens. A more holistic lens has implications for viewing the customer's business less as a purchase and more as an investment in a long-term relationship. This, in turn, invites considerations of the capabilities the company requires to fulfill the customer's needs for service, as well as to understand the customer's long-term needs for the future. Known as *relationship marketing*, this perspective is focused on the *life-time value* of a customer's revenue stream.

Many successful high-tech companies understand that the customer's purchase decision of today is potentially only a small fraction of their investments over time in a particular technology platform. Viewing the customer relationship as a long-term revenue stream allows a company to realize that the servicing of the customer account is what will further cement loyalty, allowing it to capture an increasing share of the customer's wallet. Thinking strategically about how to wrap a service component around the product being offered can be one way to allow the customer to view the company, not as a provider of a commodity (i.e. bandwidth, for example), but more as a solutions-provider who can assist the customer in managing its complete set of needs.

The field of customer relationship management and the lifetime value of customers are vital to effective high-tech marketing, and must be proactively considered.

TRENDS IN TECHNOLOGY MANAGEMENT: IMPLICATIONS FOR HIGH-TECH MARKETING

Many trends in technology management highlighted in this book have implications for the marketing of high-technology products and services. There is no question that technology development has become more complicated and expensive. Take the case of the LCD flat panel TV screen, currently a hot consumer product (overviewed previously in Box 5.1 on the 4 Ps). Building a production plant requires a $2–$3 billion investment (Ramstead, 2004). To pool resources, Philips of the Netherlands and LG Electronics of South Korea formed a joint venture, LG.Philips LCD, to manufacture this product. Competing technologies such as plasma and projection, coupled with consumer expectations of falling prices, are shortening the product life cycles of LCD displays.

Under such conditions, companies like LG have to sharpen their high-tech marketing skills at all three levels: strategic, functional, and tactical. At a strategic level, they have to decide which country markets to serve, the lifestyle and demographic profiles of their target customers in each market, and the value proposition or positioning that will appeal to each target market. At a functional level, LG has to make decisions on product, price, promotion, and placement for its LCD displays. Finally, at the tactical level of marketing, LG has to make decisions on timing of product launch, ad campaigns, press releases, and the like. All of these skills can be imparted more easily to employees in a market-oriented firm, one that truly believes in meeting the needs of customers rather than maximizing short-term gains.

Other important trends with respect to high-tech marketing specifically include the emergence of market opportunities in developing economies, and the outsourcing of information-technology services. High-tech companies are finding that they can expand their opportunities in "base of the pyramid" markets, or emerging countries with large populations – historically overlooked by companies – that have not had the purchasing power to buy discretionary products as compared to more established market economies. Technology inventions can solve problems faced by people in some of the world's poorest areas. For example, Hewlett-Packard's corporate social responsibility initiatives are bringing the benefits of technology to India, Costa Rica, and Senegal. Because customers in most developed economies are "pretty well served by information technology companies," targeting the next 4 billion provides both new revenue opportunities as well as social benefits above and beyond economic benefits (Prahalad and Hart, 2002; London and Hart, 2004; Murphy, 2002).

Another trend in the technology arena is the outsourcing of information-technology services to countries with a lower-cost infrastructure. Key among these is India. Although this trend is fueled by the need to lower costs in a competitive global economy, there are downside risks that must be addressed. These trends highlight the fact that today's market is a global market. Regulatory issues differ significantly between the US and Europe, for example, where proactive involvement by government officials in setting telecommunication standards and licensing areas of the broadcasting spectrum have given European companies (and customers) an edge over US companies. High-tech managers must be astute with respect to understanding competitive dynamics in the global landscape.

SEIZING THE HIGH-TECH MARKETING INITIATIVE

This chapter has presented issues that guide the development of a marketing philosophy in a high-tech company. To be sure, marketing is more than mere tactical development of sales campaigns. It encompasses effective cross-functional interaction between all departments in the organization, with a common focus on serving customers well. It includes the capability to make the tough decisions about which customer segments to serve – and, its corollary, which customer segments not to serve – as well as positioning the company's products relative to the competition in those segments. A market-oriented firm relies on market-based information in making its decisions. And, successful high-tech firms have a solid understanding of the customer's perceptions of the costs/benefits of adopting the new technology.

Certainly, there are *many* more topics to be considered in the world of high-tech marketing (the "chasm" companies face in the technology adoption life cycle, pricing strategies and tactics, designing effective distribution strategies, licensing considerations, protection of intellectual property, potential consumer/social backlash against technological development, setting industry standards, effectively working with partners, specific marketing strategies for markets driven by network effects, and so forth). Therefore, in addition to the topics presented here that provide the foundation for successful high-technology marketing, interested readers of this chapter are encouraged to delve more deeply into the specific tools and strategies of effective high-tech marketing with the recommended readings offered here.

Smart high-tech marketing is based on systematic consideration of critical issues in order to allow innovations to reach full commercial success. Without effective marketing of high-technology products and innovations, the benefits such innovations can yield – both to customers buying and using them and to the firm supplying and marketing them – will remain elusive.

FURTHER READING

Bruner, R., Eaker, M.R., Freeman, R.E., Teisberg, E.O., and Spekman, R.E. (1998). Marketing management: leveraging customer value. In W. D. Bygrave (Ed.), *The portable MBA* (3rd edn, pp. 103–124). New York: John Wiley & Sons. The chapter on marketing in this book provides a solid foundation of the strategic issues in marketing, including the marketing concept, steps in market segmentation, and so forth.

Mohr, J., Sengupta, S., and Slater, S. (2005). *Marketing of high technology products and innovations* (2nd edn). Upper Saddle River, NJ: Prentice Hall. This book provides detailed coverage on all aspects of high-tech marketing, from an overview of the high-tech environment, through strategic marketing planning, detailed coverage of the 4 Ps of marketing and its subcomponents (product management, pricing, place/distribution, and advertising and promotion considerations), as well as leveraging the Internet and societal/ethical considerations. www.markethightech.net.

Moore, G.A. (2002). *Crossing the chasm: Marketing and selling high-tech products to mainstream customers* (revised/updated edition). New York: HarperBusiness. This high-tech marketing classic focuses on the technology adoption life cycle (adoption and diffusion of innovations), highlighting the disruptions in this cycle for high-tech products.

Moore, G.A. (1999). *Inside the tornado: Marketing strategies from Silicon Valley's cutting edge.* New York: HarperBusiness. The follow-up to *Crossing the chasm* describes the three stages in the rapid take-off of a high-tech firm's sales following the chasm, and the specific strategies useful at each stage.

Ryan, R. (2002). *Smartups: Lessons from Rob Ryan's entrepreneur America boot camp for start-ups.* Ithaca, NY: Cornell University Press. This book is an easy-to-read guide with practical advice for high-tech start-ups.

Ryans, A., More, R., Barclay, D., and Deutscher, T. (2000). *Winning market leadership: Strategic market planning for technology-driven businesses.* New York: John Wiley & Sons. This business book presents a 10-step model on strategic market planning for executives in technology-intensive businesses, and includes many useful examples.

REFERENCES

Anderson, J. and Narus, J. (2004). *Business market management: Understanding, creating, and delivering value.* Upper Saddle River, NJ: Prentice Hall.

Booz Allen and Hamilton (1982). *New product development for the 1980s.* In-house report. Washington, DC.

Bruner, R., Eaker, M.R., Freeman, R.E., Teisberg, E.O. and Spekman, R.E. (1998). Marketing management: Leveraging customer value. In W.D. Bygrave (Ed.), *The portable MBA* (3rd edn, pp. 103–124). New York: John Wiley & Sons.

Chandy, R. and Tellis, G. (1998). Organizing for radical product innovations: The overlooked role of willingness to cannibalize. *Journal of Marketing Research,* 35, 474–487.

Chandy, R. and Tellis, G. (2000). The incumbent's curse? Incumbency, size, and radical product innovation. *Journal of Marketing,* 64(3), 1–17.

Chandy, R., Prabhu, J., and Antia, K. (2003). What will the future bring? Dominance, technology expectations, and radical innovation. *Journal of Marketing*, 67(3), 1–18.

Christensen, C. (1997). *The innovator's dilemma*. Boston, MA: Harvard Business School Press.

Dhebar, A. (1996). Speeding high-tech producer, meet the balking consumer. *Sloan Management Review*, 37(2), 37–49.

Drucker, P. (1954). *The practice of management*. New York: Harper & Row.

Dutta, S., Narasimhan, O., and Rajiv, S. (1999). Success in high-technology markets: Is marketing capability critical? *Marketing Science*, 18(4), 547–568.

Gatignon, H. and Xuereb, J.-M. (1997). Strategic orientation of the firm and new product performance. *Journal of Marketing Research*, 34 (February), 77–90.

Jaworski, B. and Kohli, A. (1993). Market orientation: Antecedents and consequences. *Journal of Marketing*, 57 (July), 1–18.

Keller, K.L. (2003). *Strategic brand management: Building, measuring, and managing brand equity*. Upper Saddle River, NJ: Prentice Hall.

Kohli, A. and Jaworski, B. (1990). Market orientation: The construct, research propositions, and managerial implications. *Journal of Marketing*, 54 (April), 1–18.

Leonard-Barton, D. (1992). Core capabilities and core rigidities: A paradox in managing new product development. *Strategic Management Journal*, 13, 111–125.

Leonard-Barton, D., Wilson, E., and Doyle, J. (1995). Commercializing technology: Understanding user needs. In V.K. Rangan *et al.* (Eds), *Business marketing strategy* (pp. 281–305). Chicago, IL: Irwin.

London, T. and Hart, S. (2004). Reinventing strategies for emerging markets: Beyond the transnational model. *Journal of International Business Studies*, 35 (September), 350–370.

Mears, J. (2003). Customer focus keeps Dell productive. *Network World*, 51 (April 21).

Mohr, J., Sengupta, S., and Slater, S. (2005). *Marketing of high-technology products and innovations* (2nd edn). Upper Saddle River, NJ: Prentice Hall. www.markethightech.net

Moore, G.A. (2002). *Crossing the chasm: Marketing and selling high-tech products to mainstream customers*. New York: HarperBusiness.

Murphy, C. (2002). The hunt for globalization that works. *Fortune* (October 28), 163–176.

Narver, J. and Slater, S. (1990). The effects of market orientation on business profitability. *Journal of Marketing*, 54 (April), 20–35.

Prahalad, C.K. and Hart, S. (2002). The fortune at the bottom of the pyramid. *Strategy+Business*, 26 (first quarter), 2–14.

Ramstead, E. (2004). I want my flat TV. Now!, *The Wall Street Journal*, May 27, p. B1.

Rogers, E. (2003). *Diffusion of innovation*. New York: The Free Press.

Slater, S. and Narver, J. (1994). Does competitive environment moderate the market orientation/performance relationship? *Journal of Marketing*, 58 (January), 46–55.

Workman, J. (1993). Marketing's limited role in new product development in one computer systems firms. *Journal of Marketing Research*, 30 (November), 405–421.

Yang, C., Moon, I., and Tasahiro, H. (2004). Commentary: Behind in broadband, *Business Week*, (September 8), 88.

Part III

The management of technology in the society

INTRODUCTION

The third part of the book focuses on the society or the environment of technology-based firms. The environment of such firms consists of all factors and actors that directly or indirectly impact the firm performance.

Several authors have modeled the external environment of a firm. Some of these models categorize relevant factors and actors in the external environment. Jain (1985), for example, describes a simple model of four sets of factors. Kotler (1991) describes a more complex model that represents different layers of the external environment. The first layer, the "core-market," represents suppliers, distributors, consumers and all actors required to produce and use products. In the second layer, the "micro-environment," are competitors, banks, and other factors that can have a direct effect on a firm without being part of the value chain. The third layer, the "macro-environment," comprises demographic, socio-cultural, physical and legal factors.

In all cases, a short analysis of the environment of a firm would reveal many actors and factors, such as suppliers, competitors, distributors, clients, labor unions, banks, insurance companies, governments, technological developments, culture, demographic developments, environmentalists, and so on. Some of these factors, e.g. demographic developments, seem to develop in a predictable trend, whereas other actors, such as governments, act on the basis of complex negotiation processes and therefore seem to be unpredictable. All factors, predictable or non-predictable, interact in complex patterns and affect the performance of a firm. Chapter 5 shows that high-tech markets are characterized by large degrees of uncertainty.

Rather than attempting to describe all relevant actors and factors in the environment of high-tech firms, the third part of the book describes five alternative perspectives on the environment. Chapter 6, "Managing the dynamics of technology in modern day society" by Karel F. Mulder discusses alternative theoretical explanations of technological change. On the one hand, technological determinists consider change as an autonomous process driven by the internal logic of the technology rather than by the environment. On the other hand, social constructionists consider the development of technology as an entirely socially determined process. In Chapter 7, "Development and diffusion of breakthrough communication technologies," J. Roland Ortt focuses on a specific theory of technological change. He considers technological change as an evolutionary process that is driven by market actors and factors as well as by characteristics of the technology. Different patterns in this process and its managerial implications will be discussed. Chapter 8 "Forecasting the market potential of new products" by David J. Langley, Nico Pals and J. Roland Ortt shows alternative approaches to infer the future market potential of new high-tech products. In addition to methods that are aimed at the extrapolation of past experiences (consumer research, data analysis and expert opinions) two alternative approaches are discussed. The first approach is the practical method, such as probe and learn. The second method is a more theoretical approach that is derived from biological evolutionary theories. In Chapter 9, "The innovating firm in a societal context," C.W.M. Naastepad and Servaas Storm focus on labor–management relations in countries and in companies. They show the effects on productivity and innovativeness and introduce two entirely different ways of managing labor–management relations with significant effects on labor productivity. In Chapter 10, "Complex decision-making in multi-actor systems," Martijn Leijten and Hans de Bruijn describe the interaction of actors in a network and the effect of these interactions on the progress and outcome of large technological projects.

REFERENCES

Jain, S.C. (1985). *Marketing Planning and Strategy*. Cincinnati, OH: South-Western Publishing.

Kotler, P. (1991). *Marketing Management. Analysis, Planning, Implementation, and Control*. (7th edn). Englewood Cliffs, NJ: Prentice Hall.

Chapter 6

Managing the dynamics of technology in modern day society

Karel F. Mulder

OVERVIEW

This chapter discusses technological change from a macroscopic viewpoint. It addresses the issue to what degree technological change is an autonomous process leading us to a utopia or dystopia, or a process of social choice, in which mankind determines the technologies it wants for its future. Technology managers should recognize where choices on technologies can be made to be able to develop strategies. It is argued that the manager of technology has an increasingly complex task because of:

- The growing complexity of technology, which creates a need for long-term visions and cooperation between various fields of expertise.
- The emancipation of consumers, workers and production site neighbors, making government licenses insufficient to deal with social issues.
- The globalization of technology creation, which makes global thinking and intercultural communication crucial.

INTRODUCTION

Popular feelings regarding modern technology are somewhat ambiguous:

- On the one hand we can observe a fear for scientists as "uncontrolled maniacs" who are able to produce monsters such as the one of Frankenstein or threaten human life by uncontrolled experiments like the one that caused the Chernobyl nuclear plant to melt down.
- On the other hand, there is often a strong belief that science and technology by itself will select the best options for progress, and lead us to a better world if we do not block their development.

If these views were correct, management of technology would consist of implementing strict control mechanisms for scientists and engineers or it would consist of just running the administration of a laboratory. However, both views are incorrect: science and technology develop in interaction with their social environment. It will be argued in this chapter that the course of technology is in part a matter of social choice. Scientists and engineers are never able to make all the choices that are involved in innovation. Technological change means choice, and therefore management of technology means not just facilitating or controlling the geniuses, but also interacting with various stakeholders in order to decide where to go. Management of technology implies developing and adjusting strategies, to create the technologies that are in legitimate demand, now and in the future. In our time, three developments make this task increasingly complex:

- the growing complexity and interconnectedness of technologies;
- the globalization of corporations, markets and R&D projects;
- the emancipation of the citizen that creates a growing need for public accountability.

Therefore, this chapter concludes that managing technological innovation is a task that needs dedicated professionals.

AUTONOMOUS TECHNOLOGY DEVELOPMENT AND TECHNOLOGICAL DETERMINISM

What drives technological innovation? This question will be dealt with in this paragraph. The vision that technology is autonomous (i.e. not influenced by external forces such as economy, law, social issues, and environmental issues) can often be recognized in popular reasoning: "The progress of technology cannot be stopped," or "If Albert Einstein had not formulated the general theory of relativity, somebody else would have done it shortly afterwards." It seems people assume that there is an invisible hand propelling technological innovation. The core of this reasoning is often that technology is seen as based on science (or is merely seen as applied science). Scientific research produces knowledge that is published and stored. The pond of scientific knowledge is ever growing, and so technology will only improve as it has more and more knowledge at its disposal. Moreover, improvements in one technology contribute to improvements in other technologies. The improved technology helps science to produce new knowledge. In this way the development of technology is seen as a self-propelling mechanism, beyond the control of anybody. Even individual scientists and engineers cannot really influence this process: if their actions are not in accordance with this mechanism, they only produce noise that can be neglected. Technological change is therefore considered to be *autonomous*. The technological artifacts are considered to be produced in a linear sequence from basic research, to applied research, to technological systems design, to design of artifacts, to marketing. In this linear model, technology is seen as the determining element of society. *Technological determinism* implies that the state of society is determined by the state of technology, not by the preferences of people. Some people reason that: "As the Internet is there, the municipalities should not distribute leaflets to their citizens, but the citizens should learn to surf the web to retrieve their information." Moreover, the availability of technology is determining demands: "Until the mid 1980s students turned in theses in handwriting or in typewriting with various pieces cut and patched. As the PC became widely available, professors no longer accepted this. Reports in nice layout with various graphics became obligatory."

Jacques Ellul is one of the most renowned philosophers who analyzed technology from a determinist viewpoint. The core theme of Ellul's work was the threat of "technologic tyranny over humanity." Ellul (1964) made a sharp distinction between traditional (pre industrial) technologies and modern technology. Traditional technology was, in his view:

- limited in application (because the technology was made for a specific task at a specific place);
- only to a small extent dependent on resources but especially depending on craftsmanship and skill;
- local in character (as it had to fit into local circumstances and local culture).

As a result, Ellul claims that people had the possibility of choice when traditional technologies were dominant. Individuals and local communities could influence the technologies that they applied. Contrary to traditional technologies, Ellul characterized modern technology by:

- *Automatism*, which means that there is automatically one type of technology that fits to one problem, wherever one is on the planet.
- *Self-replication*, which means that new technology reinforces technological growth in other areas. The result is exponential growth.
- *Indivisibility*. Technology determines our lifestyle. We have to accept it completely. For example, fully participating in society without a cell phone might create serious problems.
- *Cohesion*, which implies that technologies that are used in rather distinct applications have much in common.
- *Universalism*, which means that technology is geographically as well as qualitatively omnipresent.

For Ellul the conclusion was that technology destroys human freedom. There is no way back, and so the future prospects are grim in his vision. Part of Ellul's arguments can be recognized in the so-called Unabomber Manifesto. Unabomber, the name used by the Californian mathematician Theodore Kaczynski, committed various attacks against research institutions and airlines in the 1980s and early 1990s. A characteristic argument in his manifesto shown here:

> A technological advance that appears not to threaten freedom often turns out to threaten it very seriously later on. For example, consider motorized transport. A walking man formerly could go where he pleased, go at his own pace without observing any traffic regulations, and was independent of technological support systems. When motor vehicles were introduced they appeared to increase man's freedom. They took no freedom away from the walking man, no one had to have an automobile if he didn't want one, and anyone who did choose to buy an automobile could travel much faster than the walking man. But the introduction of motorized transport soon changed society in such a way as to restrict greatly man's freedom of locomotion. When auto-

> mobiles became numerous, it became necessary to regulate their use extensively. In a car, especially in densely populated areas, one cannot just go where one likes at one's own pace, one's movement is governed by the flow of traffic and by various traffic laws. One is tied down by various obligations: license requirements, driver test, renewing registration, insurance, maintenance required for safety, monthly payments on purchase price. Moreover, the use of motorized transport is no longer optional. Since the introduction of motorized transport the arrangement of our cities has changed in such a way that the majority of people no longer live within walking distance of their place of employment, shopping areas and recreational opportunities, so that they HAVE TO depend on the automobile for transportation. Or else they must use public transportation, in which case they have even less control over their own movement than when driving a car. Even the walker's freedom is now greatly restricted. In the city he continually has to stop and wait for traffic lights that are designed mainly to serve auto traffic. In the country, motor traffic makes it dangerous and unpleasant to walk along the highway. (Note the important point we have illustrated with the case of motorized transport: When a new item of technology is introduced as an option that an individual can accept or not as he chooses, it does not necessarily REMAIN optional. In many cases the new technology changes society in such a way that people eventually find themselves FORCED to use it.)
>
> (Unabomber Manifesto, 1995, paragraph 127)

Technological determinism is not necessarily as fatalistic as it was for Ellul. Various futurologists predict bright futures for mankind based on technological progress that is beyond anybody's control such as the abundance of energy by developing nuclear fusion technologies (Cf. Celente, 1997). Often these predictions are presented as compelling, and not as one of the options for mankind (Cf. De Wilde, 2000). Compare for example Richard Smalley's (who received the Nobel Prize for Chemistry in 1996) compelling statement for a US Congressional hearing regarding the potential of nanotechnology:

> The impact of nanotechnology on health, wealth, and lives of people will be at least the equivalent of the combined influences of microelectronics, medical imaging, computer-aided engineering, and man-made polymers developed in this century. I believe at the moment our weakness is the failure so far to identify nanotechnology for what it is: a tremendously promising new future which needs to have a flag.
>
> (Terra, 1999, p. 1)

Technologic determinism is a dubious concept on several grounds. First, it supposes one-way traffic from science to technology to society. Technology is only determined by scientific growth and self-replication. Historically, this is incorrect: indeed, quite a number of key technologies were created long before the underlying scientific principles were formulated. For example, the invention of the first steam engines (by Newcomen in 1712 and Watt in 1770) preceded the formulation of the thermodynamic principles explaining their operation by a century (Carnot cycle formulated in 1824). As L.J. Henderson wrote, penetratingly, in 1917: "Science is infinitely more indebted to the steam engine, than is the steam engine to science" (Henderson, 1917). The first aircraft flew in 1903, but the scientific explanation why man could fly on wings had to wait until Prandtl formulated his boundary layer theory in 1920. In 1895, the famous physicist Lord Kelvin, president of the Royal Society, had even claimed that it was impossible for man to fly by wings (Gibbs-Smith, 1974).

Second, historic analyses often show that technological innovation is not a process leading to a predetermined outcome, but involves processes such as social choice by various groups, feedback loops and dead end streets. For example, synthetic fibers that are now highly successful in replacing asbestos were once developed to make safer tires (in which they were not applied) (Mulder, 1992a). Unexpected inventions are called *serendipity*: making discoveries that you were not looking for (Van Andel, 1994).

Third, in their analyses, Ellul and the Unabomber fail to acknowledge the tremendous advantages that technology has brought us. The number of people living a relatively good life on planet Earth, i.e. well nourished and healthy, is historically unprecedented. Criticizing the alienation created by technology, as Ellul and the Unabomber did, is one-sided, to say the least, and therefore leads to unjustified conclusions: we sacrificed some liberties, certainly, but received new liberties in return.

SOCIAL CONSTRUCTIVISM

Technological determinism is completely rejected in the Social Construction of Technology (SCOT) model (Pinch and Bijker, 1987). In this model, technological change is analyzed as a process of social construction. SCOT aims at explaining change and stability of artifacts, i.e. man-made objects. Central to this model of technological change is the notion of *interpretative flexibility* of artifacts. An artifact is never self-evident: different people can attribute different meanings to the same artifact. The artifact "car," for example, might be especially a status symbol for its owner while being a dangerous thing to pedestrians, an object of taxation for the tax collector, or a daily necessity for the commuter.

Groups with a joint perception of an artifact are called *relevant social groups*. The way in which relevant social groups perceive an artifact determines what they see as problematic, i.e. as challenge for innovation. The success of innovation is therefore not determined by "improving on the existing technologies," as what counts as an improvement is not identical for each group. Successful innovation implies addressing the problems that at least one relevant social group has regarding this technology. Making existing cars much faster, an obvious innovation in the autonomous technology framework, is therefore not an innovation that is likely to be very successful in our society, as the social group demanding faster cars is rather small. Safety, comfort, status and environmental issues have far more support among various social groups relevant to the car, and are therefore more likely to be targets for innovation.

In the autonomous technology framework, the pace of scientific and technological change determines that artifacts are more or less stable over time. In the SCOT framework, the stability of artifacts is explained by the stability of relevant social groups and their perceptions of artifacts. The following case of the bicycle illustrates this point.

The bicycle

Nowadays we tend to think that our modern bicycle replaced the previous nineteenth-century high-wheeled bicycles in a pretty straightforward manner. However, the history is far more complicated, showing the socially determined choice processes.

The largest user group with the Penny-Farthing turned out to be young men of reasonable wealth, who possessed the courage and dexterity to handle the bikes. Besides them was a group of potential users. The Penny-Farthing riders – young brave and from the higher circles of society, radiated superiority towards their walking or horse riding brethren. For them, the Penny-Farthing was a "macho machine." For potential users such as women, long-distance cyclists or older gentlemen the Penny-Farthing was rather considered an unsafe machine. The bicycle evoked considerable resistance: "but when to words are added deeds, and stones are thrown, sticks thrust into wheels, or caps hurled into the machinery, the picture has a different aspect" (cited in Bijker, 1990, p. 47). In London, for instance, cyclists used wooden sidewalks, because the roads were otherwise unpaved. This evoked resistance with the local population, further enhanced by existing class differences. Moreover, anti-cyclists had decency problems with female cyclists.

Because the cyclist was seated almost directly over the middle of the forward wheel, with his or her legs far from the ground, every stop or bump provided the risk of falling over.

These different attributions of meaning also spawned different directions of development.

For the sportsmanlike Penny-Farthing rider enlargement of the forward wheel was the best way to increase speed. This culminated in 1892 in the Rudge Ordinary with a forward wheel diameter of about 1.4 meters. The fact that this only made the bicycle more dangerous was considered by the specific user base to be more of an advantage than a drawback.

To make the unsafe Penny-Farthing suitable for other users, experiments were done with various different models: in some bikes the wheels were reversed, in some bikes the forward wheel was made smaller and the saddle was put further backwards as in the Lawson safety bike, which used a chain to transmit the power from the pedals to the back wheel. By these changes in design, the problem of women, older gentlemen and sportsmanlike riders were all triggering innovation.

Various bikes coexisted by the end of the nineteenth century. When pneumatic tires were developed, the riding characteristics of bikes increased considerably. This could be utilized by Lawson's safety bike, as the transmission ratio could easily be adapted. As a result the safety bike developed into a faster bike. In this way, the Penny Farthing lost its sportsmanlike appeal and its fate was sealed.

However, innovation does not follow changing social demands automatically: engineers and designers are trained to solve problems in a specific way. In the SCOT model, this is called the *technological frame*. The technological frame determines the way in which engineers perceive and solve problems. Engineers are often not aware of their own technological frame as they consider it as self-evident. To show the involvement of an actor within a technological frame the term "inclusion" is used. Inclusion in different technological frames at the same time is possible. An electrical engineer has a high inclusion in the technological frame connected to his own discipline, but he may also be skilled in car repairs, i.e. he is included in another technological frame. This enables him sometimes to come up with different technological solutions. Technological frames make engineers adhere to the traditions of their discipline. Disruptive changes in technology are therefore often produced by relative outsiders, or people with low inclusion.

The changes in the preferences of the relevant social groups do not coincide with the changes in technology. New technologies might be a response to preferences of social groups that have already changed. But the preferences of Relevant Social Groups are also affected by the availability of new technologies. The result is a dynamic pattern of co-evolution of technology and society. The "Hype cycle" is a good example of such a mismatch:

> New technologies or products, that have reached a break-through, raise considerable interest. A frenzy of publicity generates over-enthusiasm and high running expectations. Later, this leads to disillusionment. Share values and popular interest dramatically fall. Some companies continue and succeed in developing practical applications. These are gradually accepted, and a productive phase is reached.
>
> (Gartner, 2004)

The debate between constructivism and determinism is not just of scholarly interest. For technological determinists, attempts to influence science and technology for commercial, ethical or political reasons are futile, producing no more than minor ripples in the pond. For social constructivists, everybody is constantly deciding upon the future shape of technologies, even the people who claim to be technological determinists.

THE FORMATION OF TRAJECTORIES: POSITIVE FEEDBACK AND LOCK IN

In the process of technological change, we sometimes get stuck in a specific technology: Microsoft Windows, VHS video recorders, or a 220 Volt/50 Hz electricity grid. The technologies in which we get stuck don't always have to be those technologies that are most efficient. The technology that wins the competition and becomes the standard, i.e. gains a major market share, doesn't have to be the best option. An example of users being caught up in an inefficient technology is the VHS video system. VHS "won" over competitors Betamax and V2000, despite the fact that it was neither the cheapest nor the technically superior system (Lardner, 1987).

The formation of technological paths, which can be inefficient, is a consequence of the occurrence of the positive feedback phenomenon, i.e. increasing returns with increasing market penetration. This means that the more a technology is adopted, the more attractive the technology becomes for others. The telephone is a good example of positive feedback: who would buy a telephone if nobody else had one? Six factors might contribute to positive feedback loops (Arthur 1990, 1996):

1 *Expectations*. The development of a certain technology can be influenced and accelerated

by the expectations people hold as to the success of the technology. Expectations contain a certain image where a future situation is sketched, connections are made and roles are described. Based upon these expectations new actions are undertaken.

2 *Familiarity*. When a technology is better known and better understood, it has an increased chance of being adopted. This is in fact "increasing returns by information."

3 *Network characteristics*. Positive feedback shows up more strongly with technologies possessing network characteristics. It is advantageous for a technology to be associated with a network of users, because this increases availability and the number of product varieties. Again, a good example here is the VHS video system. To be able to function, this technology needs a network consisting of video rental stores stocked with VHS tapes. The more users are present, the better the possibility is for users to profit from VHS-recorded products.

4 *Technological connectivity*. Feedback processes are stimulated by the occurrence of "technological connectivity." Often, a number of other sub-technologies and products get absorbed into the infrastructure of a growing technology. This gives it an advantage over technologies, which would need a partial demolition of that infrastructure to function. The QWERTY keyboard (the name of which refers to the first six keys on the top row of the keyboard) is a good example. Although alternative keyboard configurations are probably more efficient, they never replaced QWERTY as everybody has learned to type on a QWERTY keyboard. Summarized: a technology, which fits into the system of already existing technologies, has a relatively better chance to develop than a technology, which lacks those connections (David, 1985).

5 *Economies of scale*. When an increasing volume of products is produced while the costs per unit production don't increase linearly with it, the price of a product is lowered. This means that a technology can become more economical when it is applied on a larger scale.

6 *Learning processes*. Positive feedback during the development of a technology can finally take the form of a learning process, because a technology can be improved more quickly when more is learned during its use. So when a company learns a lot about using a specific technology but learns little about another, this last technology has less chance of being adopted in the future. *Interactive learning* is a specific form of learning. This kind of learning occurs when different stakeholders in the development process communicate on their mutual views on the technology.

TECHNOLOGICAL DETERMINISM AND SOCIAL CONSTRUCTIVISM AS RECURRING STAGES

The famous law of Moore (1965), stating that the number of components on a microcircuit will double every 18 months, is sometimes regarded to be a proof of autonomous technology development. However, it is clear that at the time that the law was formulated, there was only very little evidence for it. Moore, as a director of Intel, very much influenced the expectations of competitors and customers regarding the potential of microelectronics. Therefore, it could be argued that Moore's law is not an observation of autonomous technology development, but a statement that created a strong consensus among pioneering innovators, that led to such a stable pattern of development.

In the past decades, concepts have been developed that aim to integrate the determinist's main assertion, that technology is self-propelling, with that of the constructivist, that technology is the outcome of socially determined choice. These new concepts seek to account for the conditions under which technological change is propelled predominantly by social forces and the conditions under which social forces can scarcely influence that process.

Stability of technologies is to be explained by the various different environments in which technologies are embedded:

- *The socio-economic environment*: a technology must meet the demands imposed on it by all the relevant social actors in its environment (for a car these demands can refer to aspects such as the price, status, comfort, safety, appeal, or speed of the car). New technologies must, in other words, solve the problems that actors think can be solved by the artifact.
- *The physical environment*: every artifact is adapted to other artifacts, to technological infrastructure, to maintenance systems, energy sources and so on. For example, cars must fit on roadways, use available, standardized fuels, be fitted with familiar steering mechanisms and meet various performance standards before they are approved. New technologies must be compatible with these existing conditions.
- *The technological knowledge base*: technologies are based on the existing know-how, rules and accepted paths for further innovation that are accepted within a particular technological community. New technological artifacts arise from the state of knowledge and the shared beliefs regarding possible improvements within the community of practitioners designing and constructing them. More radical technological change therefore implies changing mainstream ideas within technological communities or breaking their power by creating an alternative technological community to take its place.

Since major upheavals in all three realms are rarely simultaneous, radical innovations are likewise rare. For example, an alternative for the car should be socio-economically viable (in terms of costs–performance ratio) adapted to the existing infrastructure (roads, fuel, regulation), and we should have the know-how to design it. Because of the various forces favoring technological stability, technological artifacts generally change only incrementally, over long periods of time. As in biological evolution, however, a technology may become extinct, or split into several species adapted to specific niches and circumstances.

In the large-scale technological systems of today, social institutions and technological hardware form a seamless web and any distinction between the "social" and "technological" dimensions of these systems becomes futile (Hughes, 1983). Particularly when systems fail, attempts are made to blame casualties on either "human" or "technological" factors. Such attempts are doomed to failure, though, for it is in fact impossible to distinguish the human and technological factors in any given technological system: is it the hardware that is not properly adapted to the humans operating, administering or maintaining it, or are the humans not functioning in accordance with the demands set by the hardware they are dealing with? This question cannot be answered empirically.

Technological changes are often slow and can easily lag behind the more rapid pace of change in society as a whole. However, the creativity of technologists also leads to new products and systems that revolutionize social life, such as cellular phones and computers. This is not to say that every new revolutionary technology is accepted by society. Indeed, many new technologies were not accepted, and are hardly remembered. Civil aircraft for vertical take-off and landing or soluble tablets to replace toothpaste are just some of the vast array of technologies that have been rejected. In the case of nuclear power, the issue of acceptance has still not been settled. It might therefore be argued that, insofar as technology can be distinguished from its social environment, the relationship between them is a (co-) evolutionary one: they adapt to one another, but there are mismatches. Such mismatches occur especially in times of rapid change, due either to massive breakthrough of new technologies, as in the case of IT in the 1980s, or to rapid changes in society's preferences.

The technological innovation process is not just about making the best technology, in terms of price-performance ratio as fast as possible to beat competition. It is about managing complex situations. Timing is crucial, as choices have to be made at certain moments, but cannot be made at other moments. It is also about taking the general issues in society into account as "no company or branch of industry can afford to be at war with society for prolonged periods."

SCIENCE-BASED INNOVATION

The relation between science and technology has changed dramatically during the twentieth century. In the nineteenth century the natural science departments at universities were generally dealing with phenomena that were interesting but without any practical use. Industry or trade was generally little interested in these scholarly activities. Nowadays, scientific research is often crucial for new high-tech products.

The first changes in the roles of science and technology took place in Germany. German chemists developed processes to manufacture synthetic dyes based on coal tar. Bayer AG, in Leverkusen was the first firm to develop a research laboratory directed to the scientific study of chemistry in order to create new chemical products. How to do research and how to organize it within the context of industry was an innovation as such. Competing chemical firms soon followed Bayer's example (Meyer-Thurow, 1982).

The electrical inventions of Thomas Edison fuelled the birth of the electric industry that was underway by 1890. Research soon had to supply the improvements that were needed: The General Electric Research Laboratory in Schenectady, New York, emerged and became famous due to Irving Langmuir's 1932 Nobel Prize in Chemistry. The Du Pont Corporation, Wilmington, Delaware also started a chemical research laboratory. Du Pont made a fortune during the First World War from its invention of smokeless gunpowder. Du Pont decided to continue this successful strategy by setting up even more fundamental research. Frictions between academic research and industrial research often occurred (Hounshell and Smith, 1988). The history of Nylon discovery at Du Pont serves as an interesting example of the problems in the marriage between technology and science.

The creation of Nylon

At the end of 1926, Du Pont decided that it would start a fundamental research program.

One of the main proponents of this fundamental research program, Charles M.A. Stine (1936), later explained the motives:

> Fundamental research assists one to predict the course of development of chemical industry. Pioneering applied research enables one to achieve certain objectives indicated by fundamental research. Therefore, the continued growth (as distinct from mere expansion) of chemical industry is dependent upon fundamental research. That is the basic philosophy of fundamental research.

Stine stated that fundamental research also improved industry–university interaction and created consulting specialists for applied research within the company. The main difference to university research would be that: "In university research, the discovery is a sufficient objective in itself." Du Pont recruited academic scientists who had great liberty to engage in the subjects they thought to be useful. In 1928, Wallace Hume Carothers, a chemist from Harvard University, was enlisted as head of the organic chemistry fundamental research group. Carothers had doubted this step for a long period for industrial research was not high valued by academics. At Du Pont's Experimental Station, he started research on the macromolecular concept of polymers, a subject of great interest, as it was the focal point of debate in chemistry. He made new, long chain polymers by carrying out reactions that were well understood. A real flood of publications emerged from this research project that made a massively documented case for his scientific goal: that polymers were just ordinary molecules, only longer. Carothers published the theoretical points he wanted to make. Initially Carothers did only fundamental research. Du Pont even stimulated publication of results. However, in 1930, management changed. Carothers was urged to aim his research at developing new products based on polymers. He did not resist this pressure. His main concern was his freedom of publication, which was granted to him if he was willing to give the company time for patent applications.

In September 1931, Carothers announced the possibility of obtaining useful fibers from strictly synthetic materials. On February 28, 1935, Carothers synthesized for the first time polyamide 6,6. In July 1935, this polymer was commercialized because the raw materials were comparatively cheap. At the end of 1937, polyamide 6,6 was produced in a pilot plant. The technological problems of production were enormous. The required purity of raw materials was unprecedented and the spinning process differed very much from conventional processes. About 230 engineers were working on the project and more than 200 patents were granted just for this technological work.

At the end of 1938, Du Pont launched it as "Nylon." It was an overwhelming success on the market. The spirit in which Carothers had been leading his team was circumscribed by one of the members of the group, the later Nobel Prize winner Paul Flory: "His approach to science was motivated by boundless curiosity; it was not fettered by superficial boundaries between specialties." Carothers did not live to see the success of "Nylon." He committed suicide in April 1937, deeply depressed and, although being the first industrial chemist admitted to membership of the National Academy of Science, convinced of having failed as a scientist (Mulder, 1992). Scientists like Carothers bridged the gap between academics and industrial technologists. As Harvard president James B. Conant said of Carothers' acceptance of a position at Du Pont: "he had facilities for carrying on his research on a scale that would be difficult or impossible to duplicate in most university laboratories" (Adams, 1961).

Science proved its value to industrial interests but also to military ones. During the Second World War, physicists suspected that Hitler was working on a nuclear bomb. They convinced Einstein to request President Roosevelt to start a research project to develop a nuclear bomb too (Weart and Szilard, 1978). In the Manhattan Project top physicists were gathered to build it. The first nuclear explosion ever took place on July 16, 1945, in the Alamogordo desert, New Mexico, US. The cities of Hiroshima and Nagasaki were the first targets of nuclear attacks, on August 6 and 9, 1945. Physics lost its innocence, i.e. like the other natural sciences it lost its claim to be an independent force for the progress of mankind. Physicists had not only built the bomb, but had also played a main role in the political decision to build and apply it. Physics, as the queen of science, had dirty hands as it was also subject to political and commercial forces (Rhodes, 1986; Herken, 2002).

The interwovenness of technology and science is often called techno-science. It implies that science is no longer accepted as an activity purely directed towards producing objective knowledge, as was the main legitimation for science in the nineteenth century (Van den Daele, 1978).

MAJOR CHANGES IN TECHNO-SCIENCE IN RECENT DECADES

Three main trends are increasing the challenge of management of technology: complexity of technology, globalization and the emancipation of the citizen. These trends are further analyzed below.

Complexity

To state that techno-science is increasingly complex is common sense: phenomena that were not even discovered 50 years ago are utilized in the design of new products. Especially most of our current day information and communication technologies were completely unknown half a century ago. However, technological complexity is not just growing by the development of new, often more complicated artifacts as such. The complexity of modern artifacts is often due to the use of various knowledge realms, which were unconnected before, and the interconnection to other artifacts. The car again serves as a good example. In the 1960s, the car body was only made of steel. The car design essentially only incorporated mechanical knowledge. Nowadays, various kinds of plastics have been introduced, some of them reinforced by glass fiber. Aluminum also plays an increasing role. Electronic equipment is applied to control various functions of the car. Wind tunnel experiments have lowered aerodynamic drag, while rubber research and tire design produced much more complicated, but safer tires (Cf. Prophet, 1998). The number of scientific disciplines that are used in car design has greatly increased. That makes innovation much more complicated. But that is not the only complicating factor. Redundancies in the design have been decreased and therefore every minor change in a detail of the design of a car is far more likely to have consequences for other parts of it. The effect is that innovation is far more complicated: more experts and disciplines are involved, and more adaptations are needed before a design is fit for production.

Globalization

The innovation process for products of increasing complexity requires considerably more effort and resources than it did in the past. Increasing the market could best cover these increasing innovation costs. As the innovation costs are not dependent on the volume of sales, these costs become relatively low when market sales grow. As trade barriers have decreased in recent decades, this created a strong drive for companies to market their products globally. Especially products with low transport costs can easily be marketed worldwide. Products with higher transport costs can be produced in the main regions of the world, according to the main product (and production process) design. Companies that might be unable to operate worldwide either find overseas partners, merge, or disappear. Every traveler can observe the result: some decades ago it was easy bringing home presents to the people back home. Nowadays it is much harder as the offer of presents appears to be identical in every airport of the world.

This effect can also be illustrated by the history of the television. In 1950, when only 10 percent of the US households owned a television set, 33 manufacturers sold the sets on the US market. All of these manufacturers were US companies that were not owned by foreign interests. In the year 2000, 16 manufacturers were supplying TV sets in the US. Only one manufacturer was still a US company: Curtis Mathes.

Table 6.1 *The entrance of foreign television manufacturers in the US (TV History, 2001–2004)*

Company	*Brand*	*website*	*country of origin*	*entrance US market*
Akai		http://www.akaiusa.com/profile.htm	Japan	?
Curtis Mathes		http://www.curtismathes.com/aboutus.asp	USA	1960
Hitachi		http://www.hitachi.com/index.html	Japan	1975
JVC		http://www.jvc-victor.co.jp	Japan	1976
Philips	Magnavox	http://www.consumer.philips.com	Netherlands	1976
Matsushita	Panasonic	http://www.panasonic.com	Japan	1975
Mitsubishi		http://global.mitsubishielectric.com/	Japan	1980
Thomson	RCA	http://www.thomson.net/EN/home	France	1987 (1946)
SAMPO		http://www.sampoamericas.com	Taiwan	1981
SAMSUNG		http://www.samsung.com	Korea	1989
Sanyo		http://www.sanyo.com/home.cfm	Japan	1977
SANSUI		http://www.sansui.co.jp/info/index.cfm	Japan	1987
Sharp		http://sharp-world.com/index.html	Japan	1983
Sony		http://www.sony.net/	Japan	1961
Tatung		http://www.tatung.com/	Taiwan	1979
Toshiba		http://www.toshiba.co.jp	Japan	1976
LGE	Zenith	http://www.lge.com/index.do	Korea	1999 (1948)

The reduction of the number of suppliers might be seen as remarkable, but there is a much more interesting phenomenon: in the 1950s and early 1960s, the manufacture of television sets in all industrialized countries was completely dominated by national industries. Nowadays, the television brands that one can buy in various countries are virtually the same. Except for Curtis Mathes, most brands are sold in various parts of the world. A small sample of three Internet shops in the Netherlands shows that ten TV set manufacturers also marketed their products in the US, while there were three more manufacturers (Finlux-Finland, Vestel-Turkey, Orion-Singapore) that were still working at their global expansion. This phenomenon can be observed not only for television sets. It applies to many manufactured goods such as cars, radios and washing machines. We can conclude that as consumers, we have somewhat less choice of suppliers. But far more important is that most people in the world have more and more identical choices. The world becomes more uniform, and therefore perhaps somewhat less exciting.

As corporations tend to operate more and more at a global scale, competition increases. The increased competition implies that no company can get away with products that do not meet today's technological standards. More speed of innovation is required. However, as corporate resources are limited, the company needs to focus its efforts. The effect is that companies widen their geographic area of activities while narrowing down the range of products that they make. Large conglomerates, which in the 1970s produced products as varied as coffee, coatings, drugs, steel wire, shipping and plastics, have now generally focused their activities on only one product sector.

Emancipation of the citizen

The 1960s and 1970s have formed a watershed in the political history of many Western countries. Until the end of the 1960s, the parliamentary system implied that parliaments, and the governments that were controlled by them, had almost absolute powers. The influence of citizens was limited to the elections. Formally this is still true. However, it can be observed in most countries that various groups of citizens do not accept a decision purely because of its parliamentary legitimation. Large shares of the population of industrialized countries are highly educated. They do not take government actions for granted if their own interests or convictions are at stake. Governments can no longer neglect citizens' protests, and therefore companies should not only apply for government licenses for their activities, but also seek for support within the population. Table 6.2 shows the participation in higher education in several European countries (selection of countries based on availability of data) (UNESCO Institute for Statistics, 2004).

Overall, the number of students increased by a factor of 3.26 between these years.

Before the 1960s, technology was generally regarded to be a positive sum game: the benefits of new technology always exceeded the costs. Therefore, the people who tried to stop certain innovations were barriers to progress. This image changed in the 1960s. Some technologies, such as agricultural chemicals, turned out to have far more disadvantages than advantages for society. Nuclear weapons, which were developed to make the world safer, created for the first time in human history the possibility of a total human self-destruction. It was often questionable if the sum would be positive. Moreover, the distribution of advantages and disadvantages

Table 6.2 *Participation in higher education in several European countries*

	1970	1996
Albania	25,469	34,257
Austria	59,778	293,172
Bulgaria	99,596	262,757
Finland	59,769	226,458
Iceland	1,706	7,908
Italy	687,242	1,892,542
Netherlands	231,167	468,970
Norway	50,047	185,320
Romania	151,885	411,687
Spain	224,904	1,684,445
Sweden	144,254	275,217
United Kingdom	601,300	1,891,450

was problematic: chemical producers and farmers had some benefits but the chemicals destroyed ecosystems from which, for example, fishermen earned their living.

At the end of the 1960s, the US Congress decided to start *Technology Assessment*; to create an office for assessing the (broad) effects of a new technology in terms of costs and benefits.

To assess these effects is rather hard. The direct effects, including unintended effects, are not always clear beforehand, but what about the second- and third-order effects that are caused by changed patterns of behavior because of the availability of the technology? People nowadays tend to live at larger distances from their work because of the availability of transport, but could this geographic consequence be foreseen when the car became a widely available means of transport?

The intended effect of the anti-conception pill was a more reliable and easier control of human reproduction. It permitted people to behave more promiscuously, a second-order effect. A third-order effect was an increase in sexually transmitted diseases. This caused many people to return to more traditional means of birth control.

However, it is often hard to predict these causal relations and their consequences in advance. The US Congress adopted a moderate definition:

> Technology Assessment is a form of policy research, which provides a balanced appraisal to the policymaker. Ideally, it is a system to ask the right questions and obtain correct and timely answers. It identifies policy issues, assesses the impact of alternative courses of action and presents findings. It is a method of analysis that systematically appraises the nature, significance, status, and merit of a technological program.
>
> (Daddario, 1968, p. 10)

Technology assessment also makes sense for the private sector. Companies could argue that compliance to the law is sufficient when introducing technological innovations. However, the introduction of new technologies that cause effects to third parties, although legally allowed, will often trigger the introduction of new legislation or lawsuits. At the end, this might ruin the business that was intended by introducing the innovation (Malanowski *et al.*, 2001). Therefore, the public legitimacy of corporate activities can no longer be dealt with by complying with legislation. It also requires taking societal values into account in order to be trusted. Local governments might support protests against corporate decisions. Moreover, the citizen might use its role as a consumer to put pressure on companies. The power of the consumer makes even the largest multinational vulnerable. Illustrative is what happened to Shell in the Brent Spar case:

> In 1995, the UK government allowed Shell to sink an old oil production platform in the North Atlantic. Greenpeace objected and tried to stop it but was removed from the platform. However, Shell's gas stations throughout Europe became deserted places. Various politicians joined actions against Shell. Ultimately, Shell had to change its plans and the Brent Spar was dismantled and reused in Norway
>
> (De Groot van Embden, 2001)

An important lesson from the Brent Spar affair is that the environmental protest does not necessarily have to be right: it is open for debate whether the final solution was better for the environment than dumping the platform on the ocean floor. However, what counts is that Shell's arguments were less credible in the eyes of the public than Greenpeace's arguments.

DEALING WITH THE CONSEQUENCES

The world as a global community, global issues, global cooperation

Many corporations must act on a global scale. They therefore must learn to deal with various cultures within their organization and external to it. Complying with the law in each nation is not enough: consumers and employees of the company will protest against exploitation of labor in poor countries, environmental destruction, or racial discrimination. Therefore, firms need to adapt to local circumstances but also need to develop corporate standards of behavior.

Globalization also means that firms need to be present in the main technological hot spots. Although one might argue that it might be most efficient to concentrate research and development at one location, it is often not very wise. A research laboratory acts as a node to make contact with a local research community. For a Western company it will be hardly possible to set up cooperation with a Japanese research unit without actually being present there.

Assessing the consequences of technological innovation

Technology assessment in the public sector has been carried out for more than 30 years. After the US Congress started its Office of Technology Assessment in 1972, various industrialized nations established similar institutes in the 1980s and 1990s (Smits, 1990). The changes in theorizing on technological change, that is to say the debates regarding autonomous technology or socially constructed technology, did not leave technology assessment unaffected. The technology assessment institutes gradually shifted their activities from merely advising parliaments on consequences of new technology, to strategies for steering technology and promoting active participation of citizens in discussions on new technologies (Van Eijndhoven, 1997; Rip *et al.*, 1995).

In business, technology assessments were also made (Maloney, 1982; Fleischmann and Paul, 1987; Simonse *et al.*, 1989) although these efforts were also criticized as being too narrow minded (Coates and Fabian, 1982). Technology assessments in business were often more or less extended technological forecasts. Unilever Research director, Professor Wiero Beek said in 1988:

> For me, the overriding question now is not so much to get TA even better institutionalized in scientific circles (as a scientific discipline, TA, given its subject matter, will always be weak in terms of logical conviction), but to get it better institutionalized in political and managerial circles.
>
> (Beek, 1995, p. 84)

The deep analyses were often not very effective. Therefore, companies chose instead to set up project teams of various specialists. Scenario planning became popular (Van Ginneken and Van Hulst, 1996), probably because this method allowed many people to be involved in the strategy process. The success of scenario planning was that it allowed companies to involve a large part of their human resources to participate in the creative thinking process on the future.

When facing controversial issues, most companies were very careful. Companies with interests in biotechnology took great care. In 1993, Unilever initiated consultations with various consumer and environmental organizations. Other companies joined these consultations (Anonymous, 1993). One of them, Gist Brocades, had developed a biotechnological process to manufacture chymosine, a coagulant to produce cheese. However, the dairy industry refused to use it, afraid it would lose its (export) market. Gist Brocades and three other companies joined in another project, which aimed at finding a way out of the deadlock on modern biotechnology and food. The report advised corporations to take the issue of public acceptance seriously. In practice this advice implied that public acceptance should not be left to the final R&D stages, corporations should be more open to the public, deal seriously with the public (even if it is "irrational") and should be willing to adapt products and processes. Corporations should prepare for interaction with their environment, optimize learning effects and build up a corporate memory (Jelsma and Rip, 1995).

So, technology assessment studies can be important for companies, especially if they contribute to setting up dialogues with citizens (for an overview of methods, see: Van den Ende *et al.* (1998)).

Systems innovations need inter-firm and public–private collaboration

For innovation, the globalization process creates a problem: the core business of firms is narrowed down, but the firms operate worldwide. Therefore an increasing part of innovations will require various types of technological change that are not all available within the R&D department of one firm. Senseo, the coffee machine developed by Philips Electronics and Sara Lee/DE consumer products is a good example of such an innovation: This new way of making coffee comprised a new coffee machine, a new coffee blend, and new packaging for the coffee. Inter-firm cooperation is therefore very important. However, setting up a good cooperation is an art: the main factor for success is trust between partners that recognize the values of the mutual contributions and respect mutual interests (EIRMA, 2002).

THE FUTURE OF MANAGEMENT OF TECHNOLOGY

Probably, I have discussed so many subjects that the reader must be convinced that innovation requires the ability for managing various heterogeneous subjects. Management of technology is like playing chess against some dozens of opponents simultaneously. The main message that is conveyed in this chapter is that entrepreneurial profit is, in fact, the reward for contributing to society. Profits that are acquired without a positive contribution to society are generally soon terminated by legal or consumer actions. But contributing to society is not enough. Being in time is crucial, just as communication with stakeholders is.

Innovation is a challenge for the coming decades, not just for economic growth, but also to enable high standards of living for a growing part of the world population. China and India are rapidly catching up with Western standards of living and other countries will follow. This will lead to a sharp increase in resource consumption, even if the industrialized countries could increase their efficiencies of production. Natural resources are limited and we will be confronted by these limits. Innovation will be a necessity for the future. Meanwhile it has been shown in this chapter that innovation has become increasingly complex and hard, for individual companies as well as for national authorities. Managing the innovation process is therefore a challenge of unprecedented magnitude.

FURTHER READING

Bijker, W.E. and Law, J. (1992). *Shaping Technology/Building Society: Studies in Sociotechnical Change*. Cambridge, MA: The MIT Press.

Bijker, W.E., Hughes, T.P. and Pinch, T. (eds) (1987). *The Social Construction of Technological Systems. New Directions in the Sociology and History of Technology*. Cambridge, MA: The MIT Press.

Coates, J.F., Mahaffie, J.B. and Hines, A. (1997). *2025, Scenarios of US and Global Society Reshaped by Science and Technology*. Greensboro, NC: Oakhill Press.

Cramer, J. (2003). *Learning about Corporate Social Responsibility, the Dutch Experience*. Amsterdam: IOS Press.

Malanowski, N., Kruck, C.P. and Zweck, A. (eds) (2001). *Technology Assessment und Wirtschaft, Eine Länderubersicht*. Frankfurt/New York: Campus Verlag.

Rip, A., Misa, T. and Schot, J. (1995). *Managing Technology in Society, the Approach of Constructive Technology Assessment*. London/New York: Frances Pinter.

REFERENCES

Adams, R. (1961). Wallace Hume Carothers, 1896–1937. In: Farber, Eduard, *Great Chemists*. New York/London: Interscience Publishers, 1599–1611.

Anonymous (1993). Biotechnologie reclame van Unilever, *Zeno* 1, 12.

Arthur, W.B. (1990). Positive Feedbacks in the Economy. *Scientific American*, 262 (February), 80–85.

Arthur, W.B. (1996). Increasing Returns and the New World of Business. *Harvard Business Review*, July–August, 100–109.

Beek, W.J. (1995). Technology Impact Assessment. In: Beek, W.J., *Vertoog en Ironie 2, een bundeling van publikaties van Prof. Dr. Ir. W.J. Beek (1987–1995)*. Nederlandse Unileverbedrijven, 74–88.

Bijker, W.E. (1990). *The Social Construction of Technology*. Eijsden.

Celente, G. (1997). *Trends 2000: How to Prepare for and Profit from the Changes of the 21st Century*. New York: Warner Books.

Coates, V.T. and Fabian, T. (1982). Technology Assessment in Industry, a Counterproductive Myth? *Technological Forecasting and Social Change*. 22 (3 & 4), 331–341.

Daddario, E.Q. (1968). In: Subcommittee on Science, Research and Development of the Committee on Science and Astronautics, US House of Representatives, 90th Congress, 1st Session, Ser. 1. Washington, DC: US Government Printing Office, Revised August 1968.

David, P.A. (1985). Clio and the Economics of QWERTY. *American Economic Review*, 75(2), 332–337.

De Groot van Embden, K. (2000). *Een geschenk uit de hemel, De slag om de Brent Spar*. Documentary. Hilversum: VPRO-television.

De Wilde, R. (2000). *De Voorspellers, een kritiek op de toekomstindustrie*. Amsterdam: Uitgeverij de Balie.

EIRMA (2002). *Innovation through Collaboration*. Report of special task force presented at annual meeting. Vienna, Paris: EIRMA, CD-rom.

Ellul, J. (1964, first edition 1954). *The Technological Society*. New York: Vintage Books.

Fleischmann, G. and Paul, I. (1987). *Technikfolgen Abschätzung in der Industrie der Bundesrepublik Deutschland, Ergebnisse einer empirischen Untersuchung im Auftrag BMFT*. Frankfurt am Main: Goethe Universität.

Gartner Inc. (2004). http://www.gartner.com/Init (November 30, 2004).

Gibbs-Smith, C.H. (1974). *Flight through the Ages, a Complete Illustrated Chronology from the Dreams of Early History to the Age of Space Exploration*. New York: Thomas Y. Crowell.

Henderson, L.J. (1917). *The Order of Nature*. Cambridge, MA/London: Harvard University Press.

Herken, G. (2002). *Brotherhood of the Bomb: The Tangled Lives and Loyalties of Robert Oppenheimer, Ernest Lawrence, and Edward Teller*. New York: Henry Holt.

Hounshell, D.A. and Smith, J.K. (1988). *Science and Corporate Strategy. Du Pont R&D, 1902–1980*. Cambridge: Cambridge University Press.

Hughes, T. (1983). *Networks of Power. Electrification in Western Society, 1880–1930*. Baltimore, MD/London: The Johns Hopkins University Press.

Jelsma, J. and Rip, A. (1995). *Biotechnologie in bedrijf, een bijdrage van CTA aanbiotechnologisch innoveren.* Den Haag: Rathenau Instituut. English summary available ("Biotechnology in Business").

Lardner, J. (1987). *Fast Forward: Hollywood, the Japanese and the Battle Over the VCR.* New York: W. W. Norton & Company.

Malanowski, N., Kruck, C.P. and Zweck, A. (eds) (2001). *Technology Assessment und Wirtschaft, Eine Länderubersicht.* Frankfurt/New York: Campus Verlag.

Maloney Jr, J.D. (1982). How Companies Assess Technology. *Technological Forecasting and Social Change,* 22 (3 & 4), 321–329.

Meyer-Thurow, G. (1982). The Industrialization of Invention: A Case Study from the German Chemical Industry. *ISIS* 73 (268), 363–381.

Moore, G.E. (1965). Cramming more Components onto Integrated Circuits. *Electronics* 38(8), April 19, (ftp://download.intel.com/research/silicon/moorespaper.pdf, December 1, 2004).

Mulder, K.F. (1992). Replacing Nature, the Arising of Polymer Science and Synthetic Fiber Technology. In: Gremmen (ed.), *The Interaction between Technology and Science.* Wageningen University: Studies in Technology and Science volume 3, 239–262.

Mulder, K.F. (1992a). *Choosing the Corporate Future. Technology Networks and Choice Concerning the Creation of High Performance Fiber Technology.* University of Groningen, dissertation.

Pinch, T. and Bijker, W. (1987). The Social Construction of Facts and Artifacts: Or How the Sociology of Science and the Sociology of Technology Might Benefit Each Other. In: Bijker, W., Hughes, T.P. and Pinch, T. *The Social Construction of Technological Systems. New Direction in the Sociology of Technology.* Cambridge, MA: The MIT Press, 17–50.

Prophet, G. (1998). *Cars and complexity,* http://www.edn.com/archives/(1998/090(198/18eed.htm (December 2, 2004).

Rhodes, R. (1986). *The Making of the Atomic Bomb.* New York: Simon & Schuster.

Rip, A., Misa, T. and Schot, J. (1995). *Managing Technology in Society: The Approach of Constructive Technology Assessment.* London/New York: Frances Pinter.

Simonse, A., Kerkhoff, W. and Rip, A. (eds) (1989). *Technology assessment in ondernemingen.* Deventer: Kluwer.

Smits, R.E.H.M. (1990). *State of the art of Technology Assessment in Europe,* A report to the 2nd European Congress on Technology Assessment. Milan, November 14–16.

Stine, C.M.A. (1936). The Place of Fundamental Research in an Industrial Research Organization. *Transactions of the American Institute of Chemical Engineers* 32, pp. 127–137.

Terra, R. (1999). *Congressional Hearings Favor Doubling U.S. Funding for a National Nanotechnology Initiative.* Foresight Update 37, p. 1.

TV History, 2001–2004, *Television History – The First 75 Years,* http://www.tvhistory.tv/index.html (December 2, 2004).

Unabomber (1995). *The Unabomber's Manifesto, Industrial Society and its Future.* First published on Sept. 19, 1995. The Washington Post, http://www.panix.com/~clays/Una/ (November 25, 2004).

Unesco Institute for Statistics (2004). http://www.uis.unesco.org (March 15, 2004).

Van Andel, P. (1994). Anatomy of the Unsought Finding. Serendipity: Origin, History, Domains, Traditions, Appearances, Patterns and Programmability. *British Journal of Philosophy of Science*, 45, 631–648.

Van den Daele, W. (1978). The Ambivalent Legitimacy of the Pursuit of Knowledge. In: Boeker, E. and Gibbons, M., *Proceedings of the conference on Science, Society and Education*, Amsterdam.

Van den Ende, J., Mulder, K., Knot, M., Moors, E. and Vergragt, P. (1998). Traditional and Modern Technology Assessment: Toward a Toolkit. *Technological Forecasting & Social Change*, 58(1 & 2), 5–21.

Van Eijndhoven, J.C.M. (1997). Technology Assessment: Product or Process? *Technological Forecasting and Social Change*, 54(2 & 3), 269–286.

Van Ginneken, B. and Van Hulst, W. (1996). Het scenariodenken in bedrijven, *Zeno*, no. 1, 4–7.

Weart, S.R. and Szilard, G.W. (eds) (1978). *Leo Szilard: His Version of the Facts; Selected Recollections and Correspondence*. Cambridge, MA: The MIT Press.

Chapter 7

Development and diffusion of breakthrough communication technologies

J. Roland Ortt

OVERVIEW

Previewing the diffusion pattern of breakthrough communication technologies seems relatively straightforward because empirical studies have distinguished similar S-shaped diffusion patterns for technologies such as telephony and television. However, some recent breakthrough communication technologies, such as interactive television and broadband mobile communication technology, show a more erratic diffusion pattern. Introduction of these technologies is often postponed. Once introduced, these technologies are withdrawn from the market after the first disappointing results. Several explanations for the different patterns are possible. By studying the process of development and diffusion of five breakthrough communication technologies, this chapter shows that the S-shaped curve and the more erratic curves represent subsequent phases in one pattern of development and diffusion of breakthrough communication technologies. Managerial implications of the differences between the phases are discussed. Previewing the pattern and the current position of the technology in this pattern are important since successful strategies in one phase may prove to be disastrous in another phase.

INTRODUCTION

Technological innovation is a prerequisite for the long-term viability of many companies. Therefore, considerable resources are invested in research and development of new technologies. R&D managers often have serious difficulties in establishing the proper allocation of these resources to different technologies because the (market) information to establish the long-term potential of these technologies is often not available. For the managers involved in the development and commercializing of a breakthrough technology, it is of vital importance to understand the pattern of development and diffusion because this pattern determines the short- and the long-term market potential. Many pioneers trying to introduce a breakthrough technology experience difficulties in previewing this pattern. In some cases, for instance Internet technology, the short-term market potential is underestimated. As a result, the pioneers invest too late and run the risk of losing their market position. In most cases, however, the pioneers overestimate the (short-term) market potential of their technology. As a result they invest too much, incur large losses, lose their market position, leave the market and subsequently witness that other companies, which introduce the same technology some years later, generate huge profits. Olleros (1986) refers to this phenomenon as "the burnout of the pioneers."

The chapter will illustrate patterns of development and diffusion of breakthrough technologies by investigating historical cases in the field of (tele)communication. Breakthrough technologies are characterized by a dramatic change in attainable price/performance ratios, or by new kinds of performance (Tushman and Anderson, 1986). Telegraphy, radio, fax, television, mobile communication, and Internet technology are examples of breakthrough communication technologies since they enabled new types of performance or drastically altered the price/performance ratios of existing technologies.

When looking at the actual diffusion of breakthrough communication technologies such as the telephone, radio, and television, similar S-shaped curves can be observed (Miles, 1988; Rogers, 1986; Williams *et al.*, 1988). The S-shaped curve seems to be a robust model. The similarity of the curves seems to indicate that predicting diffusion of these technologies is straightforward. Alternative theoretical perspectives on this S-shaped diffusion curve and their management implications will be described in the next section.

In practice, however, expectations turn out to be optimistic and of a dubious value for many new technologies (Schnaars, 1989; Wheeler and Shelley, 1987). In particular, some communication technologies that have been introduced since 1970, such as videoconferencing (Clarke, 1990) and interactive television (Schnaars, 1989), have been confronted with a disappointing number of adopters. The diffusion of these communication technologies cannot be captured in a simple S-shaped curve (Easingwood and Lunn, 1992).

Alternative explanations are possible for the difference in the diffusion patterns. A first explanation is that the diffusion of *successful* technologies follows an S-shaped curve, whereas the *unsuccessful* technologies tend to follow more irregular patterns. A second explanation is that different patterns of diffusion can be distinguished even among successful communication technologies. The findings in this chapter will lead to another explanation. Therefore, empirical findings regarding the development and diffusion of some breakthrough communication technologies will be discussed in the third section. A model to describe this phenomenon will be presented in the fourth section and the management implications of this model will be outlined in the fifth section. The chapter will close with some concluding remarks.

THE S-SHAPED PATTERN OF DIFFUSION: THEORETICAL PERSPECTIVES

After introducing the S-shaped diffusion curve, sociological, psychological, economic, and technological perspectives on this curve will be contrasted and the management implications of these perspectives will be described.

The S-shaped diffusion curve

Diffusion refers to the gradual adoption of an innovation in a market segment or in a society. Diffusion studies usually describe the rate at which a particular product form incorporating the new technology is adopted. The diffusion is often depicted as an S-shaped curve indicating the cumulative percentage of a population that adopts a product in the course of time. The shape illustrates the initial low number of adopters, then the rise of the adoption rate until, finally, the number of adopters approaches a maximum. The rate of adoption is related to the steepness of the diffusion curve, while the potential market is related to the maximum height of the diffusion curve (see Figure 7.1). Some products are adopted by households, e.g. the first telephones, while other products are adopted by an individual, e.g. the mobile phone. So the "adopter" may refer to different units. The potential market refers to the maximum number of adopters that can reasonably be expected.

Sociological perspective on innovation diffusion

Sociologists consider an innovation primarily as an idea. From this perspective, diffusion is the communication of this idea across a population. A simple sociological model assumes that all population members behave similarly when communicating about innovative ideas. One of the members begins to communicate the idea to other members, who, in turn, share this idea with other members. The diffusion starts off as an exponential function until, after some time, members try to communicate the idea to other people who have already heard it. So, after a while, the exponential function bends downward until the entire population is informed and the function approaches its maximum. The result looks like an S-shaped curve.

A managerial implication of this perspective is that the speed of diffusion can be increased by means of a deliberate communication strategy. Different forms of communication can be applied depending on the

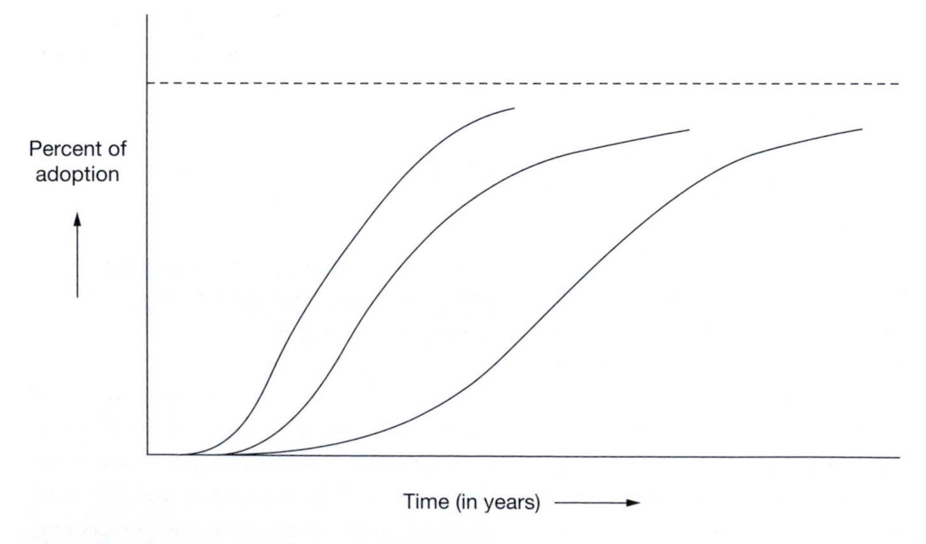

Figure 7.1 *Examples of S-shaped diffusion curves*

specific goal of the message. Mass communication, for example, can be applied to attract the attention of a large group of potential users, to raise their interest or curiosity and thereby to increase the speed of person-to-person communication. Another communication goal is to explain the basic idea of the innovation to potential users. If none of the potential users is familiar with the innovation, communication of the basic idea means that the working principle of the innovation should be explained, and the advantages of using the innovation in practice (benefits) should be communicated clearly. Lectures, discussions with groups of potential users, public demonstrations of the innovation or personal sales visits, represent communication strategies that can convey these types of messages and thereby increase the speed of diffusion of the idea. Mass communication strategies primarily increase the speed of person-to-person communication whereas other strategies, to some extent, bypass the process of person-to-person communication.

Psychological perspective on innovation diffusion

Psychology offers at least two refinements to the previous model: (1) subgroups are distinguished in a population, such as innovators, members of the early and late majority and laggards; (2) adoption of an innovation by an individual involves a process with multiple phases (see Kotler, 1991, p. 195), the last of which are "adoption" and "implementation."

The innovative members of a population hear relatively early about the communication technology and are willing to adopt and implement it relatively early. They usually form a small subgroup. A managerial implication of the idea that subgroups can be distinguished in a population and that the innovators constitute a small percentage of the potential users, is that the diffusion curve often starts slowly. If the innovation subsequently attracts the early majority then diffusion takes off. The fact that innovators adopt, however, does not necessarily imply that the majority of the population will do the same. Innovators are different from the majority of the population and tend to alter their preferences and interests relatively fast. A managerial implication is that the diffusion of some innovations takes off quickly and declines quickly as well. The consequences can be profound: diffusion can die out and a careful segmentation of the market is required to prevent this.

The managerial implication of the existence of an adoption process is that potential users may hear about the idea but do not proceed with adoption and implementation of the innovation. So, the basic idea of an innovation can be communicated rapidly among the entire population although the diffusion may not start. Diffusion requires more than communication of the idea, it also requires deliberate strategies to influence the evaluation of the idea, to stimulate the opportunity for potential consumers to try the innovation, and finally, diffusion requires deliberate strategies to stimulate adoption and implementation of the innovation.

Economic perspective on innovation diffusion

Diffusion models based on sociology and psychology tend to focus on the potential consumers or the demand side of the market. Economic models explain diffusion by looking at the way actors on the supply and the demand sides of the market interact.

A managerial implication of this notion is that introduction strategies should also consider actors on the supply side of the market. Examples of strategies focused on the supply side of the market are strategies aimed at distributors of the innovation and strategies to promote standards among producers of the product or among suppliers of complementary products.

Technological perspective on innovation diffusion

The technological perspective focuses on the development of the technology rather than its diffusion. As a result, the performance of the technology is chosen as a dependent variable rather than the cumulative adoption of innovations incorporating the technology. The increase in performance of a specific technology in the course of time also tends to follow an S-shaped curve. The relationship between both S-shaped curves can be summarized in three statements:

1. The performance pattern can start once the technology is demonstrated for the first time whereas the diffusion pattern starts after the market introduction of a product or process incorporating the technology. In practice, the market introduction can be several years after the first demonstration of a technology.
2. The performance of a technology will generally increase sharply once considerable resources are devoted to its improvement. This is more likely when the technology diffuses widely and ample resources become available to invest in this improvement. So, the large increase in performance and the large increase in diffusion (when both S-curves bend upward) are likely to coincide.
3. The attainment of the maximum level of both S-curves can have different causes and therefore will not necessarily coincide. The maximum level of the diffusion curve is attained once all potential users have adopted products or processes on the basis of the technology. The maximum level of the performance curve is attained once a theoretical maximum performance is attained (because of the laws of nature).

It is remarkable that the sociologists, psychologists and economists have long ignored the process of development of technology that precedes diffusion. Economists, for example, used to consider the process of technology development as a kind of black box.

A managerial implication of the technological perspective is that the diffusion of products and processes that incorporate a specific technology should be managed in close coordination with the efforts to improve the performance of this technology. In some cases, when the technology reaches a specific performance level or a specific price level, new applications of the technology become attainable. Airbags and ABS (automatic braking system), for example, originated from the aviation industry. When performance increased and price decreased, these technologies became available in the mass market of the automotive industry.

An overview of the four perspectives can be found in Table 7.1. The way proponents of these perspectives perceive the diffusion of innovations, the assumptions and the managerial implications of these perspectives are summarized.

EMPIRICAL FINDINGS

It will be shown that the S-curve has to be extended to capture the pattern of development and diffusion of breakthrough communication technologies such as the telegraph, telephone, fax, radio, and television technology. This extension has considerable

Table 7.1 *Different perspectives on diffusion of innovations*

Discipline/Aspects	*Sociological perspective*	*Psychological perspective*	*Economic perspective*	*Technological perspective*
Understanding of diffusion of innovation	Diffusion of an innovation is considered an idea that is communicated across a population.	Diffusion of an innovation is considered as the adoption and use of a product by subsequent groups of individuals.	Economic good (product) diffuses because of the forces of supply and demand.	Innovation incorporates technology. Technological performance develops before and during diffusion of products incorporating the technology.
Assumptions	Focus on demand side of the market. Each person communicates the idea of an innovation to multiple persons in peer group until the entire population is aware of the idea.	Focus on demand side of the market. Subgroups are distinguished within the population on the basis of their innovativeness. Adoption and use are the result of an adoption process.	Supply and demand factors interact during the diffusion of innovations.	Performance is upward bounded by theoretical limits, physical barriers or other reasons. Innovation is needed to shift the performance.
Managerial implications	Deliberate communication strategies can speed up or even by-pass peer-to-peer communication in a population.	The first target group to which innovation is communicated can determine whether an innovation diffuses widely or dies out. Diffusion requires more than communication of the idea, it also requires deliberate strategies to influence the evaluation of the idea, to stimulate trying behavior, and to stimulate adoption and implementation of the innovation among potential users.	A need on behalf of a group of consumers is not sufficient for diffusion. The diffusion of the innovation should also fulfill the needs or business goals of actors on the supply side of the market.	Diffusion of products and processes that incorporate a specific technology should be managed in close coordination with the efforts to improve the performance of this technology.

managerial implications and these will be discussed later on. Four aspects of breakthrough communication technologies are described: (1) the point in time that a breakthrough communication technology is invented, (2) the process of technical refinement and development of the technology, (3) the moment of the first application of the technology, and (4) the start of the wide-scale adoption of the technology. These aspects correspond with the four columns in Table 7.2. Before looking at the table the four aspects will be defined.

1 *The point in time that a breakthrough technology is invented*: The invention is defined as the first demonstration of a technological breakthrough. The idea that an invention can be unmistakably attributed to one inventor at a specific moment in time has been questioned by some authors (Sahal, 1981; Agarwal and Bayus, 2002). An illustration of their point of view is the fact that many things are invented independently and almost simultaneously by different persons. In such cases, the names of multiple inventors will be given. Furthermore, from an evolutionary perspective on technology development, it is hard to distinguish a single invention among a line of gradual improvements in the technology (Bassala, 2001). An invention can be defined in many ways, ranging from the moment an idea is presented, a patent is filed or the principle of a technological breakthrough is demonstrated, to the start of the first pilot application of a breakthrough technology. When possible, both the time a patent was granted, and the time a technology was first demonstrated in public are shown in Table 7.2.
2 *The process of technical refinement and development of the technology*: Some hallmarks in the development of the technology after the invention are listed in Table 7.2.
3 *The moment of the first application of the technology in the market*: The timing of the first known commercial or practical applications of the breakthrough communication technologies is described in Table 7.2. A pilot in the market without a commercial goal is not considered to be a first application of the technology.
4 *The start of the wide-scale adoption of the technology*: Although the mainstream applications for each of the breakthrough technologies are well known, it is hard to define precisely the moment of "wide-scale adoption." We will therefore indicate in which decade the diffusion of products, on the basis of the breakthrough communication technologies, increased significantly.

A first conclusion from our analysis is that the average time from invention to the first market introduction is between seven to ten years for these breakthrough communication technologies. The time from invention to the first market introduction is seven years for the telegraph and sixteen years for the fax technology. For the telephone, this time interval is either one or fourteen years, depending on the question as to whether 1863 or 1876 is considered the date for the invention of the telephone. For radio technology this time interval is about four years and for television technology it is either six or ten years, depending on the question as to whether the demonstration of a mechanical television system in 1925 or the demonstration of an electronic television system in

Table 7.2 *Invention, technical refinement and application of breakthrough communication technologies*

Diffusion/ technology	*Date of invention*	*Technical refinement*	*First applications*	*Widely used application*
Telegraph	1837: Morse in US and Steinhill in Germany demonstrate telegraph; 1837: Cook and Wheatstone get a patent in the UK.	Number of words/minute increases; 1855: the first letter print telegraph; 1874: multiple use of one channel; 1919: introduction of telex.	1844: first telegraph line in the US. Applications: railroad traffic control; transmission of news.	Public use of telegraphy. Stations in post-offices and organizations. Diffusion takes off during the 1850s. International standards after 1865.
Fax	1843: first patent granted to Bain; 1847: first image transfer.	1902: experimental fax transmission of photographs using optical scanning; speed and quality of transmission improves steadily between 1902 and now.	1863: first commercial fax system between Lyon and Paris. Applications: 1906: fax between newspaper offices; sending weather charts to ships.	After 1960 fax becomes popular in Japanese and (after 1970) in European business.
Telephone	1863: transmission of sound by Reis; 1876: telephone demonstration by Bell.	1878: improvement of the microphone; introduction of better cables and amplifiers increases the range.	1877: burglar alarm service in the US; toy for adults; internal communication in companies; local communication in cities.	More connections after WWI. After WWII, diffusion takes off.
Radio	1896: first radio by Mercian (Italy) and Popoff (Russia); 1898: first demonstration of radio.	Development of crystal detector and later electron tube; 1957: first transistor radio from Philips.	From 1900 on: communication with ships, airplanes; radio amateurs built their own radio. 1897: radios are built commercially.	1932: 1 million radios are sold. Radio becomes a mass medium.
Television (TV)	1925: mechanical TV demonstration by Jenkins (US) and Bain (UK); 1926: mechanical TV demonstration by Baird. 1929: electronic TV demonstration	1928: first trans-Atlantic transmission; 1929: demonstration of a color TV; invention of teletext; 1929–1935 experimental broadcasting in UK.	1935: regular broadcasting service in Germany; 1936: first broadcast in the Netherlands.	After WWII the diffusion of TV takes off (US). TV becomes a mass medium.

1929 is considered to be the date for the invention of television.

A second conclusion from our analysis is that it generally takes a decade or more before diffusion takes off after the first introduction of a communication technology into the market. When a technology is first introduced may be difficult to assess. Establishing when its diffusion takes off proves even more difficult. However, even rough estimates of the intervals between first introduction and a significant increase in diffusion rates reveal that these are considerable. The telegraph was first introduced in 1844, its diffusion took off in the 1850s when increasing numbers of telegraph stations were opened and telegraphy became a public service. This shows a time interval of more than six years between introduction and diffusion take off. Somewhat longer time intervals can be found for the telephone, radio and television (at least a decade each). Diffusion of the fax took off about a century after the first market introduction. Fax transmission was introduced into the market in 1863 but significant increases in diffusion rates would last until the 1960s.

A third conclusion is that most of the communication technologies are used in small-scale specific applications directly after their introduction. These applications are totally different from the more wide-scale and well-known applications. The first telephones, for example, were used as a burglar alarm, as a toy, and as an appliance for internal communication in companies. Telephony was also used by the local telegraph office to transfer telegrams to clients, rather than the telegram being delivered to the home or office. The last application seems to have paved the way for wide-scale telephony since lines to the telegraph office could be connected in pairs to establish a local area telephone conversation. In due course, the telegraph office became a telecommunication center. Similar small-scale applications can be found for the other communication technologies.

Implications of these conclusions for the pattern of development and diffusion

The fact that it takes some years after the invention of a technology before the first product is introduced in the market (conclusion 1) and the fact that it takes at least an additional decade before the diffusion of a successful communication technology takes off (conclusion 2), does not imply that the S-shaped diffusion curve is an inappropriate model. The curve generally starts about a decade after the invention of a technology and the curve is stretched at the beginning of the diffusion process. A managerial implication of this finding is that investment in breakthrough technologies requires a long-term investment.

The fact that the first small-scale applications of a communication technology often differ from the later wide-scale application (conclusion 3) also has important implications. It implies that the early stage of the diffusion process is hardly captured in a single S-shaped curve. Each of the small-scale applications can be described in a separate diffusion curve. These small-scale applications have an important role in stimulating wide-scale diffusion of the technology. A managerial implication of this finding is that, after introduction a flexible strategy has to be adopted in which the application and product can be adapted easily. In section five the managerial implications will be discussed further.

PHASES IN THE PATTERN OF DEVELOPMENT AND DIFFUSION OF BREAKTHROUGH COMMUNICATION TECHNOLOGIES

The S-shaped diffusion model aims to describe the diffusion of product forms rather than technologies (Dosi, 1982; Clark, 1985). Based on an analysis of five cases, an extended model is developed to describe the development and diffusion of a breakthrough communication technology. Three phases in this process will be distinguished, the last of which is represented by the well-known S-shaped curve (see Figure 7.2).

The beginning and the end, the average length, and the market actors and factors that generally play a major role, will be described for each phase.

The innovation phase

The first phase, which we will call the innovation phase, comprises the period from invention of a technology up to the first market introduction of a product incorporating the technology. After the invention, a technology is available in some rudimentary form. In the innovation phase this technology is transformed into a marketable "product." Several authors describe the difference between invention and innovation (e.g. Dosi, 1982; Mansfield, 1968; Utterback and Brown, 1972; Weiss and Birnbaum, 1989). The basic difference is that an invention is just an idea in some form, a sketch, a model or a kind of prototype, while an innovation is something that is actually marketable.

The length of this phase can vary considerably. Periods are found between one (for the telephone) and sixteen years (for the fax) between invention and market introduction for five breakthrough communication technologies. Mansfield (1968) claimed that the average time from invention to the start of the commercial development process is about ten to fifteen years. From the start of this process up to the market introduction, again, a couple of years elapse. Utterback and Brown (1972) estimate that, on average, this takes an additional five to eight years. So, according to these authors the period from invention to the first market introduction comprises fifteen to twenty-three years. Agarwal and Bayus (2002) found an average period of twenty-eight years between invention and commercialization for thirty breakthrough innovations from diverse industries.

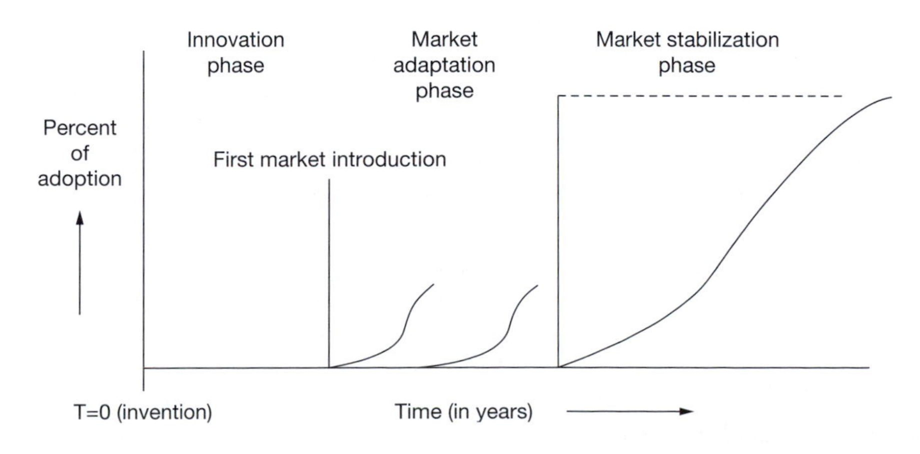

Figure 7.2 *Three phases in the diffusion process*

The large differences in these estimates can be attributed to multiple factors. First, differences can be attributed to the type of industry (Mansfield, 1968). For example, commercialization takes relatively long in the pharmaceutical industry compared to the fast-moving goods industry. Second, considerably different periods are found for technologies in one industry. The radio, for example, was introduced less than four years after its invention whereas the fax was introduced about sixteen years after its invention. Third, the length of the time interval depends on the specific definition of invention. We found time intervals from invention to commercialization for the telephone that varied from one to fourteen years depending on the question as to whether the demonstration of Reis (in 1863) or the demonstration of Bell (in 1876) is considered to be the date of invention of telephony.

During the innovation phase, organizations such as research institutes and universities that are often co-funded by the government, play a central role. To attract potential consumers, the reliability and performance of the technology often has to increase whereas the price of the technology has to decrease. Potential applications have to be found and new products and services have to be developed on the basis of the technology before it can be introduced into the market. Although in this phase no products or services based on the breakthrough technology will be introduced into the market, other types of market mechanisms can be witnessed. In this phase, a good position in the market of supply and demand for research funds and researchers is essential for success.

The market adaptation phase

The second phase, referred to as the market adaptation phase, begins after the first market introduction of a product on the basis of the breakthrough technology and ends when the diffusion of this product takes off. After the first introduction, instead of a smooth S-curve, an erratic process of diffusion may occur in practice. In this situation the market is unstable and, as Clark (1985) puts it, in a "fluid state." The diffusion is characterized by periodic introduction, decline and re-introduction of multiple products in multiple small-scale applications. Such a pattern is common for communication technologies (Carey and Moss, 1985).

We estimated that this phase comprises more than a decade for the five breakthrough communication technologies. An extended period is also found for other technologies (Mansfield, 1968; Utterback and Brown, 1972). "A review of past forecasts for video recorders and microwave ovens illustrates the length of time required for even the most successful innovations to diffuse through a mass market. . . . Both took more than twenty years to catch fire in a large market" (Schnaars, 1989, p. 120). "Most innovations, in fact, diffuse at a surprisingly slow rate" (Rogers, 1983, p. 7). Agarwal and Bayus (2002) indicate that this phase, on average, lasted 18.7 years for breakthrough technologies invented before the Second World War.

Companies try to establish a standard with their product during the market adaptation phase, and competition may intensify. The wide-scale diffusion of a breakthrough communication technology, however, requires coordination in the market among competitors, potential consumers, producers of complementary products or services and suppliers. In practice, this cooperation is hampered by the "chicken-and-egg problem." Suppliers of complementary products and services demand a critical mass of users before they consider entering the market, yet these suppliers are desperately needed to establish this critical mass of users in the first place. Finally, since the technology quickly develops during this phase, and since dominant market applications have not yet been discovered, technology standards and dominant product designs mostly still have to be established. This means that this phase tends to have a fierce and rather Darwinist character. In the struggle to produce the fittest products and services, many companies do not survive (Olleros, 1986).

The market stabilization phase

The third phase, referred to as the market stabilization phase, begins when the diffusion of a product on the basis of the breakthrough communication technology takes off, and ends when the technology is substituted. Clark (1985) refers to this phase as a more or less rigid state. The rigidity refers to the fact that dominant product designs and applications emerge from the second phase. In the third phase, diffusion of a product form may be depicted in a single diffusion curve, which mostly resembles an S-curve when cumulative adoption is depicted in the course of time. Different curves have been found (e.g. Rink and Swan, 1979; Tellis and Crawford, 1981), but in these cases no phases were distinguished in the diffusion pattern, which means that some of the divergent patterns may be attributed to the fact that the diffusion was still in the market adaptation phase.

The diffusion of the five breakthrough technologies, i.e. television, radio and telephone technology, is still going on. While for the telegraph and the fax, the period from the take off to substitution took about 100 and 30 years respectively. The length of this period, also referred to as the technology life cycle, can vary from a couple of years up to centuries (Jain, 1985).

More or less standard strategies can be pursued in the market stabilization phase. During this phase, companies strive for typical business goals such as a large market share, large profits and so on. During the market stabilization phase the technology and resulting products and services will be improved constantly, although the dominant design will essentially remain the same. For example, several features were added to the television during the market stabilization phase. Many of these features became part of the standard product in due course (Thölke, 1998). Color televisions replaced black-and-white sets, teletext was added, televisions became portable, and, to attain economies of scale yet remain flexible, modular product designs or product platforms were formed. Companies may also strive to intensify the use of a product in existing markets and thereby increase market potential. The time that the market potential for television was formed by the number of consumer households is long overdue. Currently, television sets are commonly installed in each room of a home, in cars, in caravans and boats, in hospitals and so on. The level of the market potential has shifted considerably during the market stabilization phase. Companies are also able to segment the market and offer differentiated products for each segment.

Some of the general differences between the three subsequent phases in the process of development and diffusion of breakthrough communication technologies are summarized in Table 7.3. The first row indicates the beginning and the end of each phase. The second row lists some findings regarding the length of each phase. In the third and fourth rows, the typical kind of market actors and factors as well as typical market mechanisms in each phase, are described.

Table 7.3 *Differences between the three subsequent phases in the process of development and diffusion of breakthrough communication technologies*

Phase/ characteristics	*Innovation phase*	*Market adaptation phase*	*Market stabilization phase (the S-shaped pattern)*
Begin and end of the phase	From invention of a technology up to the first market introduction of a product incorporating the technology.	Begins after the first market introduction of a product on the basis of the breakthrough technology and ends when the diffusion of this product takes off.	Begins when the diffusion of a product on the basis of the breakthrough communication technology takes off and ends when the technology is substituted.
Length of the phase	Length can vary considerably (1–30 yrs), but on average comprises 7–10 years.	Length can vary considerably, but mostly comprises a decade or more.	Length coincides with the life cycle of a product category.
Market actors and factors in the phase	Individual inventors and entrepreneurs, R&D institutes, universities, and governments (in the role of provider of research funds).	Potential competitors working on the same type of product-technology. Innovative consumers and lead users. Market actors with products and services that are complementary to the technology. Government in the role of lead user or regulator.	Early adopters up to the late majority of consumers, competitors of the same product or service, suppliers and organizations providing complementary products, and services.
Market mechanisms	Supply and demand for research funds and excellent researchers.	Substitution of alternative product technologies. Chicken-and-egg problem. Critical mass effects. Finding the best product-market combinations on the basis of the technology. Establish or reinforce standards. Supply and demand for complementary products and services.	Product life cycle mechanisms. Gradual substitution by new product technologies.

MANAGERIAL IMPLICATIONS OF THE PATTERN

In Chapter 5, general marketing approaches in high-tech markets were discussed. Many of these approaches apply to breakthrough technologies. In the current paragraph the focus will be on managerial implications of the specific development and diffusion pattern that breakthrough technologies tend to follow.

After investigating the pattern of development and diffusion of five breakthrough communication technologies, it is concluded that the well-known S-shaped diffusion curve in fact represents just one phase of this pattern. Three phases are distinguished in this pattern. First, the innovation phase covers the period from the invention of a breakthrough communication technology up to the first market introduction of a product on the basis of the technology. In this phase, which lasts about a decade, the technology is turned into a marketable product. Second, the market adaptation phase comprises the period from the first market introduction up to the point where the diffusion takes off. This phase, also lasting about a decade, often shows an erratic pattern of diffusion with the introduction, withdrawal and re-introduction of various products on the basis of the breakthrough technology. Third, the market stabilization phase begins when a dominant product design, i.e. a basic product form that turns out to be the standard for several years, emerges and the diffusion of this product takes off. The third phase ends when the product based on the breakthrough technology is substituted and sales drop.

In the introduction it is stated that the patterns of development and diffusion of several new communication technologies such as videoconferencing, and interactive television are difficult to capture in a simple S-shaped curve. These patterns diverge considerably from the S-shaped patterns of diffusion of some of the older communication technologies such as the telegraph, telephone, fax, radio, and television technology. At first sight these results imply that different patterns have to be distinguished for different types of communication technologies. This chapter shows, however, that the S-shaped pattern of diffusion of the old technologies are preceded by similar erratic patterns of diffusion. The divergent diffusion patterns of the newer communication technologies are therefore attributed to the fact that these technologies are in the market adaptation phase instead of the market stabilization phase.

The idea that the development and diffusion of breakthrough communication technologies follows a pattern with three distinct phases, rather than a single S-shaped curve, has important managerial implications.

The findings indicate that the commercialization of a breakthrough communication technology is a matter of long endurance. The time from the invention of such a technology up to the point where diffusion of the technology takes off, covers about two decades on average. An implication of this finding is that small companies, which essentially focus on one technology, may be confronted with cash-flow problems during this period. Large companies and governmentally subsidized organizations may be in a better position to survive this period.

The findings also indicate that two distinct phases can be distinguished before the diffusion takes off. The fact that these phases differ from the S-shaped diffusion curve has important managerial implications. Different actors and factors, and different market mechanisms in each phase require different strategies of the companies trying to commercialize a breakthrough communication technology. Suppose that an invention is the result of a basic research project in a large company. Such an R&D project, which is often mono-disciplinary, is confronted with two intra-company market mechanisms, i.e. the supply of, and the demand for, top researchers and research budgets. The successful demonstration of a new technological principle, in many cases, heralds a period of new funds. Instead of continuing the research activities, a switch is required to start up innovation activities. The latter type of activity usually requires multi-disciplinary cooperation among various actors outside the R&D department of a company. For smaller companies, a similar switch of activities is required. In the pharmaceutical industry, for example, many small biotechnology research companies look for an alliance with a large company to commercialize an invention or novel drug after the invention of a new molecule. After the invention, project members from other disciplines are required to develop the new drug and to organize the required safety trials before the drug is accepted for commercial use. So, after the invention a switch is required in the strategy.

Similar switches in strategy are required during the transition from the innovation to the market adaptation phase and finally to the market stabilization phase. These findings indicate that it is important to establish the position of the technology in the pattern of development and diffusion and that strategies should be tailored to this position.

The differences between the market adaptation and market stabilization phase have important managerial implications as well. One of these implications is the different scale and approach to production and marketing in the phases. When a product is introduced in the market stabilization phase, once a critical mass of users is attained, companies typically try to attain a large market share by striving for large-scale production and marketing. In this scenario, such companies try to establish economies of scale in production and marketing and try to attain a dominant position. However, when a product is introduced during the market adaptation phase, the introduction probably marks the beginning of an erratic pattern of introduction, withdrawal and subsequent re-introduction of the technology. In this scenario, a strategy of large-scale production and marketing may have dramatic effects for a company. Cooper and Smith (1997) describe an example of a company that built a large, fully automated plant to produce germanium transistors at the time when silicon transistors became the standard. This advanced plant could be closed down at the moment it was ready to produce. There is a large risk of betting on the wrong standard when a company starts large-scale production during the market adaptation phase. Rather than striving for scale, in this scenario a company should strive for a quick learning process to establish mainstream applications and dominant product designs in the market, and should try to keep pace with technological developments. A learning strategy requires small-scale and flexible ways of production and marketing that enable prompt reactions (i.e. new products) to market and technological developments (Sanchez and Sudharshan, 1992; Lynn *et al.*, 1996).

Another implication of the differences between the market adaptation and market stabilization phases is that different types of alliances should be sought in each phase. In the market adaptation phase the main concern is to establish a market for a new product category on the basis of a breakthrough technology. In many cases establishing a new market means that an existing market with well-known products, alliances among market actors, and habits among consumers, has to be changed, or even replaced, by this new market. Therefore, to establish the new market, efforts have to unite among many actors such as the potential producers of products on the basis of the breakthrough technology and companies of complementary products and services. Bijker (1992) describes an example of this type of cooperation. The electricity utilities and General Electric, an incandescent light bulb producer, formed an alliance during the 1930s in the US. In a united effort, both companies tried to develop the market for electric lighting. However, when the market for electric lighting grew, and different types of lighting were developed (e.g. the fluorescent lamp) the cooperation was stopped since the interests of the light bulb producers to develop and sell energy efficient lamps, no longer matched the interests of the electric utilities that wanted to supply more electricity. So, during the innovation and market adaptation phase many pre-competitive alliances are established in an effort to establish a new market. Yet, when a market is established, the goal is to strive for market share at the expense of direct competitors. Previous alliances are often abandoned and replaced by other types of alliances.

CONCLUDING REMARKS

Generalizability of the findings

Finally, three questions remain: (1) Do breakthrough communication technologies always diffuse in the same pattern? (2) In what type of conditions do the findings from this chapter particularly apply? (3) Are the findings in this chapter also applicable to technologies other than communication technologies?

This chapter has shown that many breakthrough communication technologies develop and diffuse in a similar pattern. In some cases, the phases of the diffusion process can have quite different lengths, or phases can even be omitted. Once a breakthrough communication technology can be applied in an existing infrastructure and can benefit from prevailing procedures, organizations and so on, it can be hypothesized that the period from invention up to wide-scale diffusion of this technology will be relatively short compared to a breakthrough technology that requires new infrastructures, procedures and organizations. While with a technological breakthrough that can build on a previous dominant product design, such as the transistor that replaced vacuum tubes in radios, it can be hypothesized that the market adaptation phase will probably be relatively short or non-existent. The three-stage pattern is, therefore, a general pattern in which numerous variations can be expected.

The findings in this chapter particularly apply under the following conditions:

- The technological change represents a technological breakthrough.
- The technology can be incorporated in multiple product forms, which can be considered major innovations from the perspective of potential consumers.
- The technology can be applied in multiple market applications.
- The diffusion has to cope with considerable externalities, such as a network, in the case of telecommunication appliances.

The conditions imply that the findings can be generalized to other technologies. The three-stage pattern is also observed for:

- Breakthrough materials such as Kevlar, Dyneema and memory metal.
- Breakthrough technologies in the automotive industry, such as turbochargers and jet-engines.

- Breakthrough technologies in medical applications, such as MRI-scan and CT-scan technology.
- In large technological systems such as the global positioning system (GPS).

In conclusion, we think that the three-stage process of development and diffusion of breakthrough technologies explains many controversies with regard to diffusion. This process has important managerial implications for companies that aim to introduce breakthrough communication technologies in the market.

Relationship of the findings with trends in technology management

In the introduction to this book the following trend is described: technology development becomes more complicated and more expensive. The trend implies that the risk of investing in new technologies also increases. A good assessment of a technology in terms of the market potential or the expected kind of pattern of development and diffusion as well as the position of this technology in this pattern will decrease the risk of following the wrong strategy.

Also, in the introduction to the book, another trend is described that has a relationship with this chapter: product life cycles are shortening. This chapter describes the pattern of development and diffusion of technologies that are incorporated in these products. If new products and different standards are introduced quickly after one another during the market adaptation phase, the resulting market turbulence may hamper the emergence of a critical mass of users. Or, to put it differently, short product life cycles in this phase may slow down the speed of diffusion of a new communication technology.

Another trend from the introduction is that more alliances are formed between organizations. This chapter shows that different types of alliances are needed in the subsequent phases of the pattern of development and diffusion of breakthrough communication technologies.

FURTHER READING

Rogers, E.M. (1983). *Diffusion of Innovations.* New York: The Free Press. Rogers' book represents the sociological and psychological perspective on the diffusion of innovations. Rogers is considered one of the founding fathers of the diffusion literature.

Sahal, D. (1981). *Patterns of Technological Innovation.* Reading, MA: Addison-Wesley. Sahal describes and contrasts technological and economic perspectives on patterns of technological innovation.

Bassala, G. (2001). *The Evolution of Technology.* Cambridge: Cambridge University Press. Bassala describes an evolutionary perspective on the development and diffusion of technology.

REFERENCES

Agarwal, R. and Bayus, L. (2002). The Market Evolution and Sales Takeoff of Product Innovations. *Management Science.* 48(8), 1024–1041.

Bassala, G. (2001). *The Evolution of Technology*. Cambridge: Cambridge University Press.

Bijker, W.E. (1992). The Social Construction of Fluorescent Lighting, or how an Artifact was invented in its Diffusion Stage. In: Bijker, W.E. and Law, J. (Eds). *Shaping Technology/Building Society*. Cambridge, MA: MIT Press, 75–102.

Carey, J. and Moss, M.L. (1985). The Diffusion of Telecommunication Technologies. *Telecommunications Policy*. 6, 145–158.

Clark, K.B. (1985). The Interaction of Design Hierarchies and Market Concepts in Technological Evolution. *Research Policy*. 14, 235–251.

Clarke, A.M. (1990). Is the Failure of Videoconferencing Uptake Due to a Lack of Human Factors or Poor Market Research? In: *Proceedings of the 13th International Symposium on Human Factors in Telecommunications*, 33–140.

Cooper, A.C. and Smith, C.G. (1997). How Established Firms Respond to Threatening Technologies. In: Tushman, M.L. and Anderson, P. (Eds). *Managing Strategic Innovation and Change*. Oxford: Oxford University Press, 141–155.

Dosi, G. (1982). Technological Paradigms and Technological Trajectories – A Suggested Interpretation of the Determinants and Directions of Technical Change. *Research Policy*. 11, 147–162.

Easingwood, C.J. and Lunn, S.O. (1992). Diffusion Paths in a High-Tech Environment: Clusters and Commonalities. *R&D Management*. 1, 69–80.

Jain, S.C. (1985). *Marketing Planning and Strategy*. Cincinnati, OH: South-Western Publishing.

Kotler, P. (1991). *Marketing Management. Analysis, Planning, Implementation, and Control*. (7th edition). Englewood Cliffs, NJ: Prentice Hall.

Lynn, G.S., Morone, J.G. and Paulson, A.S. (1996). Marketing and Discontinuous Innovation: The Probe and Learn Process. *California Management Review*. 38(3), 8–37.

Mansfield, E. (1968). *Industrial Research and Technological Innovation; An Econometric Analysis*. London: Longmans, Green & Co.

Miles, I. (1988). *Home Informatics. Information Technology and the Transformation of Everyday Life*. London: Pinter.

Olleros, F. (1986). Emerging Industries and the Burnout of Pioneers. *Journal of Product Innovation Management*. 1, 5–18.

Rink, D.R. and Swan, J.E. (1979). Product Life Cycle Research: A Literature Review. *Journal of Business Research*, 7, (September) 219–242.

Rogers, E.M. (1983). *Diffusion of Innovations*. New York: The Free Press.

Rogers, E.M. (1986). *Communication Technology; The New Media in Society*. New York: The Free Press.

Sahal, D. (1981). *Patterns of Technological Innovation*. Reading, MA: Addison-Wesley.

Sanchez, R. and Sudharshan, D. (1992). Real-Time Market Research: Learning-by-Doing in the Development of New Products. In: *Proceedings of the International Product Development Management Conference on New Approaches to Development and Engineering*, Brussels, 515–530.

Schnaars, S.P. (1989). *Megamistakes; Forecasting and the Myth of Rapid Technological Change*. New York: The Free Press.

Tellis, G.J. and Crawford, C.M. (1981). An Evolutionary Approach to Product Growth Theory. *Journal of Marketing*. 45 (Fall), 125–134.

Thölke, J.M. (1998). *Product Feature Management*. PhD Dissertation. Delft: Delft University of Technology.

Tushman, M.L. and Anderson, P. (1986). Technological Discontinuities and Organizational Environments. *Administrative Science Quarterly*. 31, 439–465.

Utterback, J.M. and Brown, J.W. (1972). Monitoring for Technological Opportunities. *Business Horizons*. 15 (October), 5–15.

Weiss, A.R. and Birnbaum, P.H. (1989). Technological Infrastructure and the Implementation of Technological Strategies. *Management Science*. 35(8), 1014–1026.

Wheeler, D.R. and Shelley, C.J. (1987). Toward More Realistic Forecasts for High-Technology Products. *The Journal of Business and Industrial Marketing*. 3 (Summer), 55–63.

Williams, F., Rice, R.E. and Rogers, E.M. (1988). *Research Methods and the New Media*. New York: The Free Press.

Chapter 8

Forecasting the market potential of new products

David. J. Langley, Nico Pals and J. Roland Ortt

OVERVIEW

This chapter addresses the broad range of forecasting methods that are applied during different stages of the product development process. The authors provide an overview of these forecasting methods and categorize them into five main groups. Three groups are based on past experience: consumer research, expert opinions and data analysis. Two other groups of methods are theoretical approaches and practical approaches. The authors argue that the latter two groups of methods can be applied particularly when past experience is not expected to provide a good basis for predicting the future, such as for major innovations. Examples of each group of methods are given, including both traditional, tried and tested methods as well as some of the latest developments in the field. The advantages and disadvantages of the methods as well as the specific conditions in which they can be applied will be described. The chapter will end with a comparison of all methods.

INTRODUCTION

One of the unique human capabilities, something that sets us apart from all other animals on earth, is being able to look into the future and prepare ourselves for what may happen. Inherent in us is this capability to make forecasts and it has given us huge advantages throughout the development of mankind. Now that we live in a technology-rich age, we can try to use this capacity for thinking ahead to help us mold the technology into forms that will be of most use to us in the near future. The aim of this chapter is to describe various approaches to forecasting the market potential of new products. The focus will be on predicting the market acceptance of high-tech products during the early stages of development.

The forecast can take different forms. It can be an estimation of the market penetration levels of specific product concepts, or an analysis of a set of product variants, each of which is based on the same technology, but with different functionalities, user interfaces, etc. Different levels of analysis can be identified:

- whole areas of technology, such as mobile phone-based Internet;

- specific technologies, such as UMTS (Universal Mobile Telecommunications System; one of the third-generation mobile phone technologies);
- applications supported by the technology, such as video services on mobile phones (such as watching clips of football matches);
- a specific service offered by one company under a particular brand name (such as pre-paid mobile telephony offered by Orange).

The ability to forecast the market potential of new products is especially relevant for technology or R&D managers who are involved in the early phases of the development. In these phases, decisions about the investment of resources are made that determine further development and subsequent diffusion patterns. In general, it is useful to make forecasts in situations with two characteristics: when there is uncertainty and when there is risk. When a business wants to develop a new technology or a new application or product, very often there is uncertainty about the likely market success of the development and there is usually a large degree of financial risk if it fails. From time to time, industries fail to make good forecasts, with disastrous effects (Olleros, 1986; Schnaars, 1989; Wheeler and Shelley, 1987). In some cases the forecasts were wildly inaccurate or perhaps the forecasts went unheeded.

The consequences of wrong forecasts are illustrated by the case of UMTS mobile telecommunication. Around the turn of the millennium, many European governments auctioned specific transmission frequencies for sending data through the air; many companies heralded new broadband mobile Internet connectivity as a technology that was about to boom. Because of the limited number of frequencies available per country, many mobile phone operators felt that their future relied on making a successful bid. Failure was not an option. Because of this attitude, not enough attention was paid to making high-quality forecasts about the likely future use of mobile Internet services. Many companies ended up paying many billions of euros for licenses when the technology was not even ready. In the past couple of years, now that the hype has died down a little, these companies realize that they have paid over the odds. Their customers would have to increase their spending on mobile phone-related services by tens or hundreds of times in order for the licenses to become cost-effective. In fact, most mobile phone operators have now written off those huge sums of money, considering the investment wasted.

The next two sections will introduce different types of forecasts. The fourth to seventh sections will describe the methods. The chapter will close with an overview of the methods and some remarks about trends.

DIFFERENT TYPES OF FORECASTING

The field of forecasting is highly diverse with different methods being applied in many areas and to many problems. This chapter does not attempt to give a complete overview of the whole field of forecasting, but concentrates on those methods used in forecasting the market potential of technological developments in the form of products and services. When describing pros and cons, it may be useful to know if we are comparing like with like. To this end, we now describe situations that can call for different types of forecasting method.

What is the source of the knowledge about the future?

In many cases, past performance is the best predictor of future performance. By referring to valuable sources of knowledge of existing products or existing usage situations, it is possible to make good forecasts about the future. These sources of knowledge are consumers, experts or databases (Armstrong, 2001; Taschner, 1999) and are coupled to the assumption that the future will not differ significantly from the past. Other methods that do not rely on this assumption, can be categorized as practical methods of trial and error and theoretical methods describing what theory says is likely to happen. These five categories are described more fully in the section on the use and applicability of forecasting methods.

How well known is the object of the forecast?

Some things are well known to us and are relatively easy to forecast whereas other things are unusual and therefore difficult to predict. We know how the number of road users has changed over the last decades and we are likely to be quite accurate when predicting road usage for next year. The knowledge we have about the current situation is highly relevant to the situation to be forecasted and our assumptions probably apply. However, it is more difficult to forecast future sales of a wholly new product, such as television via a mobile phone that is being launched in Europe as we write this chapter. The knowledge we have about current television and mobile phone usage is not really relevant to the new situation and any assumptions we make are bound to be highly debatable.

At which stage of development does the forecast take place?

During the development of a new product, prototypes are made and trials are carried out before the product is launched. Once the product is in the market, sales and usage can be monitored and market research can identify which product improvements are called for. Throughout this process, there is increasing feedback from the market, providing increasingly rich information. This information helps forecast analysts and, therefore, the later in the development process a forecast is made, the more accurate it is likely to be.

What is the time frame?

Some forecasting techniques are specifically suited to a particular time frame. For example, intention surveys, whereby consumers are asked if they would buy a certain new product, are intended to give an indication of market potential in the range of a few weeks or months. Consumers are unable to accurately say what they may do in five years' time. Other methods have a longer time frame, such as trend-based future scenarios, whereby long-term societal trends are identified and their possible influence on the product-related behaviors are sketched in a number of explorative descriptions. The choice of time frame depends upon the sort of result that is required, and, of course, we can expect decreasing forecasting accuracy with longer time frames.

What is the forecast being used for?

Different forecasting methods provide results that can be used for different purposes. A sales manager may want to estimate turnover or usage figures. A product developer may be looking for information to help improve the product design. Marketing managers use forecasting techniques to look at the likely penetration of a new product in different segments of the market and to help develop suitable communication campaigns.

Company directors may want to determine their company's future strategy by seeing a (longer term) view of the likely changes in the market and the technology.

In the remainder of the chapter the focus will be on forecasting the market potential of new and unknown products in the early stages of development, which means that long-term forecasts are required.

THE USE AND APPLICABILITY OF FORECASTING METHODS

As described in the previous paragraph there are a number of dimensions for the description of forecasting methods. For the purpose of this book and the envisaged target group the most appropriate dimension is the "source of knowledge" on which the forecast is based (Armstrong, 2001; Taschner, 1999). This leads initially to the following typology of methods.

- *Consumer Research*: Consumer researchers measure the reaction of potential customers to product concepts and estimate future demand for the final product (Greenhalgh, 1986; Moore, 1982; Page and Rosenbaum, 1992).
- *Expert opinions*: A number of methods of using expert opinions have been developed and applied (Armstrong, 2001; Rowe and Wright, 2001). For example, the Delphi technique aims at consensus between experts on the likely success of an innovation.
- *Data analysis*: Several methods of data analysis and curve fitting can be applied to predict the adoption of innovations (Armstrong, 2001). Also, econometric models are utilized to predict the sales and market shares of products (Lilien *et al.*, 1992; Leeflang and Naert, 1978; Allen and Fildes, 2001).

The three categories of methods described above, namely user, expert and data analysis, all make a link between what has happened in the past and what may happen in the future. The forecasts made when using these methods are based, one way or another, on past experience with similar products or in similar situations. We can say that the principle underlying these methods is that "past behavior is a good predictor of future behavior." If this was the whole story, forecasting would be an easy profession and the world would be a simple (if somewhat dull) place. The main difficulty comes when new behaviors are created in a population. A new product type can create a new behavior. For example, when mobile telephony became possible, suddenly people could make telephone calls while walking along the street, sitting in the car or in a restaurant. A whole range of related behavioral issues are then thrown up that are difficult to understand simply by looking at the behaviors people exhibited in the past. All of a sudden it became possible to have a phone conversation while sitting next to someone you did not know. Telephones could ring in public places, causing annoyance to others. People could find each other in busy places more easily. Many forecasts failed to predict the massive market adoption of mobile phones prior to the mid-1990s and that is because the new product type brought about a major change in usage and behavioral patterns. The forecasting methods could not cope with such new behaviors coming into existence.

Major and minor innovations

Various typologies have been proposed to distinguish the degree of newness of product innovations (Veryzer, 1998; Garcia and Calantone, 2002). One of these is the distinction between "major" and "minor" innovations. Major product innovations comprise a completely new set of attributes and form a new product category, as opposed to minor innovations, which can be seen as small-scale alterations to existing products. In this paragraph we focus on "major product innovations" because it is these innovations that cause new behaviors to emerge. For major innovations, many existing forecasting methods are not applicable and new approaches are required.

For major innovations, the object of the forecast is not well known. The knowledge we have about the current situation, similar products and similar usage situations, is limited. This is because a major innovation is, by definition, something new. So, what we know about existing products and existing usage situations is not really relevant. Our assumptions about what people will do in relation to the new product are based on what they do in existing situations and the assumptions are, therefore, flawed. When attempting to apply user, expert or data analysis to forecast the market adoption of major innovations, we know in advance that our forecasting accuracy is likely to be low.

Requirements for other methods

We can say that experience is the basis of the first three categories of forecasting methods: the experience of users themselves, of the experts who watch and analyze what happens in the real world and experience in the form of bits of data stored for analysis. When dealing with minor innovations, this experience is highly relevant and provides the necessary insight to make good forecasts. But for major innovations, this experience is actually a handicap, preventing us from seeing how the new situation will develop and how widely the behavior related to the new product will spread. When looking for new forecasting methods that can cope with the difficulties posed by major innovations, we found two other approaches that can be added to our typology:

- *Theoretical approach*: Recently Langley *et al.* (2005) introduced a new disciplinary angle for approaching the analysis of the adoption of innovations, drawn from the fields of biology and the behavioral sciences. Insights from a new theory, called memetics, are applied to indicate how likely certain types of people are to copy certain types of product-related behavior.
- *Practical approach*: The practical approach is virtually equivalent to doing nothing, at least as far as forecasting is concerned. The approach requires the adoption of very flexible production and marketing approaches. When the product is introduced in the market, the results have to be monitored carefully, and production and marketing should be changed quickly according to the market feedback. Probe and learn is the best-known example in this category (Lynn *et al.*, 1996).

In the next paragraphs the different methods will be described in more detail using an example: video on demand. Video on demand is best described as an electronic video shop service. It delivers video movies to consumers on request using broadband networks. The advantage to the consumer is that he can order from a catalogue and need not go to the video shop. The idea is that the price is in the same order of magnitude as a video or DVD in a video shop.

CONSUMER RESEARCH

Kotler (1991, p. 5) defines needs, wants and demands as follows:

> A human need is a state of felt deprivation of some basic satisfaction. . . . Wants are desires for specific satisfiers of these deeper needs. . . . Demands are wants for specific products that are backed up by an ability and willingness to buy them.

These definitions show a clear hierarchy: demands are defined in terms of wants, and wants in terms of needs.

The existence of a need is a necessary, yet insufficient condition for consumers to formulate wants. A latent need will not activate the formulation of wants. A need should be recognized first. Yet, even recognition does not suffice: "The presence of need recognition does not automatically activate some action. . . . First the recognized need must be of sufficient importance" (Engel *et al.*, 1990, p. 490). If so, consumers are motivated to formulate wants in regard to satisfiers of the need. The existence of a want, in turn, is a necessary, but not a sufficient condition for a demand. To formulate a demand, consumers must believe that the solution that they want to fulfill their need is within their means. Ortt (1998) relates the three concepts in a very simple model:

Needs → Wants → Demands

The concepts of needs, wants and demands can be linked to various levels of product specificity (Arndt, 1978). Linking the concepts of needs, wants and demands to different levels of product specificity can be relevant. Just after the invention of a new technology it is hardly possible to investigate wants and demands, since product forms are not yet established. At this stage consumer research focuses on the *needs* of potential consumers and establishes whether or not these needs can possibly be fulfilled by the technology. Consider, for example, the invention of Kevlar, a strong fiber. One of the needs that can be fulfilled by Kevlar is bullet protection. Later on, the focus in consumer research is placed upon alternative product forms, such as a bulletproof vest or an armored car, nowadays wanted by many consumers. Finally, research can focus on the relative demand for various brands of bulletproof vests.

The hierarchy in needs, wants and demands is relevant in product development processes. Knowledge of the potential consumers' needs is a prerequisite for a consumer-oriented product development process. Needs can be felt by consumers without having a specific solution in mind. Needs can be investigated without specifying a product! In other words, need assessment does not necessarily require a (concept of a) product.

Needs assessment

Approaches to investigate needs are described by several authors (Engel *et al.*, 1990; Holt *et al.*, 1984). Two categories of approaches of assessing needs will be distinguished: the approaches in which needs are directly asked for, and the approaches in which needs are inferred.

Direct approaches to assess needs

Needs can be investigated in a direct manner by asking whether respondents, in a specific situation or in regard to a particular product category, have any needs. In an entirely open procedure, respondents spontaneously mention needs. An example of this procedure is to ask respondents whether or not they have any unfulfilled communication needs during their holidays abroad. In an entirely closed procedure, respondents react to a pre-specified list of needs. An example of this procedure is to ask respondents to what degree they have the following communication needs (pre-specified list) during their holidays abroad.

Approaches to infer needs

Needs in regard to an innovation can be investigated *without specifying that innovation*. Consumers' experiences with currently available products (belonging to the same category as the innovation) or consumers' experiences in specific situations (in which the innovation can be used) can be investigated to infer unrecognized or latent needs. There are several methods to infer needs (Holt *et al.*, 1984).

One method of inferring needs is to make an inventory of the consumers' problems with specific products (Fornell and Menko, 1981). In general, problem inventories are formed in two steps. First, consumers are asked whether they have any problems with a specific product. The most important problems will be mentioned spontaneously. Subsequently, consumers are asked to react to the problems mentioned by other consumers. The result of a problem inventory for telephony is presented by Crawford (1991). An inventory of problems can be an important step in generating product ideas that create solutions and fulfill needs.

A second method for inferring needs is to focus on specific situations or future scenarios. Then, consumers are asked to indicate whether the currently available product alternatives provide the required benefits in each of these situations. In some cases, these situations can be imitated during the experiment. The consumer reactions can be observed then. If the currently available product alternatives do not provide the required benefits, and if a new product idea can be developed to provide these benefits, then an opportunity for product development has been found.

A third method for inferring needs is to find lead users (Von Hippel, 1986). Lead users have problems with, and have experienced the limitations of, available product alternatives long before other consumers. In solving their problems they often decide to develop their own solutions. An example of this situation would be a company that has built its own video communication system so that employees could work together from a distance. Lead users are also found in the consumer market. The first radio receivers and home computers, for example, were do-it-yourself kits.

Assessing wants and demands

In the early stages of a product development process, consumers' wants and demands regarding a product are often inferred from their evaluations of a concept. Concept testing refers to a variety of marketing research-based approaches employed to determine the degree of (potential) buyers' interest in the new product idea and to refine or improve the idea (Page and Rosenbaum, 1992). "The primary purpose of concept testing is to estimate consumer reactions to a product idea before committing substantial funds to it" (Moore, 1982, p. 279). It is described as a valuable consumer research technique in the early stages of the product development process (Crawford, 1991; Greenhalgh, 1986; Moore, 1982; Wind, 1982). In practice, testing a concept is a procedure in which selected respondents (mostly a sample of potential consumers) are invited to evaluate the concept (a presentation of the product) based on a number of criteria.

Assessing needs in combination with wants and demands

Some authors suggest measuring needs in combination with wants and demands (Hills, 1981; Iuso, 1975; Tauber, 1975). Measuring wants and demands alone presumes that:

1 a concept is evaluated favorably if respondents have a need that the product fulfils;
2 a concept is evaluated less favorably if respondents lack such a need.

If these presumptions are shown to be true, then there is no reason to measure needs separately or to worry about the effect that a lack of a need may have upon the test results. However, "The question raised . . . is whether a product which generates buying interest, creates favorable attitudes, and scores well in present concept tests is necessarily perceived as solving problems or unfilled needs for consumers" (Tauber, 1973, p. 62). According to Iuso (1975) and Tauber (1973) this is not necessarily true. Consumers may like various aspects of a concept and therefore evaluate it positively. Yet, they may, nevertheless, remain rather uninvolved with the basic idea since they do not need it (Iuso, 1975). As a result, it is unlikely that they will actually adopt the product. "Conversely, a consumer may be genuinely involved with the core thrust of an idea, but may intensely dislike certain particular (correctable) features, thus resulting in an overall poor rating" (Iuso, 1975, p. 229). So, combining concept evaluations with need assessment results will increase the interpretability of the results.

Example: telephone intention survey about video on demand

During the imaginary development process of our "Video on demand" service we want to assess the potential interest of consumers in this service. We decide to use the telephone interview method for this purpose. The advantage above a written or web-based questionnaire is that there is a possibility to interact with the respondent and to ask for explanations.

Research questions

Typical research questions for this kind of surveys are:

- Which percentage of the population intends to use the service and to what extent?
- What is the most interesting target group for this service?
- What are people willing to pay for the service?

Selection of respondents

For reliable and valid results it is important to select a random sample of the potential target group. In our case this group consists of all people watching television and having a fast Internet connection. A big challenge in this kind of situation is the access to a database from which you can select such a random sample. A simple but laborious way to solve this, is to use the telephone book and ask, after a short introduction, whether the respondent has a fast Internet connection. Another important aspect of the selection is the number of respondents. For most statistical analyses, the minimum number is fifty, but if you want to distinguish different target groups, based on age, gender, occupation and so on, you need many more respondents. For a thorough analysis a few hundreds are necessary.

Interview protocol

In most cases the actual interviews are subcontracted to a call center, so it is important to have a very clear and tight interview protocol. Such a protocol consists of the following parts:

- An introduction asking the respondent for cooperation, explaining the purpose of the interview and checking whether the respondent belongs to the potential target group (having a fast Internet connection).
- A number of questions about the respondent, such as age, family situation, occupation and so on.
- Questions about the respondent's related behavior to the service. In our case, that is watching television, going to the movies, hiring videos or DVDs and so on.
- Questions about the service and the respondents' intention to use it, what would be a fair price and so on.
- Closing comment and thank the respondent for his or her cooperation.

Analysis of the results

The results are analyzed using a statistical package. For the interpretation of the results it is important to realize that what people say is not necessarily what they do. Respondents declared intentions to use the service are not a guarantee that they will do so. Since current behavior is the best predictor for future behavior it is important to ask the above-mentioned questions about behavior related to the service.

Pros and cons of consumer research

The advantage of consumer research is its interactive character. It is possible to check the understanding of the concepts involved. On the other hand, some respondents find it very difficult to understand how new products will fit into their everyday lives. Another complicating factor is that intention is not always a good predictor for behavior, so people who say that they are definitely going to purchase a new product might never do so. Overall the method is quite expensive and time consuming.

EXPERT OPINIONS

Expert opinions are often used in forecasting. The idea is that experts are familiar with the trends and developments in their fields of expertise and can therefore give reasonably accurate predictions of certain developments. Experts can be content-matter experts or people who fulfill a certain role, such as, for example, an executive in an organization. Combining the input from different experts can produce a rather complete picture. The process, however, is not simply interviewing a number of experts. Armstrong (2001) describes the different stages of a forecasting process in which the opinions of experts are used:

- formulate the forecasting problem;
- choose the method;
- apply the method;
- compose and combine the forecast;
- assess uncertainty in forecasts;
- adjust forecasts;
- evaluate forecasts.

In each of these stages judgment is involved to some extent. Armstrong also gives a number of examples of ways in which expert opinions are used:

Judgmental extrapolation

This is a subjective extrapolation of time-series data. It is, in fact, a combination of expert opinions and data analysis. In most cases domain experts, using their knowledge as well as historical data, carry this out. The best results are obtained if the data are represented in a graphical form, because it is much easier to see trends in a graph than in a list of numbers. The quality of the results can be further improved by indicating a confidence interval for the forecast. This gives at least some information about the estimated reliability of the forecast.

Delphi method

The Delphi method is a method to obtain forecasts from a panel of experts aiming at consensus. To accomplish this, a number of questions are asked to a group of experts. The answers are summarized and fed back to all members of the group anonymously and a number of questions are asked again. Group members can reconsider their opinion based on the opinion of others. The same process is repeated a number of times. This kind of iteration increases the reliability of the results. The biggest progress is made during the first iteration. The optimal group size is between six and twelve people. In larger groups the quality of the results tends to decrease, because of increased noise.

Example: Delphi method applied to video on demand

Applied to the video on demand example, the Delphi method is used to assess the developments in broadband technology used in the fixed and mobile communications network. Examples of questions that might be asked are:

- Is it possible to deliver video on demand of sufficient quality over an ADSL connection in a cost-effective way? Or do we need fiber-to-the-home?
- What will be the penetration of fiber-to-the-home two years from now? Why do you think that?

After a number of iterations the results are summarized in a final report. The information can be used to limit the number of options for the technical implementation of the service and also provides important input for the business case.

Pros and cons of expert opinions

Using expert opinions is relatively cheap and quick. Another advantage is that the available information from previous product launches can be used. On the other hand, experts tend to be over confident in the reliability of their own predictions, which are often biased and inconsistent.

DATA ANALYSIS

There are a lot of data that are continually collected: figures of product adoption or usage, such as telephone call records, repeat purchases and so forth. In those data certain patterns can be seen, such as increasing market penetration or the seasonal cycle relating to ice cream sales. Forecasting methods can make use of these datasets to make predictions about the future, so long as they take into account any effects that may have a significant effect on the existing patterns. There are a variety of methods using existing data to make forecasts, the main types of which are described below.

Practitioners of data analysis usually categorize the sort of data into two types:

- cross-sectional data: data for a single time period (a snap-shot; e.g. DVD player sales in 2005);
- time-series data: data for a sequence of values through time (e.g. DVD player sales each month throughout 2005).

Dataset analogies

The object of a forecast may be related in some way to existing products such as similar products in the same market, or the same products in different markets. In these cases, sets of data from the related cases can be used to generate forecasts, to add precision to other estimates or to show when pattern changes can be expected. Quite often, one single analogy may provide an invalid forecast due to specific differences between individual cases. In such cases, and in cases where erratic patterns appear (Duncan *et al.*, 2003), it may be prudent to combine a group of analogous datasets. This method, known as pooling, reduces the effect of outlying data points (Sayrs, 1989).

Extrapolation

Extrapolation methods use existing values in a time-series dataset as the basis for a forecast (Makridakis *et al.*, 1998). For example, a computer supplier with a large catalogue may look at sales patterns of each individual item over the past few months in order to predict next month's figures. There are a number of advantages of this method. It is quick and easy to apply, which also makes it cheap. It can make a large number of forecasts very easily, as in the example just given.

Extrapolation methods can be quite objective and are replicable (Armstrong, 2001). However, there are two main problems. First, a dataset is required, which makes these methods inapplicable for new products that have yet to be launched. Second, the blind assumption that any pattern in the dataset will be continued into the future will often simply not be true. Armstrong (2001) has collated decades of experience with these methods and lists a number of principles in applying them (see also Figure 8.1).

In practice, much domain-specific knowledge of experts is usually combined with extrapolations in order to understand what caused the past trend and what may influence that trend in the future.

Econometric models

These models are similar to extrapolation models in that they use existing sets of time-series data. However, econometric models incorporate more knowledge about a variety of relationships and causal effects. As such, they are a relatively complex type of model. Initially developed by economists in order to solve their specific problems (hence the name), these models are now applied across a broad spectrum of forecasting areas.

If some conditions are uncertain, such as the relative effect of past and present marketing activity on current sales, then econometric models are not appropriate. These models are best applied to well-known situations that have been intensively studied, which is usually not the case for new product innovations.

Example: extrapolation applied to video on demand

To forecast the market potential of our video on demand service we could use a number of

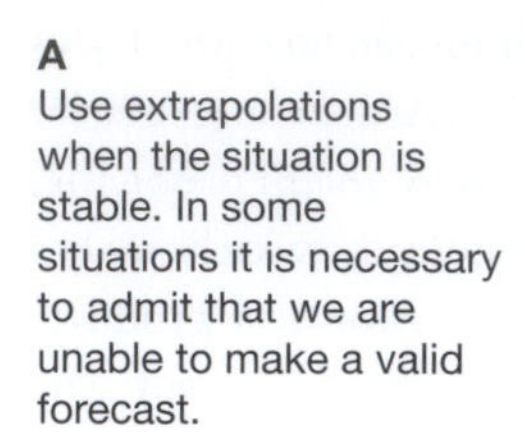

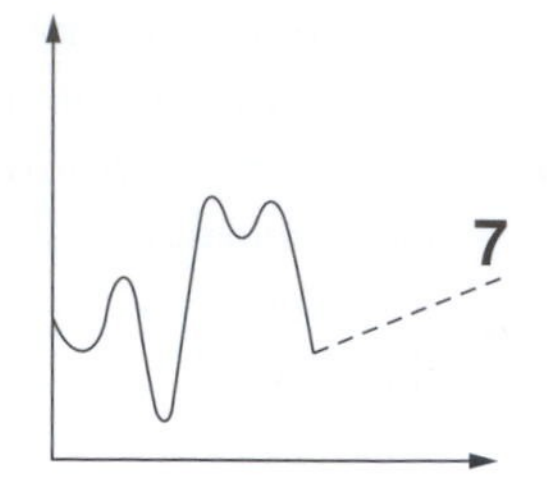

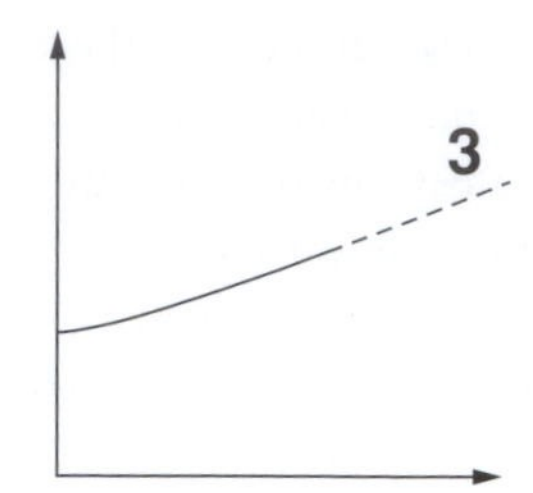

B
Use a simple representation of the trend. There may be a temptation to try to predict with a higher level of accuracy than the methods are capable of.

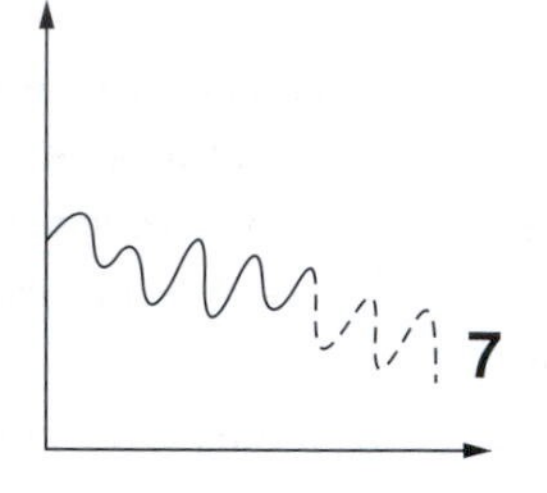

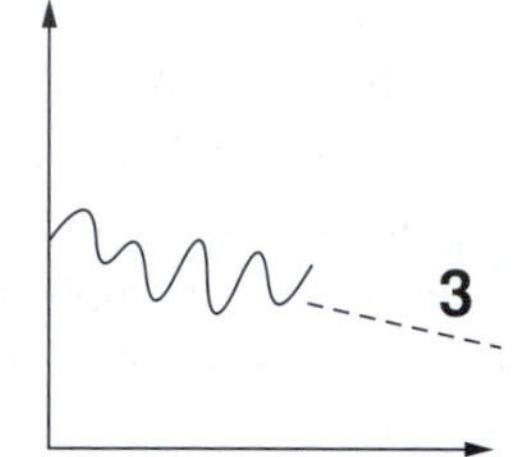

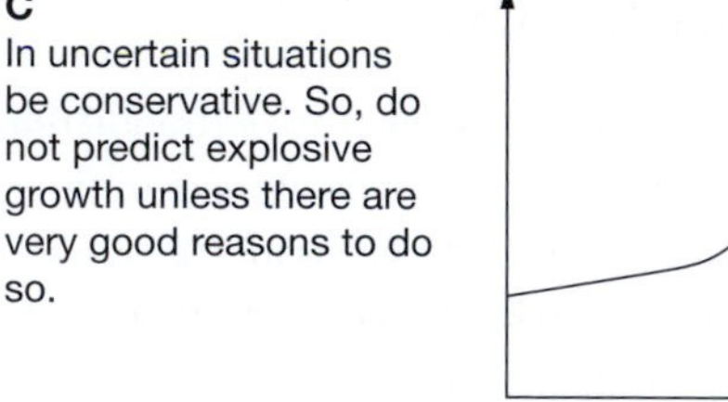

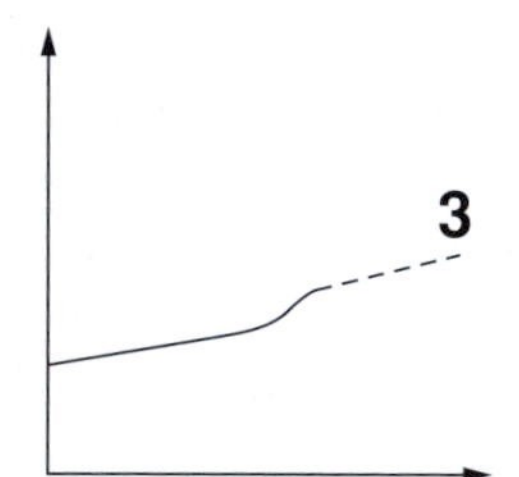

Figure 8.1 *Some principles in extrapolating patterns*
Adapted from Armstrong (2001)

data sources (if available). Interesting data are, for example:

- the penetration of broadband connections in different countries;
- the subscription of people to other broadband services;
- sales and hire out figures of DVD and video tapes.

From these three sources, the first one is used to predict the number of potential subscribers in the next few years. The second is used for a cross-sectional extrapolation estimating the number of potential users who will actually subscribe. The third is another means of estimating the number of subscribers.

Pros and cons of data analysis

Data analysis is a relatively cheap method that can be very accurate for short-term forecasts if enough data are available. It is not applicable if major changes occur, because the data do not apply to the new situation.

THEORETICAL APPROACH

When dealing with major innovations, we expect patterns of customer usage to change dramatically. Any method relying on a smooth transition from past behavior to future behavior may provide misleading results. This is because the needs and wants of users will change once new products become available, offering new functionalities, and as the market, of which they form a part, develops. User, expert and data analysis methods all rely to a large extent on experience from the adoption and use of existing products. This experience is used as a basis for making forecasts about the likely future adoption and use of the innovation in question. As such, it may be safer to assume that these methods are invalid when applied to major innovations and to look for other ways of developing forecasts.

The theoretical approach assumes that it may be possible to model what people will want in the future before they become aware of it themselves. This does not apply to each individual person, but when looking at a market as a whole, certain predictable patterns may emerge. If we are not to use specific knowledge about related situations, then we can look for new methods by applying more general knowledge, which is independent of specific experiences. One possibility is to apply general theoretical knowledge about the adoption or spread of product-related behaviors within a population. In fact, if we are to use general theoretical approaches, these do not even have to be specifically related to the use of products. There are theories that describe the reasons why people adopt some behaviors and reject others. We could apply such a theory to the adoption of product innovations and then we would be free from the problem that the experience-based methods have.

What are the approaches and theories that scientists have developed to provide an understanding of the adoption of new behaviors? Well of course there are many approaches to understanding human behavior and it is not within the scope of this chapter to describe them all. Some theories describe an "ultimate cause" of behavior, other theories describe "proximate causes." The former provide answers to the "why?" questions, such as, "Why is that person carrying out that behavior?" The latter provide an answer to the "how?" questions, such as, "How does that person do that?" When applying theories to the field of forecasting the market success of product innovations, we will perhaps be better off looking at theories that describe ultimate causes. A few such theories are briefly described:

Behaviorism: This refers to a set of theories that are based on the idea that there is a direct relationship between a stimulus (such as a reward for doing something) and the response (carrying out the required behavior). Behaviorism was developed in relation to the way people learn. It pays little or no attention to internal, cognitive processes, and instead postulates that all behaviors have an external cause. The focus of much behaviorism research has been on the reinforcement of behaviors by punishment or reward.

Attitude theory: A set of theories that relate a person's attitudes or beliefs to their actual behavior. For example, the theory of reasoned action makes a link between beliefs and intentions as a way of predicting social behavior. A major component of the theory is the "subjective norm," which describes what a person believes that other significant people believe, as a guide to select appropriate behavior (Fishbein and Azjen, 1980).

Social identity theory: A theory that describes the way that people identify themselves as members of a group and how they adopt behaviors and attitudes that are part of that group's culture. The theory describes how members of different groups relate to each other through their self-categorization as belonging to their group (the in-group) and not to the other group (the out-group) (Tajfel and Turner, 1986).

Perceived innovation attributes: An approach that measures the perceptions that potential consumers have with regard to an innovation. This perception of the innovation, for example the relative advantage, is measured in comparison to the closest alternatives of the innovation. The focus in this approach is on the perceptions that have a relationship with later adoption and implementation of the innovation (Ostlund, 1973, 1974). A review of this approach is provided by Tornatzky and Klein (1982).

Imitation theory: A body of research exploring why and how people appear to be remarkably good at imitating each other. It is proposed that only humans can understand the behavior of others in terms of the intentions of those others and the desired outcomes. This enables humans to learn through other people's experiences and allows us to have complex social interactions.

Memetics: A different theory of imitation, whereby the Darwinian evolutionary processes apply to behaviors. Elements of behavior, termed "memes," replicate through imitation and are subject to selection pressures by people choosing whether or not to exhibit those behaviors. The most successful behaviors are the ones that can trigger people, by whatever means, to copy them and to continue to carry them out.

Memetics applied to the forecasting of innovations

Langley *et al.* (2005) propose to apply the theory of memetics to the forecasting of the market adoption of products and services. They describe a number of new principles for applying this theory.

1 The behavioral component of the use of products and services, including new product innovations, can be analyzed by applying the theory of memetics. If we are to do this, we will need to assess the capacity a product-related behavior has to get itself copied in a particular population. An example of a product-related memetic process is the rapid spread, in many developed countries, of the idea of sending short text messages (SMS) from person to person regardless of location. This idea is, of course, intrinsically linked to the act of keying in and sending such messages on the physical artifact of a mobile phone.
2 A useful approach is to assess the "match" between a product-related behavior and a person or group of people. This match describes the extent to which the people are stimulated to adopt that behavior. In order for this to be possible, a relatively homogeneous group must be used in the assessment.
3 If a meme is to be suited to a particular group of people, it must combine well with the other memes already resident in their minds. The set of currently adopted memes can be reflected in the basic personality traits of a person. Therefore, one's make up in terms of those traits will determine (consciously or subconsciously) which new ideas or behaviors one is likely to adopt.

Langley *et al.* (2005) state that if we succeed in decomposing innovative concepts into a number of behavioral elements it should be possible to estimate the probability that a person with certain personality traits will copy these behavioral elements and, as a consequence, will adopt the new product. This new method should, in principle, be able to distinguish between the likely adoption of a new product by different target groups. This was realized in the SUMI instrument that will be described in the next paragraph.

Example: SUMI applied to video on demand

SUMI, which stands for Service-User Matching Instrument, is an instrument that can forecast the market adoption of major innovations, based on the general principles of the theory of memetics; i.e. that behaviors spread through a population by stimulating people to copy them (Langley *et al.*, 2005). This method can be applied to the new product type, video on demand. Although VoD is related to current television and VCR usage and, also, to some Internet-based film applications, it can be described as a major innovation, because new behaviors and usage patterns are likely to emerge if it is successful. When analyzing a product or a service with SUMI a four-step process is applied:

1. The process starts with a series of interviews in which as much information as possible is gathered about the service and about the envisaged target group. People to target during this stage are, for example, product developers, product managers, marketing directors and marketeers. Emphasis is on factual information (e.g. the market segmentation methods that are used, target group descriptions, etc.).
2. During the next phase this information is translated into the characteristics and traits used by the SUMI instrument and quantified.
3. In the next phase of the analysis the product and target group scores are fed into the instrument and the SUMI match results are calculated and analyzed.
4. The process is concluded with a workshop with stakeholders in which the results are presented and discussed. Discussion is often focused on those product characteristics that are important to the target group(s) and on possibilities to improve the product if results show a poor match for those characteristics.

In the case of VoD the first step leads to a clear description of the product concept. Which variant of all the possible ways of offering VoD would be the best? A number of assumptions must be made, such as:

- the technology is reasonably well developed;
- online video libraries are available;
- the audio-visual quality will be high (comparable with either VHS or DVD);
- recording is not possible, but pausing or rewinding/fast forwarding is;
- prices will be similar to renting a video as is now usual practice.

In order to carry out the analysis, a choice must be made of which target groups to include in the forecast. Because the VoD service is thought to be mostly of interest to young technophiles you might consider to choose a range of this sort of consumers, such as Internet hobbyists or DINKYs (Double income, no kids yet).

SUMI produces results like the fabricated results presented in Figure 8.2. According to these results VoD will be a success for most target groups.

SUMI can also provide an analysis of the strengths and weaknesses, which can be used

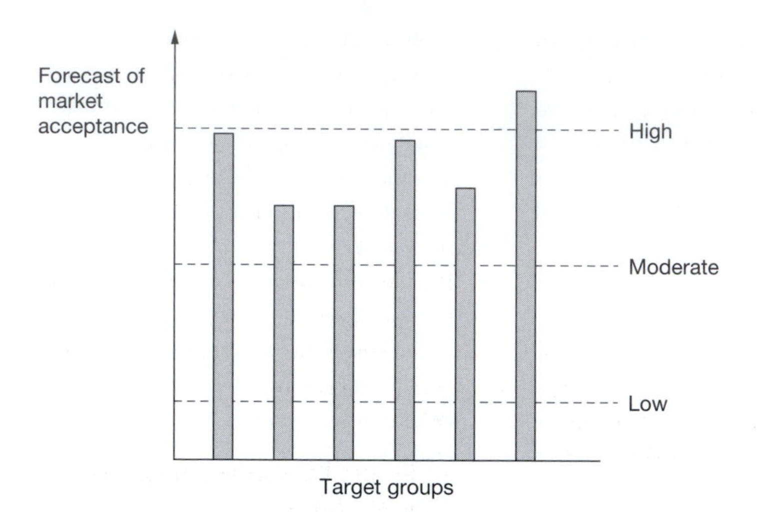

Figure 8.2 *The (fake) SUMI forecasts for the new product video on demand for a range of young technophile target groups*

to guide product development or to influence the way the product is marketed.

Pros and cons of theoretical approaches

The theoretical approach is relatively cheap and quick. The methods do not need to rely on specific experience of real-world situations, which may not be available. Furthermore, they are relatively objective and less subject to bias or inconsistencies than some experience-based methods. It is, however, not easy to incorporate relevant knowledge of real-world situations when available.

PRACTICAL APPROACH

The practical approach can only occur in the very last stage of the development process of a product, because it is almost equivalent to putting a product on the market. Several methods can be applied.

Friendly user pilot

A friendly user pilot, tests a product (prototype or alpha release) with a number of users who have a positive attitude towards the product or the company that is launching it. They are willing to take into account the possible problems and failures that may occur. Such pilots can provide very valuable information for the final adjustments to the product or to the production process.

Market pilot

As a next step a pilot can be carried out with a larger number of users in a specific target group or in a certain geographical area. Such pilots can produce information about (patterns in) user behavior, acceptance and appreciation of the product by the pilot users. A pilot can also be used to make final decisions about the pricing of a product.

Probe and learn

Probe and learn is a method to assess and to develop the market potential of major innovations by probing the market, i.e. by trying to introduce the innovation in various applications and by quickly adapting the innovation and the strategy on the basis of the first market results. Probe and learn is based on the notion that, in some cases, market analysis is not applicable to assess the market potential of a major innovation. This is typically the case when "the technology is evolving, the market is ill-defined, and the infrastructure for delivering the still-developing technology to the as-yet-undefined market is non-existent" (Lynn *et al.*, 1996, p. 10).

The principles of probe and learn

- *Quick reactions to feedback from the market after the introduction rather than relying on market analysis before introduction.* "These companies developed their products by probing potential markets with early versions of the products, learning from the probes and probing again" (Lynn *et al.*, 1996, p. 15).
- *Learning to improve quickly rather than demanding a quick return on investment.* The same authors indicate that probe and learn processes invariably require an enormous commitment in terms of time (years) and investment (usually more than US$100 million).
- *Probe and learn requires a long-term vision and commitment.* Probing makes sense particularly when a technology is of central importance to a company's mission and when companies have to learn about a technological breakthrough. Learning refers to questions such as whether it can be scaled up, which applications and potential consumers can be distinguished for the breakthrough, which applications and market segments are most receptive to various product features, and so on.

Adoption of a probe-and-learn-process implies that product development processes are organized in multiple rounds. One of the reasons is that commercializing breakthrough technologies is less "plan able" than developing minor innovations. The process is more experimental and therefore involves more interactions.

Example: probe and learn applied to video on demand

Although probe and learn is probably not the most suitable method we shall try to apply it to the video on demand example. Since video on demand is most probably a web-based application it is relatively easy to launch a first version for a certain target group or area. In the Netherlands, for example, an experiment is conducted in a district of a city where all houses will be equipped with fiber-to-the-home. Quick evaluations of the first user experiences lead to improvements of the concept, or the design, and can also provide information about up-scaling the service for larger areas. A fast method for evaluating user experiences is a group interview with a sample of the target group.

Pros and cons of the practical approach

The practical approach is by far the method with the highest validity, because it provides real feedback from real people using the real product in their real lives. It is also by far the most expensive. It sometimes may provide misleading results when the market is dynamic, for example when new products are continually being launched or when market patterns keep changing.

CONCLUDING REMARKS

The different approaches compared

This chapter described a number of methods that can be used to make forecasts of the market adoption of innovative products and services. Also, the pros and cons of the different methods were discussed. The two most important criteria for choosing a method are the quality of the results gained (validity) and the resources required to achieve these results (costs, time, etc.). The next two figures compare the methods with respect to these criteria. Figure 8.3 is an estimation of the relative validity of the methods in the case of minor and major innovations. It is obvious that for almost all methods the validity decreases with the degree of innovation as, for wholly new products, the new situation will be very different to current situations. The more drastic the innovation, the lower the validity will be. The only exception to this is the theoretical approach. Since the forecast is based on theoretical relations and not on opinions or past experiences, the validity of this approach will be roughly the same for minor and major innovations.

Figure 8.4 compares the costs of the different approaches. The practical approach is by far the most expensive. The theoretical approach and expert analyses are relatively cheap. The speed with which the different methods can be applied shows about the same pattern as the costs.

In each individual case a method can be chosen taking the described criteria into account. In some cases cost is not a problem, but validity is crucial, in other cases it can be the other way round.

Relationship of the findings with trends in technology management

In the introduction of this book several trends are described. Two of these trends have a direct relationship with the practice of forecasting the market acceptance of new products, i.e. (1) the increased complexity of technology and product development and (2) the increased client orientation.

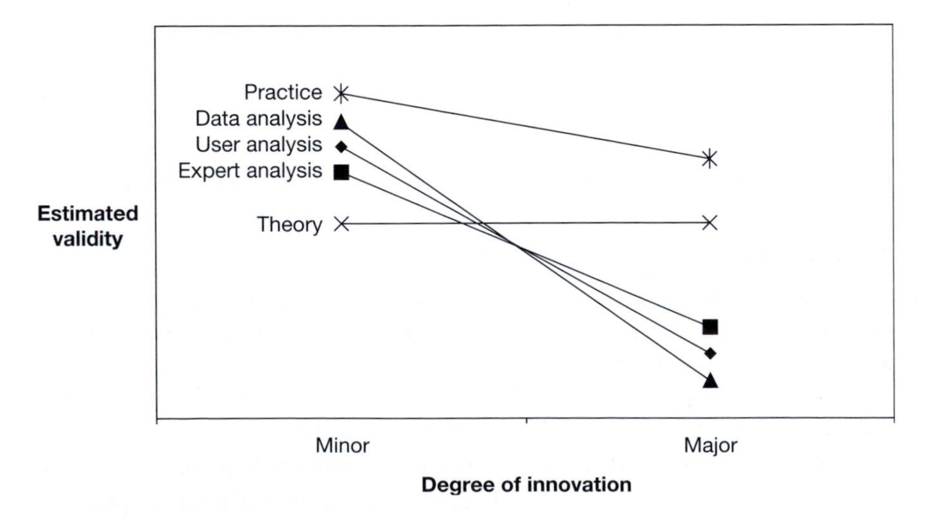

Figure 8.3 *The estimated validity of the different forecasting methods in the case of minor and major innovations*

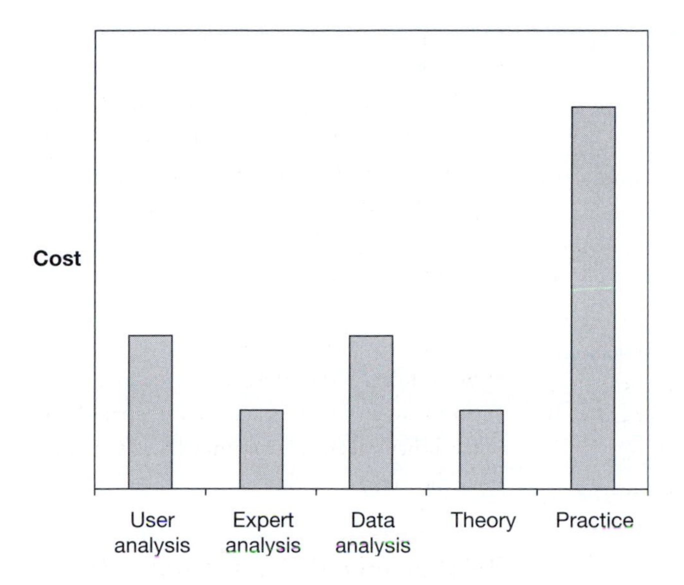

Figure 8.4 *Relative cost of the different approaches*

Because technology development becomes more complicated and expensive it is of utmost importance to evaluate the investment in the early stages. Forecasting the market acceptance is one of the approaches to assess the potential of a technology early on. The increased client orientation makes it more important to take the clients into consideration from the start of a development process onwards.

In Chapter 5, the necessity of adopting a market-oriented and customer-focused approach in high-tech markets was explained. This approach implies that management decision should be based on market and consumer considerations. Market and consumer analysis provide important information for these decisions. In the current chapter it is explained how this market- and customer-orientation can be applied in case of major innovations.

FURTHER READING

Armstrong (2001) has edited a book that provides an overview of up-to-date knowledge about market forecasting. The book begins with a classification of market forecasting techniques. The subsequent chapters, each of which describes a forecasting technique, are written by experts in the field. All chapters are built up according to a similar scheme. Managerial implications of the findings are summarized at the end of the book.

Taschner (1999) gives an overview of forecasting practices in a specific industry, the telecommunication industry.

Makridakis *et al.* (1998) is suggested for further reading on data analysis.

REFERENCES

Allen, P.G. and Fildes, R. (2001). Econometric Forecasting. In: Armstrong, J.S. (Ed.). *Principles of Forecasting: A Handbook for Researchers and Practitioners*. Dordrecht: Kluwer Academic Publishers, 303–362.

Armstrong, J.S. (Ed.) (2001). *Principles of Forecasting: A Handbook for Researchers and Practitioners*. Dordrecht: Kluwer Academic Publishers.

Arndt, J. (1978). How Broad Should the Marketing Concept Be? *Journal of Marketing*. (January), 101–103.

Crawford, C.M. (1991). *New Products Management* (3rd edition). Homewood, IL: Irwin.

Duncan, G.T., Gorr, W.L. and Szczypula, J. (2001). Forecasting Analogous Time Series. In: Armstrong, J.S. (Ed.). *Principles of Forecasting: A Handbook for Researchers and Practitioners*. Dordrecht: Kluwer Academic Publishers.

Engel, J.F., Blackwell, R.D. and Miniard, P.W. (1990). *Consumer Behavior*. Chicago, IL: The Dryden Press.

Fishbein, I. and Azjen, M. (1980). *Understanding Attitudes and Predicting Social Behavior*. Englewood Cliffs, NJ: Prentice Hall.

Fornell, C. and Menko, R.D. (1981). Problem Analysis: A Consumer-Based Methodology for the Discovery of New Product Ideas. *European Journal of Marketing*. 15(5), 61–72.

Garcia, R. and Calantone, R. (2002). A Critical Look at Technological Innovation Typology and Innovativeness Terminology: A Literature Review. *Journal of Product Innovation Management*. 19, 110–132.

Greenhalgh, C. (1986). Research for New Product Development. In: Worcester, R.M. and Downham, J. (eds). *Consumer Market Research Handbook* (3rd edition). Amsterdam: North-Holland, 425–469.

Hills, G.E. (1981). Evaluating New Ventures: A Concept Testing Methodology. *Journal of Small Business Management*. 19 (October), 29–41.

Holt, K., Geschka, H. and Peterlongo, G. (1984). *Need Assessment*. London: John Wiley & Sons.

Iuso, B. (1975). Concept Testing: An Appropriate Approach. *Journal of Marketing Research*. 12 (May), 228–231.

Kotler, P. (1991). *Marketing Management. Analysis, Planning, Implementation, and Control* (7th edition). Englewood Cliffs, NJ: Prentice Hall.

Langley, D.J., Pals, N. and Ortt, J.R. (2005). Adoption Of Behaviour: Predicting Success for Major Innovations. *European Journal of Innovation Management*. 8(1), 56–78.

Leeflang, P. and Naert, P. (1978). *Building Implementable Marketing Models*. Leiden/Boston: Nijhoff Social Sciences Division.

Lilien, G.L., Kotler, P. and Moorthy, K.S. (1992). *Marketing Models*. Englewood Cliffs, NJ: Prentice Hall.

Lynn, G., Morone, J. and Paulson, A. (1996). Marketing and Discontinuous Innovation: The Probe and Learn Process. *California Management Review*. 38(3), 8–37.

Makridakis, S.G., Wheelwright, S.C. and Hyndman, R.J. (1998). *Forecasting: Methods and Applications*. New York: John Wiley & Sons.

Moore, W.L. (1982). Concept Testing. *Journal of Business Research*. 10 (Fall), 279–294.

Olleros, F. (1986). Emerging Industries and the Burnout of Pioneers. *Journal of Product Innovation Management*. 1, 5–18.

Ortt, J.R. (1998). *Videotelephony in the Consumer Market*. PhD Dissertation. Delft: Delft University of Technology.

Ostlund, L.E. (1973). Factor Analysis Applied to Predictors of Innovative Behavior. *Decision Sciences*. 4, 92–108.

Ostlund, L.E. (1974). Perceived Innovation Attributes as Predictors of Innovativeness. *Journal of Consumer Research*. 1 (September), 23–29.

Page, A.L. and Rosenbaum, H.F. (1992). Developing an Effective Concept Testing Program for Consumer Durables. *Journal of Product Innovation Management*. 9, 267–277.

Rowe, G. and Wright, G. (2001). Expert Opinions in Forecasting: The Role of the Delphi Technique. In: Armstrong, J.S. (Ed.). *Principles of Forecasting: A Handbook for Researchers and Practitioners*. Dordrecht: Kluwer Academic Publishers, 125–146.

Sayrs, L.W. (1989). *Pooled Time Series Analysis: Quantitative Applications in the Social Sciences*. London: Sage.

Schnaars, S.P. (1989). *Megamistakes; Forecasting and the Myth of Rapid Technological Change*. New York: The Free Press.

Tajfel, H. and Turner, J.C. (1986). The Social Identity Theory of Inter-Group Behavior. In: Worchel, S. and Austin, L.W. (Eds). *Psychology of Intergroup Relations*. Chicago, IL: Nelson-Hall.

Taschner, A. (1999). Forecasting New Telecommunication Services at a Pre-Development Product Stage. In: Loomis, D.G. and Taylor, L.D. (Eds). *The Future of the Telecommunication Industry: Forecasting and Demand Analysis*. Dordrecht: Kluwer Academic Publishers, 137–165.

Tauber, E.M. (1973). Reduce New Product Failures: Measure Needs as Well as Purchase Interest. *Journal of Marketing*. 37 (July), 61–70.

Tauber, E.M. (1975). Predictive Validity in Consumer Research. *Journal of Advertising Research*. 15(5) (October), 59–64.

Tornatzky, L.G. and Klein, K.J. (1982). Innovation Characteristics and Innovation Adoption-Implementation: A Meta-Analysis of Findings. *IEEE Transactions on Engineering Management*. 29(1) (February), 28–45.

Veryzer, R.W. (1998). Key Factors Affecting Customer Evaluation of Discontinuous New Products. *Journal of Product Innovation Management*. 15, 136–150.

Von Hippel, E. (1986). Lead Users: A Source of Novel Product Concepts. *Management Science*. 32 (July), 57–71.

Wheeler, D.R. and Shelley, C.J. (1987). Toward More Realistic Forecasts for High Technology Products. *The Journal of Business and Industrial Marketing*. 3 (Summer), 55–63.

Wind, Y.J. (1982). *Product Policy: Concepts, Methods, and Strategy*. Reading, MA: Addison-Wesley.

Chapter 9

The innovating firm in a societal context

Labor–management relations and labor productivity

C.W.M. Naastepad and Servaas Storm

OVERVIEW

Based on a cross-section analysis of data from 20 OECD countries, we distinguish *conflictual* labor relations systems (featuring relatively large earnings inequality, low employment protection, weak workers' rights, and close supervision of employees) and *cooperative* systems (featuring higher employment protection, stronger workers' rights, requiring less direct supervision and smaller earnings differentials). Our regression analysis suggests that labor productivity growth and technological progress are promoted by strict employment protection legislation and by more cooperative labor relations in general. This statistical relationship is indicative of how different industrial relations systems affect productivity growth and technological change. We finally draw out the implications of our analysis for technology management at the national and the enterprise level.

INTRODUCTION

High labor productivity growth is generally considered to be a critical factor in international competitiveness and economic growth. Why is productivity growth so important? Because it reflects productive investments in the latest labor-saving technologies and process innovations. Productivity growth also creates the possibility that future generations will be able to enjoy a more comfortable standard of living. Finally, rapid productivity growth can help to reduce inflationary pressures, and, as a result, create the conditions for a looser monetary policy and, potentially, lower interest rates (Galbraith and Darity, 1994). With lower interest rates, investment by firms may increase further – and thus a virtuous cycle of rapid technological change, higher productivity growth and more investment may result. At the enterprise level, achieving high productivity growth is crucial to (manufacturing) firms that operate under the pressures of intensified international competition in three ways. First, because productivity growth leads to a reduction in production costs, it will raise the international cost competitiveness of firms. Second, productivity growth is a reflection of the firm's capacity for process innovation. Finally, productivity growth, particularly when it is due to computer-based technology, is often accompanied by increased organizational flexibility at the enterprise level; this, in turn, makes it easier for firms to respond to changes in global competitive pressures (Lorenz, 1992, 1999).

It is therefore important to understand the determinants of labor productivity growth. Based on an extensive review of the literature, this chapter argues that labor productivity growth (and technological progress in general) is affected by the national system of industrial relations and the nature of a country's labor market. Empirical evidence for this proposition is provided by an analysis of long-run productivity growth, industrial relations, and labor markets in a cross-section of 20 industrialized (OECD) countries.

The chapter is organized as follows. Section 1 reviews the literature on the productivity effects of labor-management relations *at the firm level*. It concludes that labor productivity is higher (ceteris paribus) in enterprises featuring relatively greater worker involvement in production, participation in decision-making, and profit sharing. However, many of these studies show that the productivity gains of greater worker involvement and participation depend heavily on government policies and regulations that encourage cooperation. This suggests that we may learn a lot from comparisons across countries featuring different kinds of macroeconomic labor-management systems, "however difficult or vexing those comparisons may be" (Gordon, 1996, p. 147).

In Section 2, the variation that exists across countries in important dimensions of their industrial relations systems is assessed using a comparative international data set. Our analysis and literature review suggest that it is possible to distinguish two distinct industrial relations systems: "cooperative" (or coordinated) and "conflictual" (or competitive). Section 3 investigates the relationship between national systems of industrial relations and labor productivity growth. We test the hypothesis that labor productivity growth is higher – and the rate of technological progress faster – in countries in which workers' rights are stronger, employment protection legislation is stricter, and labor-management relations are cooperative. We do find a (statistically significant) positive correlation between cooperative labor-management relations and labor productivity growth. Conclusions are presented in section 4.

THE PRODUCTIVITY EFFECTS OF LABOR–MANAGEMENT RELATIONS: "TRUST VERSUS CONTROL"

Faced with increasing (global rather than national) competitive pressures, business organizations significantly intensified their search for competitive advantage during the last two decades. To many organizations it has become clear that labor productivity, a key determinant of a firm's competitive advantage, depends strongly on the nature of the relationships between managers and workers. (As we explain in the Appendix, other important determinants of labor productivity include (a) capital intensity, or the ratio of fixed capital and hours worked; (b) the "quality" of the capital goods employed; and (c) the "quality," or average skill level, of the labor force.) The insight that labor productivity depends significantly on management–labor relationships has motivated (micro-) economists to search for organizational structures that generate efficient use of labor within the firm. Much of the resulting micro-economic analysis of the impact on productivity of management–management relations is a variant of the efficiency wage model (Akerlof and Yellen, 1986). (See Himmelweit *et al.* 2001, Chapter 14 for an exposition of efficiency wage theory and see Buchele and Christiansen 1999 for a critical examination of this theory.)

The efficiency wage model assumes that:

1 There is an inherent conflict between employers, whose aim is to extract as much work as they can from their workers for as little pay as is necessary to retain them, and employees, who aim to exert as little effort as is necessary (to avoid dismissal) for as much pay as they can obtain.
2 Employers have incomplete information about their employees' level of effort; hence, employees need to be supervised and monitored.
3 Labor productivity depends positively on worker remuneration, which is supposed to determine work effort. Here we abstract from other determinants of productivity growth such as technological change and rising capital intensity. The link between high wages and high productivity is generally attributed to higher effort by workers due to a higher cost of job loss; a worker who is paid more than could be obtained in another job, it is argued, will value the current job more highly, and work harder to avoid being dismissed.

The higher the wage, i.e. the higher the cost of dismissal to the individual employee, the more effective is the threat of dismissal and the less the need for close supervision of the worker's activities. Wages and supervision can thus be seen as substitutes, and firms are expected to trade one against another: while keeping total (wage and supervision) costs constant, a firm can choose between higher wages and less intensive supervision, on the one hand, and lower wages and more intensive supervision, on the other. Employers supposedly choose that combination of supervision intensity and wage incentives which maximizes their profits – assuming a trade-off between wage levels and the intensity of supervision at a given level of labor productivity.

If efficiency–wage theory is right, we should find an inverse relationship between the real wage rate (or its rate of growth) and the intensity of management supervision. It is not easy to find evidence of this relationship using data for just one single country, because there is not enough variation in average wage rate growth and management ratios across firms and industries, because firms and industries operate in the same socio-economic, legal and policy environment (Gordon, 1996). We will therefore investigate this relationship using comparative international data for 19 OECD countries. France is not included in this analysis, because of lack of internationally comparative data on supervision intensity. Figure 9.1 presents some basic data exploring the relationship between real wages and the intensity of management supervision, covering 19 OECD countries. The horizontal axis presents data on the percentage of the (non-agricultural) labor force working in administrative and managerial occupations during 1984–1997. This management ratio is calculated using data on employment by occupation published by the

International Labour Organisation (ILO), which, since the early 1960s, has carefully reclassified national employment surveys into standardized and commensurable categories for comparative international purposes. This management ratio (*MR*) is used as an indicator of the intensity of supervision and monitoring by management (Gordon, 1994, 1996; Buchele and Christiansen, 1999). It can be interpreted as an (negative) indicator of the extent to which management trusts employees, and of the degree of autonomy workers have in organizing and coordinating their work activities. The vertical axis represents the average annual real wage growth (per hour worked) between 1984 and 1997.

The scatter plot suggests that there indeed exists a negative relationship, or trade-off, between real wage growth and the intensity of supervision. The results of an ordinary least squares (OLS) regression of the average annual rate of growth of real wages on the management ratio *MR*, presented in the first row of Table 9.1, confirm that there exists a statistically significant (at 5 percent) and negative relationship; the explanatory power of the estimated relationship is, however, quite low (the adjusted R^2 is only 0.08).

However, closer inspection of the data in Table 9.2 and of the scatter-plot in Figure 9.1 reveals that the relationship is not robust: it strongly depends on the cases of Australia, Canada, and the US. If we exclude from the analysis any combination of two out of these three countries, the estimated coefficient of real wage growth becomes statistically insignificant. If Australia and Canada are excluded, the *t*-value becomes –0.90; if Australia and the US are left out, it becomes –1.19; and if we exclude Canada and the US, the *t*-value is –0.93. The negative association thus is a statistical artifact, because it depends on wage–supervision combinations in only two or three out of 19 countries.

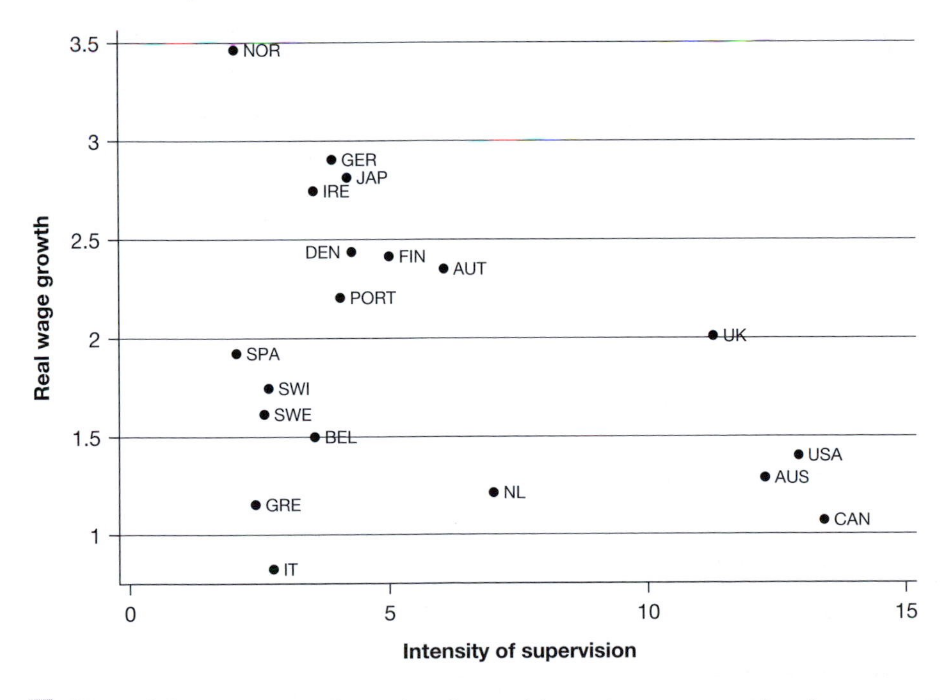

Figure 9.1 *Scatter plot of intensity of supervision and average annual real wage growth: 19 OECD countries (1984–1997)*

Table 9.1 *Regression results for 20 OECD countries, 1984–1997*

	Dependent variable	*Constant*	*Independent variable(s)*				$\bar{R}^2$	*F*	*n*	*df*
1	Real wage growth (*RWG*)	2.33 (7.44) >	−0.07 *MR* (−2.08) >				0.08	4.33 (0.05)	19	17
2A	*EPL*-index	3.37 (9.69) >	−0.22 *MR* (−6.56) >				0.52	43.06 (0.00)	19	17
2B	Workers' Rights and Cooperation Index (*WR–C*)	1.20 (3.00) **	−0.19 *MR* (−4.00) >				0.53	16.01 (0.00)	14	12
3	*EPL*-index	2.05 (14.88) >	0.91 *WR–C* (10.01) >				0.74	100.25 (0.00)	15	13
4A	Earnings inequality	3.62 (10.96) >	−0.41 *EPL* (2.94) **			+1.70 *CDP* (−2.08) >	0.35	5.24 (0.02)	17	14
4B	Earnings inequality	2.82 (17.99) >	−0.43 *WR–C* (2.71) >				0.30	7.34 (0.02)	15	13
4C	Earnings inequality	2.13 (9.38) >	0.12 MR (3.54) >				0.38	12.54 (0.00)	16	14
5A	Labor productivity growth	0.75 (0.96)		0.75 *RWG* (5.09) >	−0.02 GDPG (0.10)	1.55 *CDI* (1.77) **	0.70	15.79 (0.00)	20	16
5B	Labor productivity growth	−0.04 (−0.07)	0.25 *EPL* (2.75) >	0.70 *RWG* (5.12) >	0.10 GDPG (0.69)	1.48 *CDI* (2.74) >	0.81	21.37 (0.00)	20	15
5C	Labor productivity growth	0.82 (5.49)>	0.25 *WR–C* (3.99) >	0.68 *RWG* (7.70) >			0.80	41.74 (0.00)	15	12
6A	*TFP* growth	−1.30 (−2.44) **	0.22 *EPL* (2.03) **	0.70 GDPG (6.44) >		1.04 *CDG* (7.64) >	0.57	8.61 (0.00)	18	14
6B	*TFP* growth	−2.08 (−1.98) **	0.31 *EPL* (1.83) **	0.79 GDPG (5.60) >	0.22 *R&D* (0.95)	0.87 *CDG* (3.18) >	0.58	6.83 (0.00)	18	13

6C	*TFP* growth	−0.07 (−0.27)	−0.08 *MR* (−2.87)>	0.63 GDPG (13.49)>		0.94 *CDG* (6.42)>	0.67	11.65 (0.00)	17	13
6D	*TFP* growth	−0.49 (−1.84)**	−0.12 *MR* (−3.65)>	0.66 GDPG (18.37)>	0.38 *R&D* (2.13)**	0.49 *CDG* (1.97)** −1.04 *CDS* (2.91)>	0.74	10.01 (0.00)	17	11

Notes:
1 *MR* = the ratio of supervisors to total non-agricultural labor force; *RWG* = the average annual rate of real wage growth; *EPL* = OECD Employment Protection Index; *WR–C* = Worker's Rights and Cooperation Index; *GDPG* = the average annual rate of real GDP growth.
2 All equations include a constant. Robust *t*-statistics adjusted for heteroskedasticity of unknown form appear in parentheses. *, ** and > denote statistical significance at the 10, 5 and 1% level, respectively. Figures in parentheses in *F*-column are *p*-values.
3 Some equations control for outliers by including country-specific dummy variables: *CDG* = country dummy for Germany; *CDI* = country dummy for Ireland; *CDP* = country dummy for Portugal; and *CDS* = country dummy for Sweden.

Source: Naastepad and Storm (2004).

Table 9.2 *Productivity growth, capital–labor ratio, GDP and wage growth, and labor relations*

Country	Labor productivity growth[a] 1984–97	TFP growth[a] 1984–97	R&D expenditure (% GDP) 1985–97	Capital intensity growth[a] 1984–97	Real GDP growth[a] 1984–97	Real wage growth[a] 1984–97	Earnings dispersion D9/D1 1984–95	Ratio of supervisory to total workers[a] 1984–97	Strictness of employment protection legislation[b] EPL-index late 1980s	EPL-index late 1990s	EPL-index average 1989–99	Index of workers' rights & co-operation[c] 1990s
1 Australia	1.40	1.29	1.47	1.9	3.60	1.28	2.84	12.34	0.9	0.9	0.9	−0.412
2 Austria	2.20	1.00	1.50	3.4	2.35	2.35	3.54	6.05	2.2	2.1	2.2	+0.829
3 Belgium	2.11	0.77	1.75	3.0	2.30	1.50	2.34	3.54	3.1	2.1	2.6	+0.653
4 Canada	1.09	0.05	1.62	1.3	2.83	1.07	4.28	13.51	0.6	0.6	0.6	−1.388
5 Switzerland	1.29	n.a.	2.80	2.4	1.55	1.74	2.69	2.66	1.0	1.0	1.0	n.a.
6 Germany	3.16	1.85	2.49	3.4	2.30	2.90	2.52	3.90	3.2	2.5	2.9	+0.654
7 Denmark	2.01	0.83	1.55	2.6	2.17	2.43	2.18	4.27	2.1	1.2	1.7	−0.486
8 Spain	2.18	1.15	0.76	3.6	2.90	1.92	n.a.	2.04	3.7	3.1	3.4	n.a.
9 Finland	2.95	1.72	1.97	3.4	2.18	2.41	2.44	4.98	2.3	2.0	2.2	+0.180
10 France	1.86	0.58	2.29	3.3	1.93	1.29	3.22	n.a.	2.7	3.0	2.9	+0.215
11 UK	2.19	1.06	2.06	3.1	2.69	2.01	3.24	11.36	0.5	0.5	0.5	−0.756
12 Greece	1.27	0.81	0.41	1.6	1.65	1.15	n.a.	2.40	3.6	3.6	3.6	n.a.
13 Ireland	4.21	3.50	1.06	1.8	5.26	2.74	n.a.	3.54	0.9	0.9	0.9	n.a.
14 Italy	2.12	1.03	1.03	2.4	2.11	0.82	2.47	2.73	4.1	3.3	3.7	+2.084
15 Japan	2.94	1.98	2.77	5.0	3.25	2.81	3.11	4.18	2.7	2.4	2.6	−0.479
16 Netherlands	1.58	0.79	1.91	0.8	2.87	1.21	1.94	7.02	2.7	2.1	2.4	+0.142
17 Norway	2.98	n.a.	1.68	1.2	3.48	3.46	2.07	1.99	3.0	2.6	2.8	+0.059
18 Portugal	3.20	1.78	0.53	4.0	3.25	2.21	3.73	4.03	4.1	3.7	3.9	n.a.
19 Sweden	1.78	0.74	2.97	2.9	1.87	1.61	2.08	2.58	3.5	2.2	2.9	+0.773
20 USA	1.38	0.88	2.69	1.2	3.38	1.39	4.00	12.99	0.2	0.2	0.2	−2.069
Mean	2.19	1.09	1.77	2.6	2.74	1.92	2.86	5.30	2.4	2.0	2.2	0.000
St. deviation	0.82	0.81	0.76	1.1	0.94	0.73	0.71	4.02	1.3	1.1	1.1	1.000

Sources: a Naastepad and Storm (2004); b Nicoletti *et al.* (2000); c Buchele and Christiansen (1999), Table 3b.

Looking again at Figure 9.1, we may conceive of the 19 countries as falling into three broad groupings – rather than being arranged along a continuum. Four countries, including Australia, Canada, the UK, and the US, cluster in the lower right corner of the graph, sharing the features of low real wage growth and very high management ratios. We may call these the "low-trust" economies. The second group, located in upper left corner of Figure 9.1, consists of Austria, Denmark, Finland, Germany, Ireland, Japan, Norway, and Portugal. These countries are characterized by relatively high rates of real wage growth and low management ratios; we may call these the "high-trust" economies. While the pattern of wage growth and supervision intensity exhibited by these two groups of countries is – in principle – consistent with the predictions of efficiency–wage theory, the pattern exhibited by the third group of countries, located in the lower left corner of the graph and including Belgium, Greece, Italy, the Netherlands, Spain, Sweden, and Switzerland, cannot be explained by it. (Note that the Netherlands falls half way between the first and the third group, because of its relatively higher management ratio.) These countries can be regarded as "high-trust" in view of their low supervision intensities, but at the same time they display relatively low rates of real wage growth. This group of countries is our big question mark, because for them the trade-off between the "carrot" of wage incentives and the "stick" of intensive management supervision implied by efficiency–wage theory may not exist. In fact, using the data in Table 9.2, it can be shown that the average rates of labor productivity growth in the first and the third group of countries are of equal size (in a statistical sense). This suggests that since labor productivity in the third group of countries is not different from that in the first countries – it may depend on neither wage incentives nor intensive supervision – and, accordingly, there must be other factors, significantly more important than wages and supervision, determining worker motivation and labor productivity.

Note that the trade-off between wage incentives and supervision may also be non-existent in our group of "high-trust" countries (characterized by low supervision intensity and high real wage growth). In these countries (located in the upper left corner of Figure 9.1), we may well find that an increase in supervision intensity (with given wages) may cause a drop in workers' effort and productivity (Drago and Perlman, 1989). The reason is twofold: (a) employees associate low supervision with an understanding that their employer trusts them; and (b) they have a notion of a fair level of effort for their wage.

In these circumstances, a rise in supervision intensity is regarded as an indicator of a decline in trust. Workers' motivation, effort, and productivity decline as a result. There exists, as a result, no trade-off between wages and supervision as is assumed in efficiency–wage theory. In efficiency–wage theory, workers' motivation is treated as exogenous to the firm and the industrial relations system (note that worker motivation is assumed to depend solely on the real wage rate). From the lack of trade-off between wages and supervision it follows that workers' motivation must be treated as *endogenous to the nature of labor–management relations*, because it is heavily influenced by the wider social environment in which workers operate, and within which notions of trust and fairness are defined. That is: we must not only look at the relationship between managers and workers, but also at the organizational structures that mediate their exchange – i.e. the social relations of production (Gordon, 1996; Buchele and Christiansen, 1999).

There is another, related, reason why the analysis must be broadened so as to include work organization, namely that *individual* effort is not a decisive determinant of overall average firm-level labor productivity, as is assumed in the efficiency–wage model. As Buchele and Christiansen (1999, p. 91) argue, continuous improvements in productivity depend, not on individual efforts, but on "the effective *interaction* among workers (teamwork), among work *groups* or departments (coordination), and between labor and management (cooperation). . . . If workers' efforts are not appropriately organized and coordinated, they may exert increased efforts without increasing value added," and hence overall average productivity will not increase. In Chapter 3 of this book different HRM practices are described that serve as a mechanism to organize and coordinate. While this problem of organization and coordination – in Buchele and Christiansen's words – "has been largely ignored by . . . economists," it does receive considerable attention in the industrial relations and management literature.

For example, in a major survey of the literature on the impact of work organization on firm performance, Levine and D'Andrea Tyson (1990) conclude that "[i]n most reported cases the introduction of substantive shop floor participation leads to some combination of an increase in satisfaction, commitment, quality and productivity, and a reduction in [labor] turnover and absenteeism." Likewise, using a sample of nearly 1,000 US firms, Huselid (1995) finds that firms with cooperative and participatory labor relations had lower cost and higher productivity than did firms using adversarial labor relations practices. Michie and Sheehan (2003), using evidence from an original survey of about 240 UK firms, find that "low trust" practices, including the use of short-term and temporary employment contracts and a lack of employer commitment to job security, are statistically significantly and negatively correlated with process innovation (and productivity), while, in contrast, "high-trust" work practices are positively correlated with innovation. Finally, in an important analysis for 44 plants in three US manufacturing industries (the industries concerned are steel, apparel, and medical electronic instruments and imaging), Appelbaum *et al.* (2000) find positive effects of "high-trust" work practices on plant performance. In sum, labor productivity is higher (ceteris paribus) in "high-trust, high-wage, low supervision" enterprises that feature relatively greater worker involvement in production, participation in decision-making, and profit sharing.

The studies mentioned identify at least four high-trust practices as having a (statistically) significant positive impact on productivity (Gordon, 1996). Importantly, these studies point out that labor productivity gains cannot be achieved through piecemeal organizational changes, but require simultaneous and coordinated change in all four dimensions.

The four dimensions are:

- real sharing of productivity gains with workers (i.e. a strong commitment to real wage growth);
- the absence of "in-your-face" status differentials and extreme wage differentials between workers and management;
- significant employment security (so that employees do not have to worry that process innovations will result in lay-offs) and protection of workers' rights in general; and

- substantial organizational changes to build group involvement, not just individual participation (since much of workers' contribution to production depends on group efforts and coordination).

It must be emphasized that workers' willingness to give up the protection offered by rigid work rules, disclose their proprietary (tacit) knowledge, and initiate changes in the production process that raise labor productivity and the firm's capacity for innovation, depends, to a large extent, on the trustworthiness of management in honoring its commitments to "high-trust" work practices (see Buchele and Christiansen, 1999). The most solid foundation for this kind of trust, as Lorenz (1992) has argued, is that labor is able to *enforce* those commitments. This, in turn, requires an industrial relations system that offers legal protections to workers' rights and in which labor is organized so as to give workers an effective and safe say and stake in how they do their jobs and how their firms are run. We emphasize that Lorenz does not see labor-management cooperation as an automatic consequence of legislative constraint. As Lorenz (1999) makes clear: legislation provides a procedural framework within which labor and management *can* interact and learn about the likely behavior of their partners when confronted with contingencies. There is therefore no guarantee that cooperation will succeed, even when the legal circumstances appear to promise mutual gain. Legal protection of workers' rights thus must be regarded as a necessary but not a sufficient condition of labor–management cooperation.

The above suggests that labor productivity growth is higher in countries in which workers' rights are stronger, employment protection legislation is stricter, and labor–management relations are more cooperative. To test this hypothesis, we first evaluate the nature of the industrial relations systems in our cross-section of OECD countries to check whether there exists significant cross-country variation in labor–management relations. This will be done in the next section.

A CLASSIFICATION OF NATIONAL INDUSTRIAL RELATIONS SYSTEMS

Table 9.2 presents average annual labor productivity growth and so-called total factor productivity (*TFP*) growth over the period 1984–1997, as well as four indicators of the nature of labor–management relations, for our sample of 20 OECD countries. The concept of *TFP* growth is defined and explained in the Appendix. Unfortunately, important qualitative aspects of labor–management relations are hard to quantify; however, we believe that these unquantifiable aspects are correlated with the four quantitative indicators presented, including:

1. Supervision intensity or the percentage of the (non-agricultural) labor force working in administrative and managerial occupations during 1984–1997. We interpret this indicator as explained above (in our discussion of Figure 9.1).
2. The ratio of top (10 percent) to bottom (10 percent) earnings, which is taken as an (negative) indicator of how fairly compensated employees are likely to feel. The lower this ratio, the fairer workers will perceive their share of earnings to be (Akerlof and Yellen, 1984; Buchele and Christiansen, 1999).
3. An employment protection legislation (*EPL*) index developed by the OECD (1999) (see Nicoletti *et al.*, 2000) and

designed as a multi-dimensional indicator of the strictness of legal protection against dismissals for permanent as well as temporary workers. The *EPL*-index reflects (i) procedural inconveniences which the employer faces when trying to dismiss employees; (ii) notice and severance pay provisions; and (iii) prevailing standards of and penalties for unfair dismissal. See OECD (1999). The higher the *EPL*, the more restricted is a country's employment protection regulation. Table 9.2 presents the *EPL*-index for 1989 and for 1999 and also includes the average figure for 1989–1999.

4 An index of workers' rights and cooperation (*WR–C*), constructed by Buchele and Christiansen (1999) on the basis of data on employment protection, collective bargaining coverage, the ratio of supervisory to production workers, the ratio of top to bottom earnings, and public expenditure on social protection (as a percentage of Gross Domestic Product). The higher the workers' rights and cooperation index, the greater the job and income security of workers is and the relatively more coordinated labor–management relations are.

Table 9.2 shows that there exists considerable variation among countries in these four variables. For instance, the average intensity of supervision during 1984–97 ranges from a low of 2 percent in Norway to a high of 13 percent in the US and Canada. The relative size of the Canadian and US management bureaucracy is more than three times the size of that of Germany, Italy, Belgium, Denmark, and Sweden. Note that this difference between the US and the other countries is not due to differences in sectoral structure of the economies concerned, as Gordon (1996) shows. Similarly, the variation in earnings dispersion is large, ranging from 1.9 in the Netherlands to 4 and 4.3 in the US and Canada, respectively. In terms of the *EPL*-index, the US, the UK, Canada, and Australia are the least regulated countries, while employment protection is highest in Portugal, Italy, France, Germany, Sweden, and Norway. The workers' rights and cooperation index is highest in Italy, Austria, Germany, and Belgium, and lowest in the US, Canada, the UK, and Australia. In line with comparative studies of labor–management relations in the industrialized (OECD) countries (e.g. Gordon (1994, 1996); Buchele and Christiansen (1999); and Lorenz (1992, 1999), these findings substantiate our distinction between two types of industrial relations systems: cooperative (coordinated) versus adversarial (or competitive).

Cooperative industrial relations systems are found in Belgium, France, Germany, Italy, the Netherlands, and the Scandinavian countries. As observed by Buchele and Christiansen (1999), these countries have legally mandated work councils, which are typically under a legal obligation to seek cooperation with the employer and to resolve disputes by negotiation rather than by conflict. Employers are typically required to consult with the council on matters of work reorganization, new technology, outsourcing, overtime scheduling, and health and safety issues. Wage bargaining is mostly centralized, which reduces employer resistance to wage increases, because these are equal for all firms and hence play no role in inter-firm competition. Centralized wage bargaining also facilitates labor–management cooperation at the workplace by removing conflict over wages from the local level. In this system, employees are motivated by job security, wage growth, and effective participation in firm decision-making – hence, supervision intensity and earnings dispersion are relatively low.

Industrial relations in the US, Canada, Australia, and the UK, in contrast, are more "conflictual" and competitive: earnings inequality is greater, employment protection is low, and workers' rights are significantly less well established. The wage bargaining process is highly decentralized. As a corollary, supervision intensity in these economies is high.

Many of the characteristics of economy-wide labor–management systems thus tend to vary together. If a system features low *EPL*, it is likely to display low workers' rights and cooperation index *WR–C*, relatively high earnings inequality, and high supervision intensity. These interconnections can be brought out more explicitly by econometric analysis, using the data in Table 9.2. The first association tested (for only 19 countries because we have no comparative data on supervision intensity for France) is between the *EPL*-index (we use the average *EPL*-index for 1989–1999) and the intensity of supervision (or *MR*, the management ratio). The regression results appear in row 2A of Table 9.1. As anticipated, we find an inverse (and highly statistically significant, at 1 percent) association between employment protection and supervision intensity: the higher is the employment protection for workers, the lower is the intensity of management supervision. Likewise, as is shown in row 2B of Table 9.1, a (statistically significant) negative association exists between the workers' rights and cooperation index, *WR–C*, and *MR*. These findings support Gordon's (1996) conclusion: "More cooperation, fewer bosses."

Because of the small size of our sample (of countries), our results could be significantly affected by one or a few individual countries. In order to assess the robustness of the estimation results to variation in country coverage, we performed a sensitivity analysis for each regression equation by eliminating one country from the sample at a time and re-estimating the equation. In the case of the estimated equations appearing in rows 2A and 2B of Table 9.1, it was found that none of the individual countries (or a combination of two countries) affected the statistical significance of the estimated impact of *MR* on *EPL* or *WR–C*.

Table 9.1, next, reports (in row 3) a close association between the *EPL*-index and the *WR–C* index: the estimated coefficient is statistically significant at 0.1 percent and positive, and the adjusted R^2 is 0.74. This suggests that these two independently developed indices do, to a large extent, capture the same phenomena – though it must be emphasized that the *WR–C* index is a broader indicator, as it also reflects earnings inequality, social protection, and coordination in wage bargaining.

There also is strong empirical evidence of a negative relationship between earnings inequality on the one hand, and either the *EPL*-index or the workers' rights and coordination index on the other – as can be seen from rows 4A and 4B. Again, the findings are statistically robust in the sense that they are not affected by variations in country coverage. This is also true for the estimation results appearing in row 4C, which show that there exists a statistically significant (at 1 percent) and positive association between earnings inequality and the management ratio *MR*. In other words: the higher the ratio of supervisors to the (non-agricultural) labor force, the higher is the dispersion in earnings.

Finally, we do not find a statistically significant (positive) correlation between the *EPL*-index (or alternatively, the workers' rights and cooperation index *WR–C*) and real wage growth (these regression results are not included in Table 9.1). That is, we cannot conclude for our sample of 20 OECD countries that real wage growth is higher in the more cooperative countries than in the more conflictual ones. Why this is so can be seen from Figure 9.2, which plots a country's *EPL*-index against its average annual rate of real wage growth. As was the case in Figure 9.1, we can distinguish three broad groupings of countries. The first group – including Australia, Canada, the UK, the US, and Switzerland – features low employment protection and low wage growth. Note that in contrast to the other countries in this group (and as can be seen from Figure 9.1), Switzerland has a relatively low management ratio *MR*. The second group, consisting of Austria, Denmark, Finland, Germany, Japan, Norway, and Portugal, exhibits high employment protection as well as high real wage growth. Note that Ireland, where wage growth is high but the *EPL*-index is low, falls in between these two groups. The third group includes Belgium, France, Greece, Italy, the Netherlands, Spain, and Sweden; these seven countries feature relatively strict employment protection legislation, but at the same time relatively low real wage growth. It is because of this third group of countries that we do not find a statistically significant positive correlation between real wage growth and the *EPL*-index. In all countries of the third group, collective bargaining coverage is high: the wages of between 80 to 90 percent of the labor force in these countries are set through collective, centralized bargaining. The location of these countries in the lower right corner of Figure 9.2 thus shows that labor unions in these countries have pursued a high employment protection

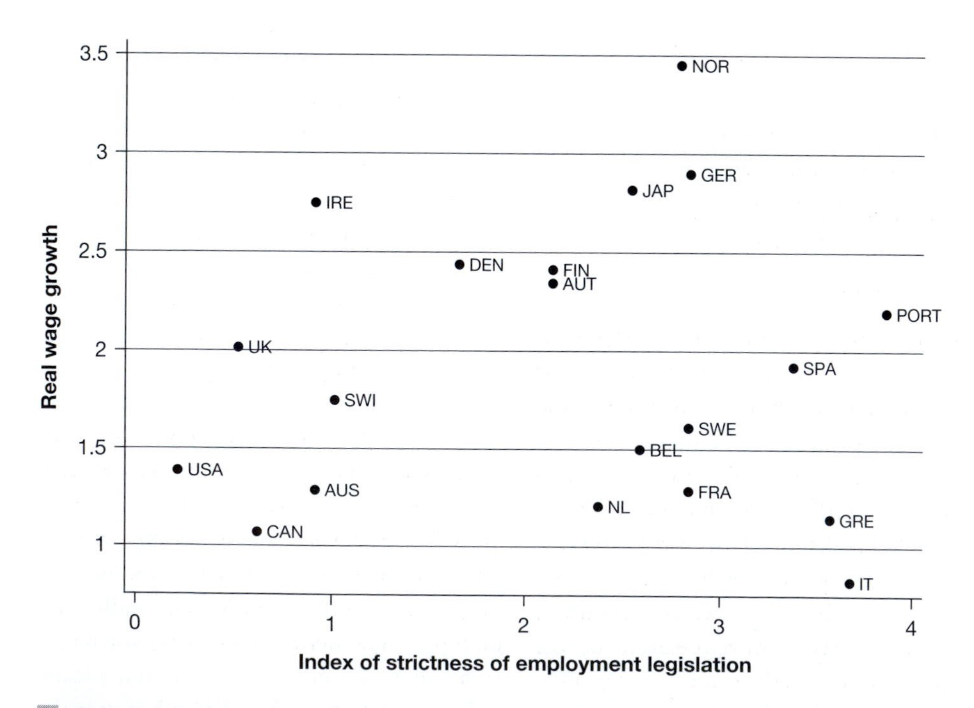

Figure 9.2 *Scatter plot of* EPL*-index and average annual real wage growth: 20 OECD countries (1984–1997)*

strategy rather than simply pressing for high real wage growth. The Dutch experience is a case in point: while real wage growth was voluntarily moderated after 1982 (in exchange for a promise of a rise in employment growth), labor unions opposed policies to deregulate the labor market (i.e. to reduce *EPL* and workers' rights).

It thus appears that there are substantial cross-national variations among the 20 industrialized countries in our sample in terms of major characteristics of their industrial relations systems. It further appears that characteristics of industrial relations systems co-vary. In particular, we find that greater employment protection goes together with stronger workers' rights and more coordinated labor–management relations, a lower management ratio, and lower earnings inequality. The crucial question for our purposes is: do the considerable cross-country differences in labor regimes affect aggregate labor productivity growth and technological change?

PRODUCTIVITY, LABOUR RELATIONS, AND THE REAL WAGE: A COMPARATIVE ANALYSIS FOR 20 OECD COUNTRIES

We examine the effect of labor–management relations on labor productivity growth in our cross-section of 20 OECD countries over the period 1984–1997. To do so, we use the notion of a "Productivity Regime" developed by Naastepad (2005), and extend the analysis by including labor relations indicators. A "Productivity Regime" specifies the relationship that exists between, on the one hand, labor productivity growth (denoted by $\hat{\lambda}$), and, on the other hand, (i) real wage growth ($\hat{w}/p$), and (ii) the growth of real Gross Domestic Product (or output) $\hat{x}$:

$$\hat{\lambda} = \beta_0 + \beta_1 \frac{\hat{w}}{p} + \beta_2 \hat{x}\,,\ \beta_1, \beta_2 > 0. \qquad (1)$$

As shown in the Appendix, the Productivity Regime (1) can be derived from an increasing-returns-to-scale CES production function.

Coefficient β_1 in equation (1) is envisaged to capture the following (static and dynamic) effects on productivity growth of real wage growth:

- *Factor substitution*: an increase in the growth of the real wage (vis-à-vis the growth of the price of capital) induces cost-minimizing firms to substitute the relatively cheapened factor capital for labor. As a result, the technique of production becomes more capital-intensive and consequently, labor productivity rises.
- *Vintage effects*: in the face of a rise in real wage growth, it is efficient for firms to invest in new, labor-saving vintages of capital, to replace older, less productive, capital goods (Naastepad and Kleinknecht, 2004).
- *Induced labor-saving technological change*: an increase in real wage growth makes it profitable for firms to invest in R&D directed at labor-saving process innovations; as a result, labor productivity will increase (Foley and Michl, 1999; Funk, 2002).

Coefficient β_2 is the so-called Verdoorn coefficient, which captures the positive impact on productivity growth of output growth (Naastepad, 2005). Output growth raises productivity growth because it leads to greater specialization in production and to new investments embodying the latest technology.

We estimated equation (1) using the average annual growth rates during 1984–1997 for the group of 20 OECD countries given in Table 9.2. The regression results appear in row 5A of Table 9.1. The adjusted R^2 is 0.7. As expected, the estimated coefficient β_1 is positive and statistically significant (at 1 percent). The estimated real-wage coefficient is 0.75, which suggests that real wage changes have a considerable impact on labor productivity growth. The estimated Verdoorn coefficient β_2 is not statistically significantly different from zero.

To estimate the impact on productivity growth of the industrial relations system, we include the *EPL*-index in equation (1):

$$\hat{\lambda} = \beta_0 + \beta_1 \frac{\hat{w}}{p} + \beta_2 \hat{x} + \beta_3\, EPL, \qquad (2)$$

$$\beta_1, \beta_2 > 0; \beta_3 \neq 0.$$

The estimated coefficient β_3 (reported in row 5B, Table 9.1) is statistically significant (at 1 percent) and *positive*: we thus find that *higher* employment protection is associated with *higher* labor productivity growth. The statistical significance of coefficient β_3 does not depend on one particular country. Due to the inclusion of the *EPL*-index, the real wage coefficient (while remaining statistically significant at 1 percent) becomes 0.7. The Verdoorn coefficient remains statistically insignificant. The adjusted R^2 rises to 0.81. We obtain a similar statistically significant and positive impact of cooperative labor relations on productivity growth, when we estimate equation (2) using the workers' rights and cooperation (*WR-C*) index rather than the *EPL*-index (see row 5C, Table 9.1); again, this finding does not depend on one specific country. Hence, our findings appear to be robust across labor-relations indicators.

In addition to analyzing the impact of labor–management relations on the growth of *labor* productivity, we investigate its possible influence on the growth of *total factor productivity* (*TFP*). We do this, because *TFP* growth is often used as an indicator of the rate of (neutral) technological progress. However, as explained in the Appendix, it is not exactly clear what *TFP* growth represents. The reason is that *TFP* growth, being determined as a residual (see the Appendix), measures not only neutral technological change, but it also includes all other sources of labor productivity change that are not taken into account by the growth rates of capital and labor (including errors of measurement). For this reason, *TFP* growth has been appropriately called a "measure of our ignorance." Another drawback to *TFP* growth as an indicator of technological change is that it depends heavily on the assumptions of (i) a constant-returns-to-scale production function and (ii) equilibrium in factor markets (Foley and Michl, 1999).

Acknowledging these limitations, Table 9.2 presents the rates of *TFP* growth during 1984–97 for 18 countries in our sample, as estimated by the OECD; for Norway and Switzerland no official figures are published. The average annual rate of *TFP* growth during 1984–97 in our sample is 1.1 percent, which is equivalent to about half of the average annual rate of labor productivity growth in the same period. *TFP* growth has been significantly above the group-average in Germany, Japan, and Ireland; it was substantially below average in France and, particularly, in Canada.

In their search for the source of total factor productivity (*TFP*) growth, recent theories of economic growth have mainly focused on R&D expenditure, (increasing-returns-to-scale) technology, and various government policies as independent variables. Keeping in line with these theoretical insights, we estimated the following two relations between average *TFP* growth (1984–97) on the one hand, and labor-relations indicators, R&D

intensity, and the average growth rate of real Gross Domestic Product (GDP) on the other:

$$\hat{TFP} = \phi_0 + \gamma_0\, EPL + \xi_0\, (R\,\&D) + \theta_0\, \hat{x}, \quad \gamma_0, \xi_0, \theta_0, > 0 \tag{3}$$

$$\hat{TFP} = \phi_1 + \gamma_1\, MR + \xi_0\, (R\,\&D) + \theta_1\, \hat{x}, \quad \gamma_0 < 0; \xi_0, \theta_1 > 0 \tag{4}$$

In equation (3), the *EPL*-index is the labor-relations indicator used, while in equation (4) we test for a positive association between the intensity of supervision (*MR*) and *TFP* growth. Following McCombie and Thirlwall (1994), we include real GDP growth in both equations to test for the presence of a Verdoorn effect of output growth on technological change, which (if found to be significant) may be taken to reflect the existence of increasing returns to scale in production. The third independent variable, R&D, is the average gross domestic expenditure on research and development as a percentage of Gross Domestic Product in each of the 18 countries during 1985–97 (see Table 9.2). R&D intensity varies considerably – between a low of 0.4 and 0.5 percent in Greece and Portugal to a high of 2.7–3 percent in the US, Japan, Switzerland, and Sweden.

The estimation results appear in Table 9.1. We first estimated equations (3) and (4) without the variable R&D (i.e. $\xi_0 = \xi_1 = 0$); the results appear in rows 6A and 6C, respectively. The Verdoorn coefficients θ_0 and θ_1 have a value of about 0.6 and are statistically significant (at 1 percent), suggesting a strong relationship between total factor productivity growth and output growth (see McCombie and Thirlwall (1994) for similar findings). The coefficient γ_0 is positive and statistically significant (at 5 percent), providing evidence that the rate of technological progress is positively influenced by a country's employment protection legislation. As expected, the intensity of supervision is negatively associated with *TFP* growth: the coefficient γ_1 is negative and statistically significant (at 1 percent). The higher the intensity of supervision, that is, the larger is the management bureaucracy relative to the total labor force, the lower will be the rate of *TFP* growth, and, hence, the slower the rate of technological progress.

These results do not change when we include the variable R&D in the regressions – as can be seen from rows 6B and 6D. The coefficient of *EPL* (row 6B) remains positive and statistically significant (at 5 percent). What is surprising, however, is that, unlike the strictness of employment protection legislation, R&D intensity is not a statistically significant determinant of *TFP* growth; we have checked that this finding does not depend on variations in country coverage. R&D intensity does appear to have a statistically significant (at 5 percent) impact on *TFP* growth in equation (4), presented in row 6D (Table 9.1), but this result is found to depend solely on the case of Japan, where both R&D intensity and *TFP* growth are relatively high (see Table 9.2). When Japan is excluded from the sample of countries, the *t*-statistic of the R&D coefficient is 0.88; the elimination of Japan from the sample reduces the statistical significance of the *MR* coefficient to below the 1 percent confidence level, but it remains well above the 5 percent level ($t = -2.34$). We conclude that labor–management relations are a statistically significant determinant of *TFP* growth – and more important than R&D intensity.

CONCLUSIONS AND IMPLICATIONS

Our cross-country data (Table 9.2) confirm the presence of significant differences in industrial relations among the industrialized economies of the OECD, especially in the extent of employment protection legislation, the intensity of management supervision, earnings dispersion, and workers' rights and the degree of coordination. We find – in line with the literature – that these characteristics tend to vary together, which makes it possible to distinguish, broadly, two types of labor relations systems: conflictual or competitive systems, featuring relatively large earnings inequality, lower employment protection, weaker workers' rights, and close supervision of employees; and cooperative or coordinated systems, featuring higher employment protection and stronger workers' rights, which require less direct supervision and result in smaller earnings differentials.

Our cross-country regression analysis suggests that strong workers' rights and more cooperative labor relations promote both labor productivity growth and technological progress. However, the suggested relationship should not be interpreted in a mechanical way, but must be regarded as indicative of how different industrial relations systems affect productivity growth and technological change. A change in only one of the dimensions of such a system is unlikely to have, by itself, a strong effect on productivity growth; it is only when the system is transformed in all (or most) of its dimensions that there will be an impact on productivity growth.

Our results are important for technology management in (at least) the following two ways. First, at the national level, our findings are important in light of two major, and interrelated, trends, namely, the deregulation of labor markets in (many) OECD countries and the intensification of global competitive pressures on firms. OECD countries were urged to deregulate their labor markets by the international institutions, including the European Commission (2003) and the OECD (2003), in the belief that increased labor market flexibility – that is: a lower "employment protection legislation" (*EPL*-) index and weaker workers' rights – would lead to higher productivity growth in the following two ways:

1 by making it less costly for firms to adjust their workforce in the face of changes in (world) demand or exogenous technology shocks, and
2 by raising the share of profits and the returns on investment, which was expected to lead to higher investment in R&D and modern capital goods.

As a result of labor market deregulation, the industrial relations systems in the OECD countries have become less cooperative and more conflictual during the 1990s.

Evidence that there is some convergence towards a more uniform and lower level of employment protection is given in Table 9.2. It can be seen that, between 1989 and 1999, the *EPL*-index (measuring the strictness of employment protection legislation) has significantly declined in the Netherlands and Spain (by –0.6 points), Germany (–0.7 points), Italy (–0.8), Denmark (–0.9), Belgium (–1.0), and Sweden (–1.3 points) – which are all "high-trust" cooperative countries. There was no change in the *EPL*-index of the "low-trust" conflictual countries Australia, Canada, the UK, and the US. As a result, the average value of the *EPL*-index (for 20 countries) as well as its standard deviation declined between 1989 and 1999 – indicating that industrial relations in the OECD are becoming more adversarial. Likewise, in most countries, earnings inequality and management ratios have increased (Naastepad and Storm, 2004). Our findings should caution against further deregulation and flexibilization of OECD labor markets. The proposed move to a "low-trust" labor relations regime is likely to lead to a deteriorated productivity performance, because it fails to effectuate the contribution that workers can make to the process of organizational and technological innovation that raise labor productivity.

There is a second important lesson for technology management – one that applies at the enterprise level. Because "high-trust" labor–management relations have an internal logic, a structural coherence, firms cannot achieve productivity gains through limited and piecemeal reforms. Productivity gains at the firm level require – what Gordon (1996, p. 91) called – a *transformation of the workplace*, involving simultaneous and coordinated changes in managerial practices, work practices, and labor–management relations towards sharing power, authority, responsibility, and decision-making (see also Chapter 3). This final major conclusion is in line with many firm-level studies (Levine and D'Andrea Tyson, 1990; Huselid, 1995; Gordon, 1996; Appelbaum *et al.*, 2000), which suggest that firms should change the basic structure of labor relations if they want results.

APPENDIX

DERIVATION OF THE PRODUCTIVITY REGIME EQUATION

Consider the following increasing-returns-to-scale CES production function (Chung, 1994):

$$x = a[\delta l^{-\rho} + (1 - \delta)\, k^{-\rho}]^{(h/\rho)}, \qquad \text{(A.1)}$$

with $-1 < \rho < \infty,\ \rho \neq 0,$
$0 < \delta < 1,\ h, \gamma > 0;$

Where x = Gross Domestic Product (GDP) of the economy under consideration (measured at constant prices), k = the economy's fixed capital stock (at constant prices), l = the number of hours worked (in a year) by the labor force, a = an "efficiency" parameter, γ = the distribution parameter, ρ = the substitution parameter, and h = the returns-to-scale parameter ($h > 1$ corresponding with increasing returns to scale). Denoting the price of capital by π, the elasticity of capital–labour substitution σ is defined as

$$\sigma = \frac{\partial\,(k/l)}{\partial\,(w/\pi)} = \frac{1}{1+\rho}. \qquad \text{(A.2)}$$

From the first-order condition, $\partial x / \partial l = w/p$ where w = the (nominal) wage rate and p = the GDP deflator, and using definition (A.2), it follows that labor productivity, i.e. λ, is equal to:

$$\lambda = \left[\frac{x}{l}\right] = (h\delta)^{-\sigma}\, a^{\sigma\rho/h} \left[\frac{w}{p}\right]^{\sigma} x^{(h-1)\rho/h(\rho+1)} \qquad \text{(A.3)}$$

Log differentiating (A.3) and dividing through by λ gives us an expression for the proportional growth rate of labor productivity (we assume that δ and h are constant):

$$\lambda = \left[\frac{\sigma\rho}{h}\right]\hat{a} + \sigma\left[\frac{\hat{w}}{p}\right] + \left[\frac{(h-1)\rho}{h(\rho+1)}\right]\hat{x}, \qquad \text{(A.4)}$$

where a superscript "hat" (^) indicates the relative rate of change. Equation (A.4) is the Productivity Regime Equation (1) used in the text. Note that (i) $\sigma = \beta_1$ is the coefficient of wage-led technological change, and (ii) $(h-1)\rho/h(\rho+1) = \beta_2$ is the Verdoorn elasticity, which is economically meaningful only if $h > 1$.

TOTAL FACTOR PRODUCTIVITY (*TFP*) GROWTH: DETERMINATION AND INTERPRETATION

TFP growth is conventionally defined as the growth of labor productivity that *cannot* be explained by capital-intensity growth with a constant-returns-to-scale production function. See Chung (1994) for an examination of the Cobb-Douglas production function and its properties. A general discussion of the growth accounting framework, and its limitations, is available in Himmelweit *et al.* (2001). Consider the following constant-returns-to-scale Cobb-Douglas production function (to simplify the algebra, we assume that technology is Hicks neutral (or output augmenting) rather than Harrod neutral (or labour augmenting)).

$$x = ak^{\alpha} l^{1-\alpha}, \quad 0 < \alpha < 1 \tag{A.5}$$

The variables x, k, l, and a are defined as above. Dividing both sides of (A.5) by l, we get the following expression for labor productivity λ:

$$\lambda = \left[\frac{x}{l}\right] = a\left[\frac{k}{l}\right]^{\alpha} \tag{A.6}$$

The aggregate productivity function (A.6) is shown in Figure 9.3. It can be seen that λ increases:

- As a result of a rise in capital intensity, caused by a rise in the real wage. Assuming cost minimization, firms will substitute the relatively cheapened factor capital for labor, which has become relatively more expensive; as a result, the technique of production becomes more capital-intensive (*technical change*) and labor productivity rises. This process of factor substitution can be represented as a movement along the production function. For example, suppose that the real wage increases and that, as a result, capital intensity rises from $(k/l)_1$ to $(k/l)_2$ in Figure 9.3; consequently, labor productivity will increase from λ_1 to λ_2.
- As a result of a rise in total factor productivity (*TFP*), i.e. a rise in a, which, in Figure 9.3, results in an upward shift of the production function from $(x/l)_1$ to $(x/l)_2$. This represents (neutral) *technological change*, because a higher level of labor productivity (λ_3 rather than λ_1) is now possible at an unchanged capital intensity.

To measure the contribution of technological change to aggregate productivity growth, we next log-differentiate (A.6) and divide through by λ; this gives us an expression for the proportional growth rate of labor productivity:

$$\hat{\lambda} = \hat{a} + \alpha\left[\frac{\hat{k}}{l}\right] \tag{A.7}$$

Using Equation (A.7), we can measure the contribution to an economy's labor productivity growth of (i) capital-intensity growth (or technical change), and (ii) *TFP* growth or technological change. Empirically, *TFP* growth is generally calculated as a residual (as was done to obtain the country-wise *TFP* growth figures in Table 9.2):

$$\hat{a} = \hat{\lambda} - \alpha\left[\frac{\hat{k}}{l}\right], \tag{A.8}$$

That is: as the growth of labor productivity that *cannot* be explained by capital-intensity growth with a constant-returns-to-scale production function. $\hat{a}$ is an unexplained residual and is called a "measure of our ignorance."

If we compare our estimates of *TFP* growth and labor productivity growth in

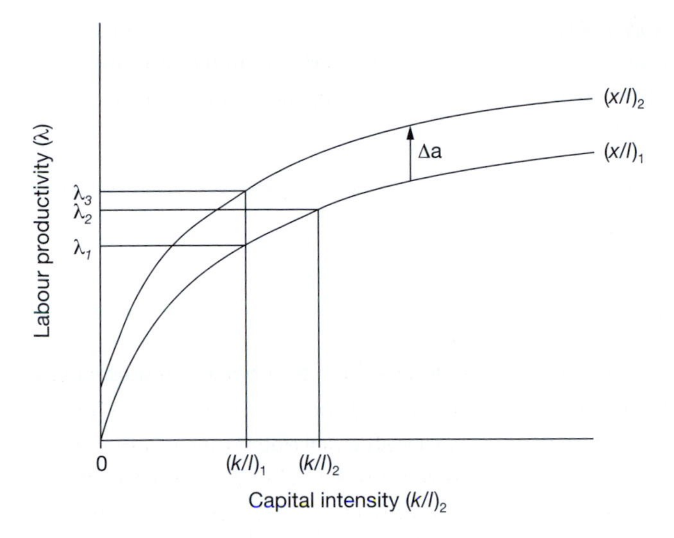

Figure 9.3 *Technical change versus technological change as determinants of labor productivity*

Table 9.2, we find that, on average, as much as 47.1 percent of OECD labor productivity growth during 1984–97 cannot be attributed to capital intensity growth – the unexplained residual $\hat{a}$ is large! It is important that this *TFP* growth be explained, because a theory of productivity growth that attributes about 47 percent of productivity growth to an unexplained residual is – as Nelson (1964) observed – "not much of a theory."

In general, if inputs are being correctly measured and the production function exhibits constant returns to scale, then the residual would be due to technological progress. This technological progress, in turn, would be a result of:

1 Improvements in the quality of the labor force (due to an increase in its average level of education).
2 Improvements in capital goods: assuming that newer vintages of machines are more efficient than old ones, an increase in investment will – by adding more efficient machines to

Table 9.3 TFP *growth as a percentage of labor productivity growth (1984–1997)*

Country	%
Australia	92.3
Austria	45.3
Belgium	36.6
Canada	4.3
Germany	58.7
Denmark	41.4
Finland	58.3
France	31.0
Greece	63.8
Ireland	83.2
Italy	48.3
Japan	67.4
Netherlands	50.1
Portugal	55.6
Sweden	41.3
UK	48.2
USA	63.8
Average	**47.1**

the existing capital stock – lead to a rise in the average efficiency of capital, thus raising the rate of productivity growth beyond the increase associated with a simple increase in capital intensity.

3 Improvements in organization.

FURTHER READING

Gordon, D.M. (1996). *Fat and Mean. The Corporate Squeeze of Working Americans and the Myth of Managerial "Downsizing."* New York: The Free Press. This book is an original, insightful and provocative analysis of labor–management relations in the US and of their impact on productivity growth, real wage growth and income distribution.

An accessible, critical and evaluative introduction to intermediate microeconomics is offered by Himmelweit, S., Simonetti, R., and Trigg, A. (2001). *Microeconomics: Neoclassical and Institutionalist Perspectives on Economic Behaviour.* London: Thomson Learning. Their book encourages readers to think critically about economics as a discipline in which there are competing perspectives rather than one single "right" approach to economic issues.

Lorenz, E.H. (1992). Trust and the flexible firm: international comparisons. *Industrial Relations.* 31(3). The article provides a very interesting historical comparison of how the introduction of computer-based technology was affected by the nature of labor–management relations in Britain, France, Germany, and Japan; examines why "high-trust" labor relations take hold in one place and not in another.

And, finally, Ichniowski, C., Shaw, K., and Prennushi, G. (1997). The effects of human resource management practices on productivity: a study of steel finishing lines. *American Economic Review.* 87(3), 291–313. The article describes an econometric analysis using data from a sample of 36 homogeneous steel production lines showing that "high-trust" work practices (including low supervision and employment security) are associated with higher levels of productivity.

REFERENCES

Akerlof, G. and Yellen, J. (Eds) (1986). *Efficiency Wage Models of the Labor Market.* New York: Cambridge University Press.

Appelbaum, E., Bailey, T., Berg, P., and Kalleberg, A.L. (2000). *Manufacturing Advantage. Why High-Performance Work Systems Pay Off.* Ithaca, NY: Cornell University Press.

Buchele, R. and Christiansen, J. (1999). Labor relations and productivity growth in advanced capitalist economies. *Review of Radical Political Economics.* 31(1), 87–110.

Chung, J.W. (1994). *Utility and Production Functions: Theory and Applications.* Oxford: Blackwell.

Drago, R. and Perlman, R. (1989). Supervision and high wages as competing incentives: a basis for labour segmentation theory. In: Drago, R. and Perlman, R. (Eds). *Microeconomic Issues in Labour Economics: New Approaches.* New York: Harvester Press.

European Commission (2003). Choosing to grow: knowledge, innovation and jobs in a cohesive society. *Report to the Spring European Council.* Brussels.

Foley, D.K. and Michl, T.R. (1999). *Growth and Distribution*. Cambridge, MA: Harvard University Press.

Funk, P. (2002). Induced innovation revisited. *Economica*. 69(273), 155–171.

Galbraith, J.K. and Darity Jr, W. (1994). *Macroeconomics*. Boston, MA: Houghton Mifflin.

Gordon, D.M. (1994). Bosses of different stripes: a cross-national perspective on monitoring and supervision. *The American Economic Review*. 84(2), 375–379.

Gordon, D.M. (1996). *Fat and Mean. The Corporate Squeeze of Working Americans and the Myth of Managerial "Downsizing."* New York: The Free Press.

Himmelweit, S., Simonetti, R., and Trigg, A. (2001). *Microeconomics: Neoclassical and Institutionalist Perspectives on Economic Behaviour*. London: Thomson Learning.

Huselid, M.A. (1995). The impact of human resource management practices on turnover, productivity, and corporate financial performance. *The Academy of Management Journal*. 38(3), 635–672.

Levine, D.I. and D'Andrea Tyson, L. (1990). Participation, productivity and the firm's environment. In: Blinder, A.S. (Ed.). *Paying for Productivity: A Look at the Evidence*. Washington, DC: Brookings Institution.

Lorenz, E.H. (1992). Trust and the flexible firm: international comparisons. *Industrial Relations*. 31(3), 455–472.

Lorenz, E.H. (1999). Trust, contract and economic cooperation. *Cambridge Journal of Economics*. 23(3), 301–316.

McCombie, J.S.L. and Thirlwall, A.P. (1994). *Economic Growth and the Balance-of-Payments Constraint*. London: St Martin's Press.

Michie, J. and Sheehan, M. (2003). Labour market deregulation, "flexibility" and innovation. *Cambridge Journal of Economics*. 27(1), 123–143.

Naastepad, C.W.M. (2005). Technology, demand and distribution: a cumulative growth model with an application to the Dutch productivity growth slowdown. *Cambridge Journal of Economics*, forthcoming.

Naastepad, C.W.M. and Kleinknecht, A. (2004). The Dutch productivity slowdown: the culprit at last. *Structural Change and Economic Dynamics*. 15(1), 137–163.

Naastepad, C.W.M. and Storm, S. (2004). *Productivity Growth and Labour Relations in OECD Countries*. Mimeo. Delft University of Technology.

Nelson, R.R. (1964). Aggregate production functions. *The American Economic Review*. 54(5), 575–606.

Nicoletti, G., Scarpetta, S., and Boyland, O. (2000). Summary Indicators of Product Market Regulation with an extension to Employment Protection Legislation. *Economics Department Working Paper No. 226*. Paris: OECD.

OECD (1999). Employment Outlook 1999. Paris: OECD.

OECD (2003). *The Sources of Economic Growth in OECD Countries*. Paris: OECD.

Chapter 10

Complex decision-making in multi-actor systems

Martijn Leijten and Hans de Bruijn

OVERVIEW

In this chapter two entirely different approaches to managing projects will be contrasted. The focus will be on complex technological projects. Decision-making often takes place in a network of many actors. Organizations faced with a multi-actor system tend to fall into a standard hierarchical pattern, but when the problems or tasks are complex this pattern is often out of step with the nature of the organizational structure (which is network-like), the problems (which are unstructured) and the context (which is dynamic). An approach is needed that is better able to respond to these characteristics, one that revolves around the process the actors are engaged in rather than the content of the problem or task. In this approach decisions are made on the basis of consultation and negotiation, and decision-making is an ongoing process of adaptation. Complex infrastructure projects provide a valuable illustration of the phenomena facing multi-actor systems in a complex network.

INTRODUCTION

Organizations with a technology-intensive primary process are often large and complex networks, and on top of this they often have to operate in a complex and dynamic environment. The complexity and dynamism is even greater if the organizations need to operate in a broader context – a multi-actor system – and thus in turn these organizations form part of a network. A multi-actor system is a constellation of actors (companies, stakeholders etc.) who share an interest in something, e.g. a project. In communities of this kind the technology-intensive primary process is not the only concern: another one is the process of reaching the right decisions together, and this generally means that decision-making in such organizations and constellations is far from simple. To give just a few examples:

- Decision-making often calls for in-depth expertise on technology, but managers are hardly ever likely to have

sufficient expertise of this kind, so they are highly dependent on professionals in the organization or from outside. In spite of this they have to take decisions.
- Good decision-making should ideally be preceded by good information. The reality is usually different: the manager has to make decisions with only limited information at his disposal (many investment decisions, for instance, are surrounded by uncertainty).
- The environment of an organization or multi-actor system is often in a state of flux: the company's competitors do not stand still, and the state of the art can change. This calls for adaptive decision-making which is able to cope with constantly changing circumstances.

This chapter discusses decision-making in these and similar situations. We indicate first why decision-making in a multi-actor system may be so complex, and then show how a traditional organization sets about it. We shall use an example to show that while project management can be an effective strategy in many engineering projects, it is not when a manager has to take decisions in a complex environment. This requires a different vision and approach from this manager and an understanding of when project management is adequate and when process management is a more effective method. We then indicate what strategies there are for managers to make their decision-making effective, backing this up with examples from the practice of decision-making in complex infrastructure projects.

THE COMPLEXITY OF DECISION-MAKING

Why is decision-making often such a complex affair? Table 10.1 lists three characteristics of organizations that generate this complexity.

Table 10.1 *Different characteristics of organizations that make decision-making relatively simple or complex*

Decision-making relatively simple	*Decision-making relatively complex*
Structure is hierarchical	Structure is network-like
Problems are structured	Problems are unstructured
Context is stable	Context is dynamic

Where there is a hierarchical structure there are clear relationships of superiority and subordination. In any given situation there is someone in charge who can take decisions and unilaterally impose them on other people. The predominant style is one of command and control. In many cases the leader of the organization is also the figurehead. As will be demonstrated later on in this chapter, this is not a structure that suits the complexity of many technical problems. Being so dependent on, e.g. professionals, the manager cannot take a purely hierarchical approach (Hanf and Scharpf, 1978). The units in a multi-actor system can be autonomous for various reasons:

- They have specialist technical expertise that makes top-down management difficult.
- They have to operate in a market that has special characteristics that also make top-down management difficult.
- Sometimes a multi-actor system has such a variety of units that it is impossible to run it from a central point: the solution, of course, is for the units to become autonomous.
- Autonomy of units may be an element in the governance of a company or a

multi-actor system: it is relatively easy to make autonomous units accountable for results.

When the units have a lot of autonomy, the organization has a network structure (Mandell, 1988; Scharpf, 1993, 1994; Marin and Mayntz, 1991; Mayntz, 1993). The companies or multi-actor systems are made up of these autonomous units, limiting the scope for top-down management. This conflicts with the hierarchical structure: there are interdependent relationships between the units themselves and between the top and the units. It can be the case in a multi-actor system with a pronounced network structure that the unit administrators are more powerful than the system leader.

The multi-actor community itself is, in turn, embedded in a network, of government bodies, suppliers, customers and so on. Hardly any organization is in charge in a network of this kind; the organizations are dependent on one another. Command and control has little chance of success in a network, as management does not have a strong enough position of power. On top of this, different actors may have completely different ideas on what decisions are desirable: what is an attractive decision for unit X may not be for unit Y, and unit Y is likely to try to obstruct it. Instead of command and control, a different management style predominates in networks, that of consultation and negotiation (De Bruijn and ten Heuvelhof, 2000).

Another characteristic of complexity is that problems are unstructured (Douglas and Wildavsky, 1983; Hisschemöller and Hoppe, 1996). A structured problem has one right solution. An example of a structured problem is calculating the sum of 1 + 1: there is but one right answer. In the case of an unstructured problem there is no one correct solution, for various reasons:

- There is no correct information: the effects of a new technology cannot be ascertained objectively; they require an assessment that is surrounded with uncertainty. A phrase often heard in connection with technology-intensive decision-making is "unruly technology," technology that has not developed to the point where unequivocal information is available (Wynne, 1988). (In Chapter 8 methods are described to gather market and consumer information in these situations).
- Weighing up the various information components correctly is not an objective process. Decision-making almost always requires weighing up competing values: long-term versus short-term factors, return versus investment, economics versus safety, and so on. Any specific decision may involve a large number of such values. Weighing them up is often a subjective business and therefore open to question (Johnson, 1996; De Bruijn *et al.*, 2002).

Say an unstructured problem requires a decision, and this has to be made in a network. Different actors may have widely divergent opinions on the quality and significance of certain information, or on the relative importance of particular values. If problems are structured, the right decisions can be taken using management by expertise: the technical expert makes the right decision. This does not work in the case of unstructured problems, as any opinion on the right decision can be open to question. Here again, there is only one option to take the right decision through consultation and negotiation: negotiated knowledge (Jasanoff, 1990).

The third characteristic of complex decision-making is dynamism, and this applies both to the actors involved in the decision-making and the content of the problem.

- A common phenomenon is that attention to, and involvement in, decision-making on the part of particular actors can vary during the course of the process. Our unit Y is likely not to show much involvement at the start of the process. When it becomes clear that the decision is going in a direction that unit Y does not like, it will become more active, perhaps mobilizing the support of other divisions.
- The content of a problem can also change over time. The role of the "technological dynamic" is often referred to in the case of technology-intensive decision-making: a technology that is dynamic is constantly creating fresh opportunities and threats. Decision-making needs to be able to cope with this dynamic, e.g. by preventing insufficient advantage being taken of technological innovation, also by preventing too many risks being taken with technology that is not yet tried and tested.

In a stable situation, decision-making is a one-off event: once the decision has been made, no further changes are expected, so there is no need to modify the decision. In a dynamic situation, decision-making is an ongoing process: changes can still take place, hence decision-making is a continual process of adaptation to prevent insufficient advantage being taken of the dynamic opportunities or to prevent the risks of dynamism becoming reality (De Bruijn and ten Heuvelhof, 2000; De Bruijn *et al.*, 2002).

TECHNOLOGY AND COMPLEX NETWORKS IN PRACTICE

To illustrate the notions in this chapter let us consider the construction of underground infrastructure in urban areas. In many urban areas the solution to congestion problems and pressure on space is sought in building tunnels and other sub-surface structures. This is far from simple, however, as they have to be constructed in densely built-up and inhabited areas, making for complex projects. Those involved in projects of this kind are faced with an environment that is vulnerable both physically and socially: they have to deal with, e.g. the risk of local subsidence, ground water pressure, people using the surrounding area, and so on. This makes the problem an unstructured one in various ways. A lot of information is required, for example, in particular on technical matters; there is uncertainty about the information that is available (e.g. how representative is a soil survey of the whole project area?); the factors are not very susceptible to objective interpretation; there is no one right solution, and so on. On top of this there are large numbers of actors involved in the project (government, designers, builders, researchers, future users and investors), not to mention those who consider they are involved in some way (representatives from a wide range of interest groups).

On top of this internal complexity, projects of this kind also take place in a complex environment in many cases. For example, ongoing technological innovation – e.g. the development of new methods of construction – can enrich a project, but it can also make it vulnerable if the new technology is not yet tried and tested. Other developments external to the project can also have an effect. Complex underground projects in highly urbanized areas are a perfect example. The city does not stand still while the project is being built. Traffic systems have to be modified and the physical environment changes, but there may also be changes in laws and the political situation. Projects that take a long time to complete are particularly subject to changes in political power and regulations. The complexity features – the unstructured nature of the problem, the network structure and the dynamics – are very pronounced here too, then.

An interesting example of a complex problem of this kind is the design of the Souterrain project in The Hague in the Netherlands (a new tram tunnel-cum-underground car park in the busy city center). The project had a complex physical and social environment. The existing buildings were very close to the site where the tunnel was to be excavated, and it was not really feasible to close the street to traffic. There was also an extensive network of actors with conflicting interests, revolving around the client, the Municipality of The Hague (see Figure 10.1). Once a preliminary design had been drawn up for the scheme, the municipal Building Inspectorate warned of local subsidence, requiring an additional element in the design. The Municipality was faced with the problem that it could not allow the budget to rise, however, so in consultation with the engineering designer it adopted an innovative technical system that would not involve any additional cost as the solution. This involved combining the system for preventing subsidence with the system for sealing the excavation to prevent ground water entering, which had to be done anyway. The technical constraints for these two systems were different, however, which is why a few of the actors in the network warned that this would affect the quality of the design. Despite these warnings and the complex technical deliberations (with which the Municipality was not entirely au fait), the designer and the Municipality went ahead, and the other actors did not obstruct the process. During

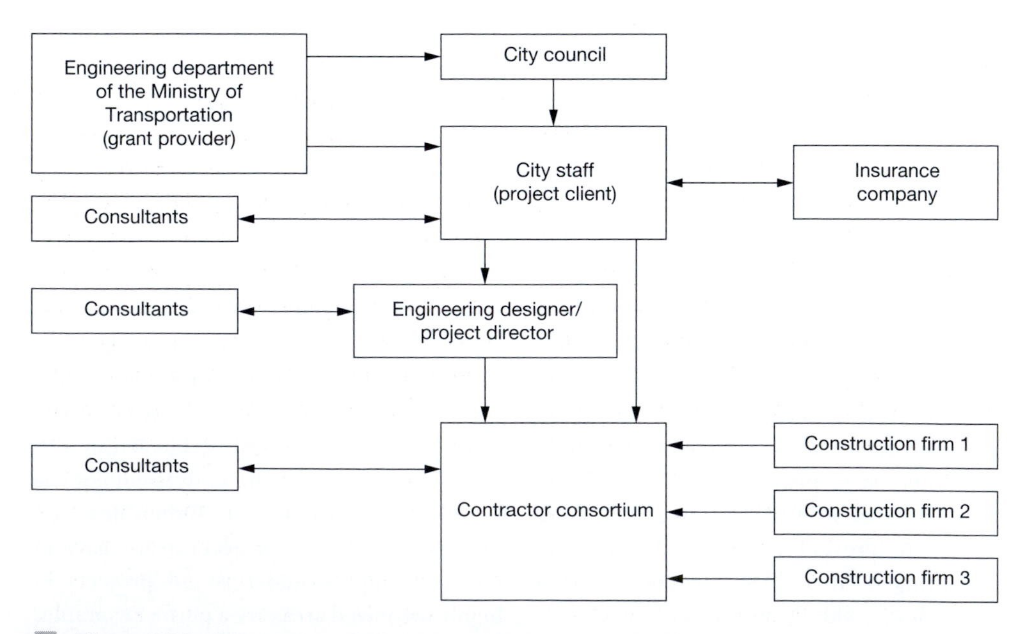

Figure 10.1 *Organogram of the Souterrain project in The Hague*

construction the innovatory technical system developed a leak (which was precisely what some of the actors had feared).

This project had a lot of the features that make problems unstructured: there was uncertainty about the information, especially on the technical system, which became critical as a result of applying innovatory technology. On top of this there was the need to weigh up various information components, setting robustness against cost – two clearly competing values. Then there were many actors with different opinions on quality and, moreover, different levels of information and ability to assess the information (to make technical judgments). In other words, there was no unequivocal or authoritative solution to the problem. The contractor, for instance, had a lot more technical knowledge than the Municipality, whereas the latter – the client – ultimately took the decisions.

The network structure also made this project complex. Networks comprise large numbers of autonomous units, in most cases individual companies or government agencies: the contractor, the engineering designer, the insurance company, various consultancies, the grant-giving body (the national Public Works agency), the municipal City Management Department, the municipal Building Inspectorate, the chain stores along the street under which the tunnel was to be built, and so on. Despite the fact that these were clearly autonomous actors, there were also many interdependencies between them: the Municipality – the client – was clearly dependent on the engineering designer and the grant-giving body and the cooperation of the chain stores; the builder was dependent on the engineering designer's design; and so on.

Finally, the project had a certain dynamic. Not only did the importance of the actors change over time (initially the engineering designer and consultancies were strongly in the ascendant, later on it was the contractor); there was a certain dynamic in the design itself, due to changing requirements and criteria: the project started out solely as a tram tunnel, then the two-level underground car park was added, as this would solve part of the parking problem in the city center (a major wish on the part of the chain stores). This made designing the project a good deal more complex. Then there was the additional demand by the Building Inspectorate to deal with the risk of subsidence.

The project involved a network of actors who probably did have enough knowledge between them to bring the design job to a satisfactory conclusion. Even when it came to the quality and reliability of the innovatory system, it ultimately turned out that various parties had made the right judgment (i.e. that the system was too risky). And yet the tunnel developed a leak while it was under construction. Evidently the structure of the network of actors was such that the right information did not arrive at the right place at the right time. The Municipality of The Hague (the client and only decision-maker) and its engineering designer made critical interventions in the design, while the subsequent contractor, the central government grant-giving body and the insurance company's experts – the parties with possibly the most knowledge and information – were not directly involved. The actors that were not directly involved subsequently expressed doubts about the interventions but did not have the incentive to obstruct the process. They simply made sure not to carry any responsibility and went ahead. There was a hierarchical relationship between the client (the decision-maker) and the contractors, who were subordinate. The decisions made in this case show that the client was not the party with the correct information, while the hierarchical structure applied here was based on that assumption. The ultimate failure might have been prevented by organizing the interests and interdependencies correctly – of necessity in a network, in this case – as a network structure is much more responsive to a situation where the client does not have enough information (or power) to manage effectively. Figure 10.2 shows this: the diagram of decision-making competence is an exact negative of the diagram of technical knowledge. We look at decision-making in problems of this kind in more detail in the next section.

DECISION-MAKING IN PRACTICE

How does decision-making on dynamic, unstructured problems work in a network? To answer this question, let us compare this type of decision-making with project-based decision-making, often the predominant style in engineering projects. Project management has certain characteristics (De Bruijn *et al.*, 2002):

- The decision-making proceeds in a number of stages, e.g. formulating the problem – setting the goals – structuring the problem – collecting information – making the decision – implementation – evaluation.
- At each of these stages "focus," "precision" and "completeness" are important: the problem needs to be defined as precisely as possible, the goals need to be as focused as possible, and the information needs to be as complete as possible.
- The stages take place consecutively: decision-making is inconceivable, for instance, unless a goal has already been set.

Does this approach work in a network? The answer is no. An actor that formulates a goal very precisely then has to negotiate with other actors to see whether they identify with that goal sufficiently. The more precisely a goal is defined, the less room for maneuver the other actors have, and the less chance that they will support it. Something that is effective in a hierarchical, project-based context is not effective in a network, where decision-making is a process of negotiation. Hence the term "process management" (De Bruijn *et al.*, 2002).

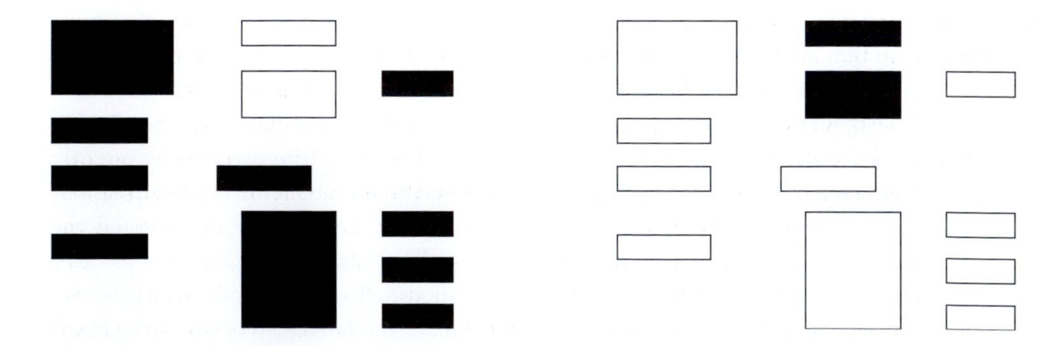

Figure 10.2 *The discrepancy between technical competence and decision-making competence: on the left a schematic representation of Figure 10.1, showing the actors with the essential technical competences in black; on the right the same figure, but showing the actors holding the decision-making competences*

Table 10.2 *Different types of decision-making in project management and process management*

Decision-making as project management	*Decision-making as process management*
Precise problem definition	Open-ended problem definition
Clear goal	Broad goal
Structuring serves to reduce complexity	Structuring serves to increase complexity
Information is objective	Information is negotiated knowledge
Decision-making modifies and requires detailing	Decision-making codifies and requires openness

In this section we shall now make a broad comparison between the essence of decision-making as project management and decision-making as process management (see Table 10.2). We then go on to discuss the points and illustrate them with cases of complex underground projects in urban areas.

Problem definition

A project-based approach requires a problem definition that is as precise as possible. The more clearly the problem is set out, the more guidance it gives towards its solution. Thus a good deal of energy can go into analyzing and identifying the right problem definition in a project-based approach.

In a process-based approach a clear problem definition can be counterproductive for at least two reasons. We have already mentioned the first one: the more precisely a problem is defined, the less freedom of movement actors have to negotiate and agree on a problem definition. This is essential: only if other actors accept co-ownership of the problem is it worthwhile for them to cooperate in solving it. The second reason is that precise problem definitions can result in fixation: if someone defines and fixes a problem, he cannot learn, and learning is vital in a decision-making process. The problem definition is closely related to the goals of the project, so we illustrate the influence of the problem definition under that heading.

Goals

In a project-based approach the goal is the vital point of reference in the decision-making process, guiding the actors' actions. In a process-based approach the goals are formulated broadly. The broader the goal, the more chance that other actors will see something in it to identify with and will negotiate with one another on the problem that needs to be solved. The need is said to be to identify "problem complexes": the more problems are on the agenda, the more chance an actor will accept co-ownership of the problem. Actors in a network often ask themselves the simple question, "What's in it for me?" The more broadly the goals are formulated, the more chance a critical mass of actors will be able to identify with them. Goals, then, do not relate only to the problem definition, they also relate to the stakeholders. The tool of "goal-stretching" can be used here: this means "stretching" the goals so that enough stakeholders identify with them. A comparable tool is "goal intertwinement": finding a solution that does not necessitate the exchanges of objectives, but that manages to simultaneously achieve the varying objectives of parties (Klijn and Koppenjan, 2000; Koppenjan and Klijn, 2004, p. 63).

An interesting example is the decision-making process on the Central Artery/Tunnel Project in the American city of Boston (Leijten, 2004a). In the 1970s and early 1980s there were two separate plans for improving the city's traffic system: (a) a third tunnel under the harbor to the airport (goal: to improve access to the airport) and (b) moving the raised Central Artery freeway (which wound its way through the city center, where it met with serious capacity problems and had an adverse effect on quality of life) entirely underground (goals: to increase capacity and improve quality of life). There was plenty of support in the community for the plans for the third harbor tunnel, whereas opinions on the Central Artery plans were very divided, certainly among the politicians. The two existing tunnels under the harbor to the airport connected up with the Central Artery in the city center, where they caused some of the congestion on the freeway. This created the possibility of linking problem definitions and goals, and this linkage (and the concomitant broadening of the scheme) enabled the advocates of the Central Artery plan to get it adopted after all, taking advantage of the broad support for the third tunnel. The advocates of the third tunnel accepted the incorporation of "their" project in the broader plan once they realized that this automatically ensured that the tunnel would be built. The advocates of the Central Artery scheme were thus able to take advantage of the goal-intertwinement.

Structuring

In a project-based approach there is a need for clearly structured constraints, as these are regarded as reducing complexity. Once the constraints are explicit the problem can be clearly defined (and thus solved) and it is clear in what situations the chosen solution does and does not apply. In a process-based approach, setting constraints causes problems, as it suggests that there is one party able to impose its constraints on the others. It can also restrict the "decision-making space," thus inhibiting the actors' learning processes, as well as reducing the opportunities for "package deals." Constraints, then, can only play a part in process management if the parties agree on them; if there is disagreement they can only be the result of negotiation. On top of this, if an initiator (e.g. the client) has a problem and formulates clear constraints he is not such an attractive partner for the others, who do not see enough scope for solving their own problems.

In a project-based approach, attempts are generally made to reduce the complexity still further by splitting the problem up into sub-problems, each to be solved by itself. The danger here is that this can give rise to single-issue yes/no situations for each problem or solution, which is not conducive to good and efficient decision-making in a process-based approach, where increasing the complexity can, in fact, be worthwhile: the more problems and solutions are involved in a decision-making process, the easier it is to link and unlink them. It is also easier to make a "package deal."

Two examples will illustrate this, one showing the value of using limited constraints and one showing the value of deliberately increasing the complexity. The first example relates to the toll tunnels currently being built by the private sector in Germany.

Here the client lays down only broad constraints in many cases. A tunnel is being built under the River Trave in the North German city of Lübeck, for instance, to replace a bascule bridge that had to be opened every time a large vessel passed through. The only basic constraints laid down by the city authorities were that it should be a fixed link such that the intersecting flows of traffic (shipping and road traffic) did not get in each other's way, and if it were to be a tunnel, care was to be taken to avoid the two ground water aquifers, one containing polluted water and the other drinking water, coming into contact. The details were left to a private-sector consortium, allowing it to achieve its own interests: not only would it eventually build the tunnel, it would also be given a thirty-year franchise to operate it and thus recoup its investment (including a percentage for profit). This also ensured that this private-sector consortium would carry out maintenance over the entire period. The builders have a clear incentive to deliver good-quality work, as they will be stuck with the results for the next thirty years and are themselves responsible for any cost overrun. The precondition, then, is flexibility on the part of the client.

The second example is of the increase in complexity in the Souterrain project in The Hague, discussed earlier. This was originally intended to be solely a tram tunnel, but during the decision-making process a two-level underground car park was added, increasing the complexity of the project and the process of building it. The car park would solve part of the parking problem in the city center, a clear-cut wish on the part of the chain stores along the street under which the scheme was to be built. The client, the Municipality of The Hague, needed their cooperation, as the scheme would have far-reaching effects on the surrounding area. Adding a car park, which could also provide funding for the construction from parking revenue, ensured a cooperative stance from an important group of actors in this case. Increasing the complexity by linking issues thus facilitated the process.

Information

In a project-based approach, objective or inter-subjective information is used as far as possible, and only the information needed for an actor to solve the problem is regarded as relevant. In a process-based approach, information gathering and provision is far more actor-centered, but this often makes the authority of information problematic. It cannot be taken for granted that there is objective information, because of the parties' conflicting interests: each actor may have its own perception of certain information, for instance. This is certainly the case if the problems are unstructured. Often normative decisions have to be made that cannot be objective. Another problem is that information is an important weapon, which parties can use strategically, e.g. by withholding information, making access difficult, allowing only partial access, or blowing it up or trivializing it. Consequently, inter-subjective information only comes about if the parties negotiate on it. What data are to be used? What assumptions are and are not acceptable? What methods are to be used to obtain the information? And so on and so forth. Thus the information that is functional in a process-based approach has the status of negotiated knowledge (Jasanoff, 1990; Koppenjan and Klijn, 2004).

A good example of this is the decision-making process on a restructuring project in the North Dutch city of Groningen (Leijten, 2004b). The plans included building an underground car park beneath the city's central square. The process attracted various pressure groups that had not been included in it and were vehemently against a new underground car park, mainly because they were against cars in the city center. To strengthen their argument, however, they broke into the process and pointed out that there was a potentially vulnerable picturesque church tower near the square that might be damaged by the underground work. The Municipality (the client) tried to refute this argument by bringing in a technical consultancy, which calculated that the risk of damage was less than one in a thousand. The Municipality was convinced that the figures were adequate, but it was mistaken. First, perceptions of what outcome was and was not acceptable differed sharply among the various parties, and, second, the Municipality had neglected to reach agreement on this with the pressure groups. Third, the question was whether the Municipality would be able to reach an understanding on this subject at all with the pressure groups based on this knowledge and information, as it was operating purely rationally, using objective criteria (research findings), whereas the pressure groups' arguments were based on sentiment, which does not usually translate into objective terms. Last, it may well be that the vulnerability of the church tower was merely a strategic argument on the part of the opponents, chosen because it would mobilize far more opposition than the original arguments against an underground car park, which arose mainly from local politics. These problems might have been avoided if the pressure groups had been actively involved in the process and the various actors, on the basis of negotiation, had reached prior agreement.

Decision-making

In a project-based approach there is usually a decision-making process that concludes with a single central decision. In a process-based approach the decision-making does not end after a decision has been taken; fresh rounds then take place, presenting fresh opportunities. In a broad-based process involving an extensive network of actors there are often several topics under consideration at once. When the parties negotiate on a number of topics, a link can be established between one decision and another: for example, it is conceivable that the losers on one decision may be compensated by another. The decision-making on the one topic may then appear to be closed, but in reality it influences the next decision. As a result, the role of formal decisions in a process-based approach is relative, whereas a project-based approach revolves around decisions. Thus, in the course of a process-based approach the parties may have reached an advanced stage in linking and unlinking problems and solutions; the formal decision that is taken is then merely a question of "ticking off" agreements that have already been reached. Also, decision-making that leaves scope for future developments in a process-based approach is much more attractive, as all the parties still see opportunities ahead of them, as well as being an inducement to cooperation. Last, decision-making that takes place in rounds also has advantages when it comes to the content of decisions: the idea that a single all-embracing decision can be made on a complex topic is based on the assumption that the decision-maker possesses all the relevant information at the time, and in practice this is usually an illusion.

Organizations often find it difficult to leave decision-making on infrastructure projects open-ended, as the processes by which such projects come about are, par excellence, processes leading towards a single goal, even an end-state. People think that if a closed decision is not taken at some point, the project will never be built. In projects of this kind, however, open-ended decision-making in a number of rounds can be valuable (cf. the restructuring project in Groningen again (Leijten, 2004b)). The Municipality defended its plans for an underground car park by promising to remove an equal number of parking spaces from other parts of the city (the Municipality was mainly concerned with the investment value of the new properties in the plan, which would be increased by the car park), and by promising that the approach to the car park would go underground at a considerable distance from the square, so that cars would not penetrate deep into the city center. These were all bargaining tools. The opponents, however, felt that there was no need for them to enter into negotiation, as they were not officially part of the process anyway, plus they had meanwhile gained the strong support of the community in its perception of the project. They felt they had more power outside the process, as they did not then have to make allowances for the interests of others. The argument was finally settled by a referendum that was forced by the opponents. This, in fact, is the ultimate example of a single central decision. The circumstances were no longer ideal for the Municipality: although the vote was about the restructuring project as a whole, because of the opponents' fierce opposition and strong lobbying, perceptions focused on the disputed underground car park. The vote went against it by a huge majority (partly as a result of the opportunistic argument regarding the vulnerability of the church tower, as mentioned above).

As an alternative strategy the Municipality could have entered into negotiation with all the actors, including the opponents. This being the case it might, for instance, have got its way on the building of the underground car park in one round of decision-making and met the opponents' objections regarding the traffic system (removing parking spaces elsewhere, routing the approach to the car park underground) in a subsequent decision. These were undertakings made by the Municipality in an attempt to win over the opponents (which they failed to do, of course). The vote against the underground car park also automatically meant the cancellation of the entire plan. In a process with more than one round of decision-making the rest of the plan could have remained, even if the Municipality had lost the battle for the car park, as this would have been just one – lost, as it happened – round. As there was only one round of decision-making (the referendum) there could only be one winner, and that was the opponents of the project.

CONCLUSIONS

We have tried to show in this chapter that technological issues are generally very complex and traditional project management is not adequate to handle this degree of complexity. The problems are often unstructured (there is uncertainty about information and no right solution) and there are usually large numbers of interdependent actors, each with its own values and interests. Also, the circumstances in which decision-making processes take place have pronounced internal and external dynamics. A hierarchical structure is not suitable here: it can alienate actors – on whom the decision-maker is dependent and who have important values and, often, a fair degree of autonomy – from the process, and it conflicts with the inevitably multi-faceted nature of complex problems.

An alternative is process management. This approach responds to the networking of actors in a complex technical project and makes for greater flexibility in problem definitions and goals, a different way of organizing decision-making (looser constraints and in some cases even increased complexity), negotiation on information and its interpretation, and open-ended decision-making, enabling a number of rounds of decision-making to take place. In this way characteristics such as lack of structure, interdependency and dynamism, which can cause insuperable problems of excessive complexity and thus uncontrollability, can be tackled and sometimes even turned into advantages, e.g. network-wide satisfaction and high-quality results.

Trends

The trend towards increasing importance of networks can also be seen as a shift towards greater importance of alliances between organizations. The actor in charge can no longer use its own information and power to control other actors; it must enter into a process with them by which a network comes into being. In the multi-actor system that is this network, organizations generally retain their autonomy but embark on a joint process, in which each has its own interests but they all pursue a particular goal. This can also be seen as a trend towards decentralization: decision-making power shifts from a – centralized – actor in charge to a number of relatively autonomous – decentralized – actors.

ACKNOWLEDGMENT

The research for this chapter was conducted at the research centre for Sustainable Urban Areas of Delft University of Technology.

FURTHER READING

Bruijn, J.A. de, Heuvelhof, E.F. ten and Veld, R. in't (2002). *Process Management; Why Project Management Fails in Complex Decision Making Processes*. Boston, MA: Kluwer. More extended explanation of the process management concept (as an alternative for project management).

Koppenjan, J. and Klijn, E.H. (2004). *Managing Uncertainties in Networks; A Network Approach to Problem Solving and Decision Making*. London: Routledge. Goes deeper into the feature of uncertainty and how network management can deal with it.

Quinn, R.E. (1998). *Beyond Rational Management; Mastering the Paradoxes and Competing Demands of High Performance*. San Francisco, CA: Jossey-Bass. Makes sense of the paradoxes, competing demands and contradictions of organizational life.

Scharpf, F.W. (Ed.) (1993). *Games in Hierarchies and Networks; Analytical and Empirical Approaches to the Study of Governance Institutions*. Frankfurt am Main/Boulder, CO: Campus/Westview. Authoritative work on hierarchies and networks.

REFERENCES

Bruijn, J.A. de and Heuvelhof, E.F. ten (2000). *Decision Making in Networks*. Utrecht: Lemma.

Bruijn, J.A. de, Heuvelhof, E.F. ten and Veld, R. in't (2002). *Process Management; Why Project Management Fails in Complex Decision Making Processes*. Boston, MA: Kluwer.

Douglas, M. and Wildavsky, A. (1983). *Risk and Culture*. Berkeley, CA: University of California Press.

Hanf, K. and Scharpf, F.W. (1978). *Interorganizational Policy Making; Limits to Coordination and Central Control*. London: Sage.

Hisschemöller, M. and Hoppe, R. (1996). Coping with untractable controversies; the case for problem structuring in policy design and analysis. Knowledge and Policy. *The International Journal of Knowledge Transfer and Utilization*. 4(8), 40–60.

Jasanoff, S. (1990). *The Fifth Branche; Science Advisers as Policy Managers*. Cambridge, MA: Harvard University Press.

Johnson, B. (1996). *Polarity Management; Identifying and Managing Unsolvable Problems*. Amherst, MA: HRD.

Klijn, E.H. and Koppenjan, J. (2000). Public management and policy networks; foundations of a network approach to governance. *Public Management*. 2(2), 135–158.

Koppenjan, J. and Klijn, E.H. (2004). *Managing Uncertainties in Networks; A Network Approach to Problem Solving and Decision Making*. London: Routledge.

Leijten, M. (2004a). Big Dig; een halve eeuw Central Artery in Boston. In: De Bruijn, H., Teisman, G.R., Edelenbos, J. and Veeneman, W. *Meervoudig ruimtegebruik en het management van meerstemmige processen*. Utrecht: Lemma, 151–177.

Leijten, M. (2004b). Grote Markt Groningen; down to earth. In: De Bruijn, H., Teisman, G.R., Edelenbos, J. and Veeneman, W. *Meervoudig ruimtegebruik en het management van meerstemmige processen*. Utrecht: Lemma, 203–222.

Mandell, M.P. (1988). Intergovernmental management in interorganizational networks: a revised perspective. *International Journal of Public Administration*. 11(4), 393–416.

Marin, B. and Mayntz, R. (Eds) (1991). *Policy Networks; Empirical Evidence and Theoretical Considerations*. Frankfurt am Main/Boulder, CO: Campus/Westview.

Mayntz, R. (1993). Modernization and the Logic of Interorganizational Networks. In: Child, J., Crozier, M., Mayntz, R. *et al*. *Societal Change between Market and Organization*. Aldershot: Avebury, 3–18.

Quinn, R.E. (1998). *Beyond Rational Management; Mastering the Paradoxes and Competing Demands of High Performance*. San Francisco, CA: Jossey-Bass.

Scharpf, F.W. (1993). Coordination in Hierarchies and Networks. Games. In: Scharpf, F.W. (Ed.). *Games in Hierarchies and Networks; Analytical and Empirical Approaches to the Study of Governance Institutions*. Frankfurt am Main/Boulder, CO: Campus/Westview.

Scharpf, F.W. (1994). Games real actors could play; positive and negative co-ordination in embedded negotiations. *Journal of Theoretical Politics*. 6(1), 27–53.

Wynne, B. (1988). Unruly technology: practical rules, impractical discourses and public understanding. *Social Studies of Science*. 18, 147–167.

Part IV

Managing innovation

INTRODUCTION

Management of innovation refers to managing the innovation process and the implementation of its results in organizations. The results of innovation processes can refer to new products, new production systems, and new organizations. In practice, minor product innovations are possible without changes in the production system or organization. However, in case of major product innovations this is hardly the case. Major product innovations consist of entirely new combinations of product attributes or incorporate new technologies. New technologies will often require a new production system. An example is provided by the transistor radio, a radio that uses a transistor instead of a vacuum tube. The transistor is a technology that enables the amplification of small electrical signals. This technology requires an entirely different system of production than its predecessor, the vacuum tube. As a result of the transistor technology, radios have become smaller and cheaper and require less energy. All these changes enabled the emergence of the portable radio, the marketing of which required considerable changes in radio producing and selling organizations. The case of the transistor radio also illustrates that mastering new technologies can be a vital competence for organizations. Technology management is one of the pillars of innovation management.

Chapter 11, "Corporate strategy and technology" by Marc Zegveld, describes four alternative perspectives on strategy. A central issue in this chapter is the way in which these perspectives perceive technology. The strategy of an organization has a large impact on R&D management practices. The strategy, for example, determines whether an organization is an imitator, follower or leader, and that, in turn, determines the importance of R&D and the kind of R&D management practices in an organization. The next chapter, "Innovation in context: from R&D management to innovation networks" by Patrick Van der Duin and his colleagues, describes subsequent mainstream practices of R&D management in organizations. Four generations of R&D management are distinguished. An example of a fourth-generation approach will be elaborated. Chapter 13 "Operation management with system dynamics" by Zofia Verwater-Lukszo, focuses on systems of production. In general, production and R&D management are different functions in an organization that have completely different perspectives on innovation. R&D management refers to innovation in terms of product innovations. Production on the other hand refers to innovation as production process renewal. This chapter will describe how systems of production can be made more flexible using a system dynamics approach. Chapter 14 by Erik Andriessen deals with "Managing knowledge processes." Explicit knowledge management practices are required when knowledge flows across disciplinary and departmental boundaries in an organization. An example of these boundary-crossing flows of knowledge occurs between the production and the R&D function in an organization when new products are developed (that require adaptations in the production system) or when new production machines are installed (that require adaptations in the product parts).

Chapter 11

Corporate strategy and technology

Marc A. Zegveld

OVERVIEW

Technology and strategy are, especially in this era of rapid technological developments, interrelated. Several studies are known wherein technology induces the development of corporate strategy, however the number of studies that focus on how different strategy models approach technology is limited. This chapter presents four generic strategy perspectives and how these different perspectives define technology. In the first paragraph the arena of corporate strategy is defined. Four generic strategy models are defined as the dominant perspectives of corporate strategy. In paragraphs two to five the basic aspects and processes of these four perspectives, industry-based strategic management; game theory; resource-based strategic management and dynamic capabilities are presented. In the final paragraph the differences are compared in relation to technology.

A perspective follows specific assumptions and develops a line of reasoning on the basis of these assumptions. As a result there is no right or wrong between perspectives. Consistency is, however, mandatory. Within strategy research but also within the field of business strategy itself one cannot use different perspectives simultaneously or one cannot use techniques of a single perspective within the framework of another perspective. As a result definitions, analyses and their findings only exist within the context of the chosen perspective. The concept of a perspective has several similarities with the definition of 'light'. Light is, depending on the research question, defined as 'waves' or as 'particles'. Analysis of light as particles will not fit within the perspective of light as waves and vice versa.

A large number of definitions on strategy are nowadays available (e.g. Burgelman, 1983; Chaffee, 1985; Mintzberg, 1978). Most of these definitions have in common that strategy concerns survival and deals with changing environments (Romme, 1992). The survival of companies is based on how successful companies interact with their environment. Strategy is concerned with the processes and content of these interactions and the creation of long-term wealth as a result of these interactions. Therefore, strategy is concerned with:

- The level of health or fitness, which is related to the presence and successful deployment of resources and competences, of the company. A successful deployment of strategy implies that the company generates sufficient financial value to remain competitive.
- The existence of a vision as a long-term view of how the company should develop. A vision or long-term view has to do with the positioning of the business and understanding external and internal developments and how these will affect the company. From a dynamic viewpoint the company also influences its environment, which might affect the process of positioning as part of a complex interactive process where the company and the environment interact and influence each other.
- A company should have the ability to identify and interpret these changes and respond. Identifying, interpreting and responding are the key elements of interaction and are the fundamental aspects of adaptive systems and the process of learning.

The literature on strategic management provides a wide range of different views, perspectives and related assumptions. By aggregating several strategy frameworks three reference points, namely internal, external and time, are defined as the fundamental dimensions of strategic management (Fiegenbaum *et al.*, 1996). Two original strategy frameworks, i.e. the industrial organisations school and the resource framework, use 'extern' and 'intern' respectively as their reference points. Strategy frameworks that use 'time' as a reference point stress the impact of adaptation or learning. These strategy frameworks, such as 'game theory' and 'dynamic capabilities' can be defined as dynamic versions of the industrial organisation and the resource school, respectively. Other strategy schools that use 'time' as a reference point frequently use metaphors such as 'evolution' and 'ecology' which are derived from biology.

The internal strategic reference point uses internal variables to define the success of a company. Companies set targets for strategic inputs such as cost reduction, quality improvements and new product development and evaluate their performance based on these goals. Strategic inputs can be defined around value-added activities which form the firm's central axis or the driving force of managerial concern. Simultaneously, firms also define strategic outputs such as sales and profitability and hold managers responsible for performance against these targets. As a result, self-reflection becomes very important. The external strategic reference point uses external benchmarks, such as competitors (industrial organisation), suppliers and customers (resource dependence) or society in general (institutional theory). By using time as a reference point 'past' and 'future' can be recognised as critical dimensions. Accumulating knowledge over time can be used as a source of competitive advantage and provides a reference point to spur continuous achievements. By defining a vision and related strategic intent the future can also be detected as a reference point. Primarily based on three reference points (internal, external and time), Figure 11.1 provides an overview of the four main generic perspectives of strategic management (Zegveld, 2000).

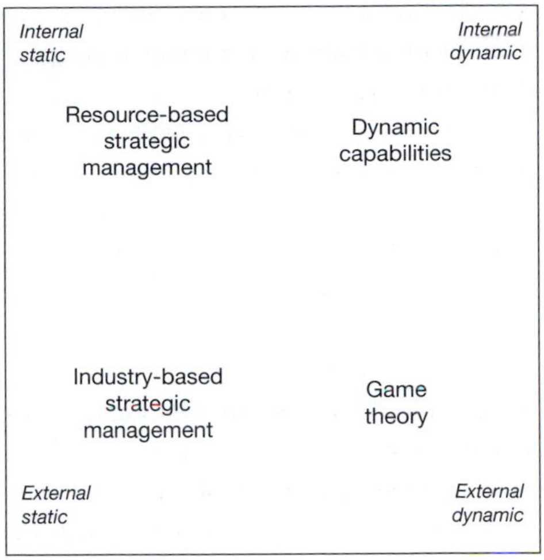

Internal static

Internal dynamic

Resource-based strategic management

Dynamic capabilities

Industry-based strategic management

Game theory

External static

External dynamic

Figure 11.1 *Positioning the four generic models of strategic management*

INDUSTRY-BASED STRATEGIC MANAGEMENT

Industrial organisation theories study specific markets, and consider a range of market structures from monopoly to oligopoly. Individual firms are large and their strategic interaction is emphasised. Firms are represented by their set of available strategic actions in relation to their cost functions. Company-related industrial organisation theories view the market structure in terms of numbers and sizes of firms and focus on the competition between firms. The principal contribution of this focus is the recognition that firms seek competitive strategies in response to the strategies of rival firms. Firms are aware of the effect of their actions on demand. According to Porter (1985, pp. 1–2):

> Two central questions underlie the concept of competitive strategy. The first is the attractiveness of industries for long-term profitability and the factors that determine it. Not all industries offer equal opportunities for sustained profitability, and the inherent profitability of its industry is one essential ingredient in determining the profitability of a firm. The second central question in a competitive strategy is the determinants of relative competitive position within the industry.

Industry-based strategic management is directly related to the Harvard industrial organization theory. Based on case studies by Mason, the Bain-Mason paradigm or structure-conduct-performance paradigm (Bain, 1968) was developed, which states that industry structure determines the conduct of firms whose joint conduct determines their collective performance (see Figure 11.2; Baaij, 1996).

Porter imported the Harvard industrial organization structure-conduct-performance paradigm into the SWOT framework (analysis of strength, weaknesses, opportunities and threats) of strategic management (Porter, 1980). The performance of the firm is a result of the industry's attractiveness and the firm's competitive position within that industry. As a result, the essence of strategy is positioning the firm in its industry environment. Porter developed his competitive forces framework for industry analysis by identifying five basic competitive forces (see Figure 11.3; Porter, 1985). As a result:

> Industry profitability is not a function of what the product looks like, or whether it embodies high or low technology but of industry structure. Some very mundane industries such as postage meters and grain trading are extremely profitable, while some more glamorous high technology industries such as personal computers and cable television are not profitable for many participants.
>
> (Porter, 1985: p. 5)

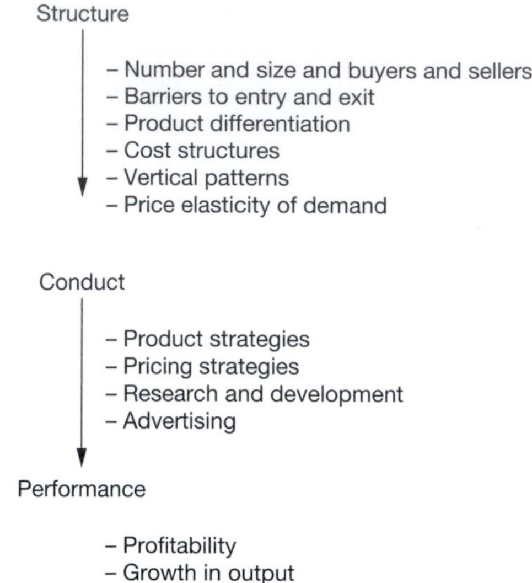

Figure 11.2 *The structure-conduct-performance paradigm*

The strength of each of the five forces is a function of the industry structure or the underlying economic and technical characteristics of an industry. The collective strengths of the competitive forces determine the intensity of competition and the ultimate profit potential of the industry. The rules of competition are embodied in five competitive forces. The collective strength of these five forces determines the ability of firms in an industry to earn, on average, rates of return on investment in excess of the cost of capital. The five forces that drive industry competition are: the rivalry among existing competitors; the bargaining power of suppliers; the threat of new entrants; the threat of substitute products and the bargaining power of buyers.

The competitive forces framework provides an analysis of industry opportunities and threats and, therefore, the industry's attractiveness, and determines the strategy of a firm. Porter defines strategy as the analytical selection of an attractive industry and subsequent selection of a competitive position within the industry.

A firm is usually not a prisoner of its industry structure. Firms, through their strategies, can influence the five forces. If a firm can shape structure it can fundamentally change industry attractiveness for better or for worse. In addition, Porter identified three generic competitive strategies to obtain these competitive positions (see Figure 11.4; Porter, 1985).

Cost leadership is, perhaps, the clearest of the three generic strategies. In it, a firm sets out to become the low-cost producer in its industry. A low-cost producer must find and exploit all sources of cost advantage. Low-cost producers particularly sell a standard, or no frills, product and place considerable emphasis on reaping scale or absolute cost advantages from all sources. A cost leader must achieve parity or proximity in the bases of differentiation relative to its competitors to be an above average performer, even though it relies on cost leadership for its competitive advantage. Parity in the basis of differentiation allows a cost leader to translate its cost advantage directly into higher profits than competitors.

The second generic strategy is differentiation. In a differentiation strategy a firm seeks to be unique in its industry or on some dimension that is widely valued by buyers. It selects one or more attributes that many buyers in an industry perceive as important, and uniquely positions it to meet those needs. It is rewarded for uniqueness with a premium price. The firm must truly be unique at something, or be perceived as unique, if it is to expect a premium price. In contrast to cost leadership, however, there can be more than one successful differentiation strategy in an industry if there are a number of attributes that are widely valued by buyers.

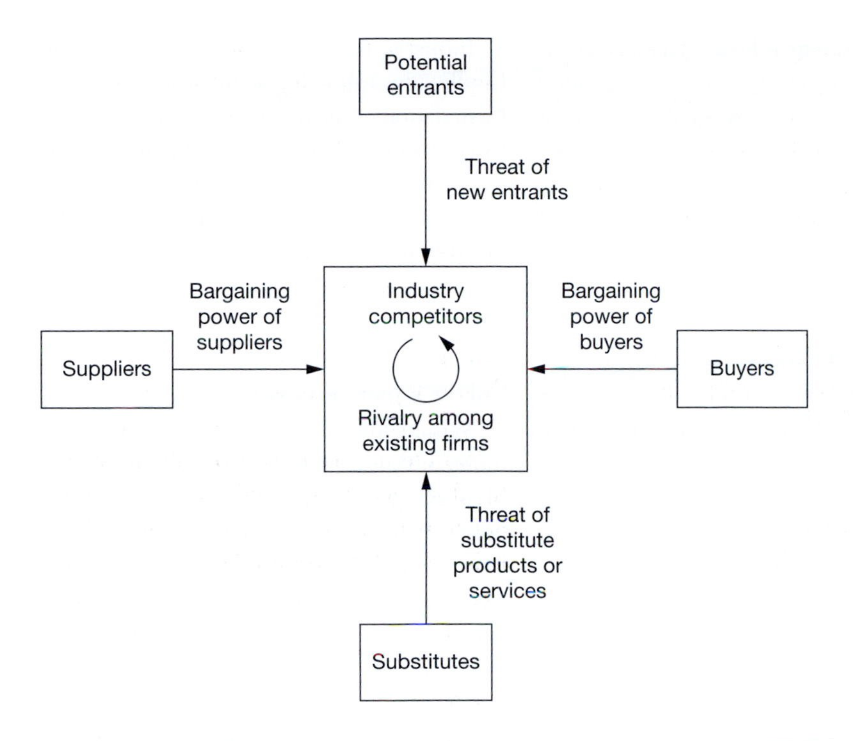

Figure 11.3 *Competitive forces framework*
(Porter, 1985)

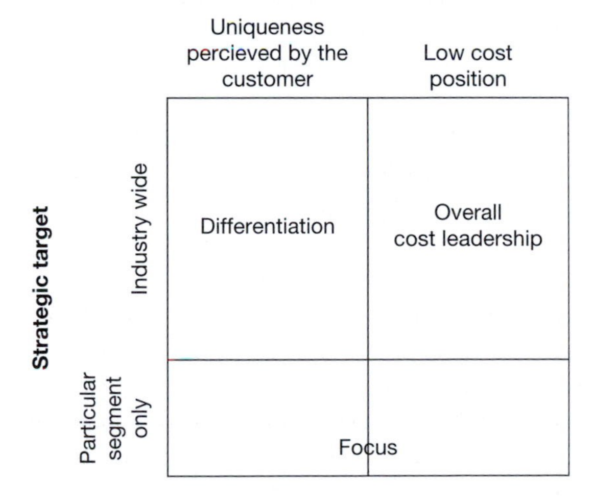

Figure 11.4 *Generic competitive strategies*
(Porter, 1985)

The third strategy is focus. This strategy is quite different from the other two generic strategies because it rests on the choice of a narrow competitive scope within the industry. The company that deploys a focus-strategy selects a segment or group of segments within the industry and tailors its strategy to serving down to the exclusion of others. By optimising its strategy for the target segment the focuser seeks to achieve a competitive advantage in this target segment even though it does possess a competitive advantage overall.

Competitive position and competitive advantage are based on a firm's ability to cope with competitive forces better than its rivals. The basis of competitive advantage is the distinctive ability of a firm to align itself with the industry environment. By developing the concept of the value chain a firm is able to analyse what activities should be related to what products in order to create a sustainable strategic advantage. Subsequently, Porter developed a theory of industry-based strategic management and used an outside-in strategy process (see Figure 11.5; Zegveld, 2000).

Industry-based strategic management handles the following sequences of strategy formulation: domain selection; competitive forces framework; domain navigation; by using the generic competitive strategies framework a firm can select a competitive position within the industry; activity configuration and after-strategy formulation, by using the value chain, a firm can translate strategy into action by selecting resources and the required activities.

Within industry-based strategic management, technology itself is not a determinant of a successful strategy. The competitive position of a defined industry and the position a firm holds within the industry are the two main criteria for the success of companies. As a result technology may influence the attractiveness of the industry or the firm's position in the industry.

Following industry-based strategic management, Philips frequently analyses the attractiveness of different industries and the most attractive position of Philips within these industries. In order to remain competitive Philips developed a strong technology

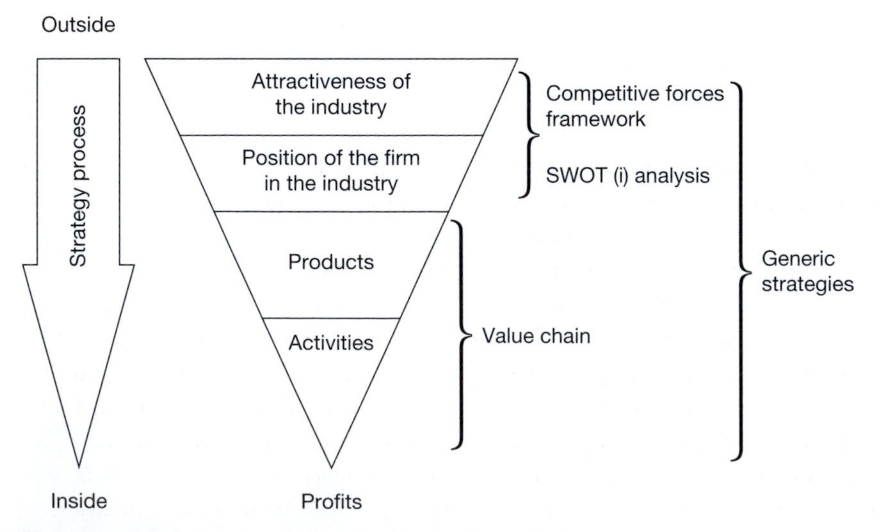

Figure 11.5 *Strategy process of industry-based strategic management*

portfolio. Philips defines technology as a very important factor that may induce change within the attractiveness of the industry or the firm's position in the industry. As a result technology research is always related to business opportunities, which improves the competitiveness of the company.

Following industry-based strategic management the success of Honda is the result of investments in new and attractive industry segments and in the position the company finally established within these industries. Within the industry-based strategic management perspective the success of technology is judged on how technology improves the attractiveness of the industry or the firm's position in the industry. The existence of Honda within different industries such as cars, motorbikes, lawnmowers, etc. reflects the analysis of Honda that these industries are attractive as well as the potential Honda defined to gain a competitive position within these defined industries.

GAME THEORY

Business language is in most cases described as a zero-sum game while business is not about winning or losing but in finding win-win strategies. According to Brandenburger and Nalebuff (1995):

> Looking for win-win strategies has several advantages. First, because the approach is relatively unexplored, there is a greater potential for finding new opportunities. Second, because others are not being forced to give up ground, they may offer less resistance to win-win moves, making them easier to implement. Third because win-win moves don't force other players to retaliate, the new game is more sustainable. And finally imitation of a win-win move is beneficial, not harmful. To encourage thinking about both cooperative and competitive ways to change the game, we suggest the term coopetition. It means looking for win-win as well as win-lose opportunities.
>
> (p. 57)

Non-cooperative game theory equilibrium can be applied to a wide variety of actions: pricing, contract terms, production, investment, advertising, R&D, product quality and other product characteristics, and is based on the seminal work of van Neumann and Morgenstern, who published their book *Theory of Games and Economic Behaviour* (Neumann and Morgenstern, 1944). In their work Neumann and Morgenstern provided a systematic way to understand the behaviour of entities, players and agents in the system where the pay-off or fortune of the interacting agents or players are interdependent. Neumann and Morgenstern distinguish between games in which a group of players interact according to a great set of rules, and games in which players interact without any constraint except for a constraint by their own individual free will. Fifty years later it was Harsanyi, Nash and Selten who received the Economics Nobel Prize for their pioneering analysis on the theory of non-cooperative games. Within business strategy it was Brandenburger and Nalebuff (1995) who initially translated the main findings on game theory towards corporate strategy:

> For rule-based gamers, game theory offers the principle; to every action there is a reaction. To analyze how other players will react to your move, you need to play out all the reactions (including yours) to their actions as far ahead as possible. You have to look backward far into the game and then reason backward to figure out which of today's actions will

lead to where you want to end up. For freewheeling games, game theory offers the principle, you cannot take away from the game more than you bring to it. In business, what does a particular player bring to the game? To find the answer, look at the value created when everyone is in the game and then pluck that player out and see how much value the remaining players can create. The difference is the removed player's added value. In unstructured interactions you cannot take away more than your added value.

(p. 62)

Assume two companies both have the possibility to choose between two strategies which are interdependent. Within the matrix in Figure 11.6 the benefits of these interdependent strategies for each of these two companies are presented.

It can be concluded that a collective optimum can be reached when company I follows strategy A and company II follows strategy 1. Besides this well-known prisoner's dilemma, a number of different structures can be visualised. In this structure the individual pay-off is important to the individual agents, however the aggregate pay-off is important to the collective. In the prisoner's dilemma, the optimal collective solution is that both players choose their first alternative. However, if player II chooses strategy 2 where the other chooses A, company I will have a negative pay-off and company II will receive a full pay-off. In the non-sequential prisoner's dilemma, lack of mutual trust will drive players to the worst possible collective solution which means that both players will choose their second alternative. Contrary to the above described prisoner's dilemma, within the sequential prisoner's dilemma learning can take place and trust might be built up. It was Axelrod (1984) who, through an experimental tournament, concluded that the strategy 'tit for tat' (tft) per-

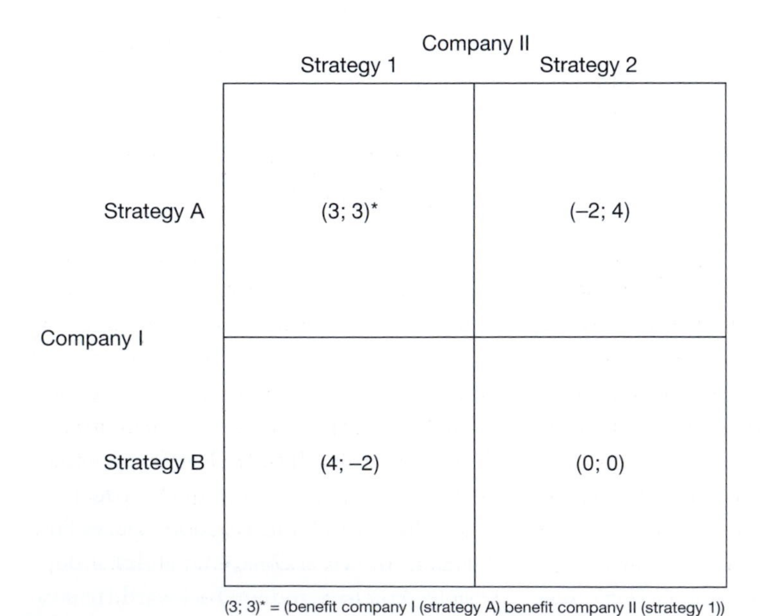

Figure 11.6 *Prisoner's dilemma*

formed best. In this strategy cooperation starts at the first move; after that a player does what the other player did on the previous move. The tft-strategy has four major advantages: avoidance of unnecessary conflict by cooperating as long as the other player does; clarity in communicating the rules of the game; retaliation in the face of a defector and forgiveness after responding to an occasional provocation. The tft-strategy is both competitive and cooperative which probably explains its robust success.

A simple example concerns the introduction of comfort class in 1993 by Trans World Airline (TWA). By removing 5 to 40 seats per plane, in order to give passengers more leg room, TWA increased passenger satisfaction and simultaneously increased the TWA occupation rate. If other companies followed this 'less chairs'-policy, the chance of a price war could be reduced due to the fact that less seats were available. If other companies induced a price war at least TWA would stand out due to its high customer satisfaction.

Brandenburger and Nalebuff defined two major steps in analysing competitive strategies. At first the value net of the company has to be analysed. Comparable to the five forces framework of Porter, Brandenburger and Nalebuff developed a model in order to recognise all players in the game. Besides customers and suppliers they recognise substitutors and complementors. The value net explores the players of the game as well as the interdependencies in the game. The second step focuses on the actual game in terms of: players; added values; rules; tactics and scope.

In order to understand the full impact of game theory one should not only analyse the impact of a coopetitive strategy on a single organisation but also on the players of the value net. This external orientation has several similarities with the industry perspective. Contrary to the industry perspective, game theory follows a dynamic interactive approach. Within game theory technology is, in most cases, defined as instrumental and is relevant in executing defined strategies. The success of 'Windows' by Microsoft can be understood by the use of game theory; the more people use Windows the more valuable Windows will be as it emerges to be the standard in communicating with others. Both first mover advantage and a continuous adding of new features within 'Windows' not only boosted the competitive position of Microsoft, it also created an operating standard wherein other companies were able to launch new products with a higher penetration than it would occur otherwise.

With the use of game theory one can understand that, e.g. the introduction of motorbikes within the US by Honda not only limited the competitive pressure within the home market of Honda, Japan, but also enlarged the market of its competitors. Game theory might have provided relevant information towards Honda on how to deal with its Japanese competitors during its expansion in the US.

RESOURCE-BASED STRATEGIC MANAGEMENT

Contrary to industry-based strategic management, resource-based strategic management states that firm-specific factors are the major determinants of performance differences between firms in the same sector (Cool and Schendel, 1988). As a result the firm itself is the main domain of study. The firm is organised as a means of mitigating the effects of uncertainty regarding production and final demand. As a result agency relationships within the firm are the subject of analysis as well as the relative efficiencies of information processing in market relationships versus organisational relationships. Activities will take place within the firm when relationships within the organisation handle information more effectively than market contracts. Asymmetric information and bounded rationality (Cyert and March, 1963) cause information imperfections and lead to opportunities (Williamson, 1975).

The resource perspective has a long history, starting with Marshall using three resources, i.e. land, labour and capital. It was Selznick, who developed a resource-based view of the firm in the 1950s, stressing that competences are the source of competitive advantage (Collis and Montgomery, 1995). Later, Penrose defined a company as a 'bundle of competencies' (Penrose, 1959) or as a configuration of technology and organisation (Foss and Knudsen, 1996; Penrose, 1959). The firm is made up of a number of resources, consisting of assets, competences and positional advantages embodied in various forms of capital (financial, human, social, commercial). Contrary to assets, competences and positional advantages are not subject to ownership and contracts. Competences refer to abilities and knowledge in the sense of know-how and are merely a group of skills, experiences and technologies rather than a single, discrete skill or technology (Nelson and Winter, 1982). Competences are firm-specific and unique and cannot be bought outside the company but have to be developed. They are the result of combining different organisational resources and their related knowledge and experiences. Besides competences, capabilities can also be identified as essential to the resource perspective (Stalk *et al.*, 1992). Capabilities can be related to product and process development, technology transfer, intellectual property, manufacturing, human resources and organisational learning. A distinction is made between core competences and core capabilities. Core competences are defined as emphasising technological and production expertise at specific points along the value chain. Core capabilities are more broadly based, encompassing the entire value chain. In this respect core capabilities are visible to the customer in a way core competences rarely are. As a result, the existence and development of core capabilities make it possible for the firm to gain benefits based on its positional advantages such as strong brand loyalty and a good reputation. The positional advantages are related to competitors, customers, employees, shareholders and partners (see Figure 11.7; Stoelhorst, 1997).

Hamel and Prahalad (1994) contributed a great deal to the current popularity of a resource-based view of the firm by developing the concept of core competences. The source of competitive advantage is the management's ability to consolidate corporate-wide technologies and production skills into competences that enable individual businesses to adapt quickly to changing opportunities. According to Prahalad (1991) companies have to bridge the performance gap (related to operational aspects and restructuring) as well as the opportunity gap (related to strategic direction and revitalisation) to create value (see Figure 11.8; Hamel and Prahalad, 1993).

> Strategic intent is a way of creating an obsession with winning that encompasses the total organisation. It is a shared competitive agenda, sustained over a long period of time, for global leadership. Extraordinary accomplishment is often based on a clearly articulated strategic intent. The US has experienced the power of a clear strategic intent. Consider the Apollo program 'man on the moon by the end of the decade' was a 'stretch' target. It meant global leadership and a domination of space. The goal was competitively

focussed; Russians were the enemy. It was very clear. While the goal was clear, managers of the project had to discover the means, and a lot of new technologies had to be developed in an enormous time pressure. Why is the spirit of the Apollo program not replicable in firms?

(Prahalad, 1991, p. 4)

The strategy orientation is therefore a mixture of leveraging and stretching the corporate resources to such a level that new business can be created (Hamel and Prahalad, 1993). Management can leverage its resources in five basic ways: by concentrating more effectively on key strategic goals; by accumulating them more efficiently; by complementing one kind of resource with another to create a higher order value; by conserving resources wherever possible; and by recovering them from the market place in the shortest possible time (Hamel and Prahalad, 1993). Besides a certain framework for the company's possibilities, a strategic

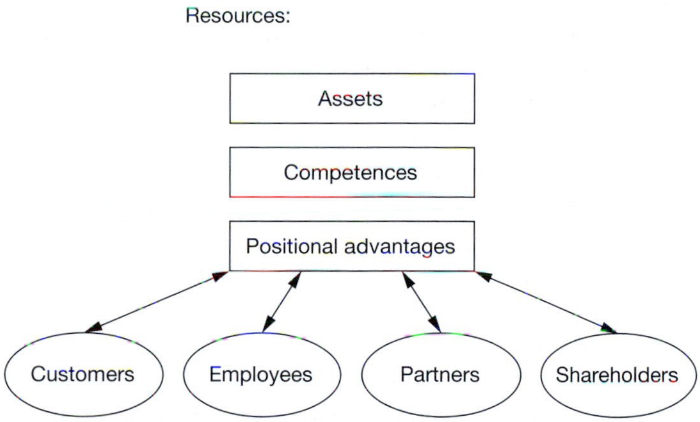

Figure 11.7 *Definition of resources*

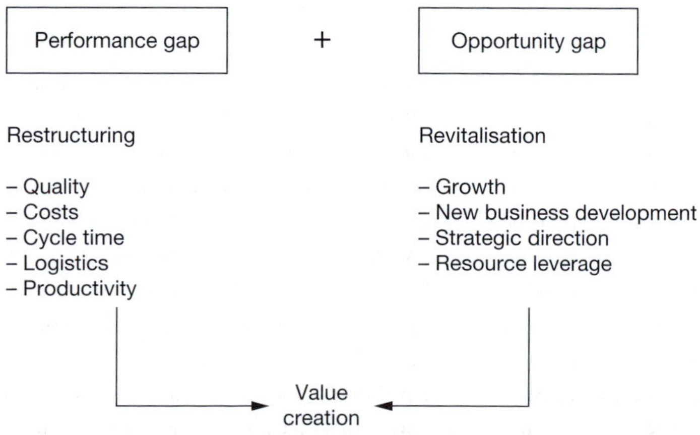

Figure 11.8 *Performance and opportunity gap*
(Hamel and Prahalad, 1993)

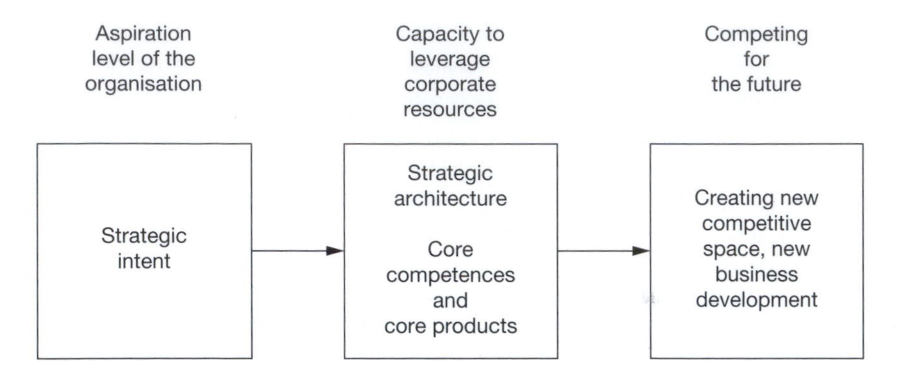

Figure 11.9 *From strategic intent to competing for the future* (Hamel and Prahalad, 1993)

intent has to be developed to create a path leading to the defined new business space (Figure 11.9). Resources are the factors for production and dictate the supply or the transformation process. This means that resources can be tangible as well as intangible, human as well as physical.

The strategy process from a resource-based strategic management perspective therefore has an 'inside-out' orientation (see Figure 11.10). Once a shared level of aspiration is defined a framework for leveraging corporate resources that is consistent with the strategic intent has to be developed. Prahalad and Hamel start with a strategic architecture, which is a way of developing a point of view regarding evolution of an industry. How will the interface with customers change? What are the new technological possibilities? How are our current and future competitors positioning themselves to approach this industry? Strategic architecture is a distillation of a wide variety of information that allows managers to identify what core competences the company has and what is needed. The concept of core competences is frequently confused with core technologies and/or capabilities. Core technologies are a component part of core competences. Core competence results when firms learn to harmonise multiple technologies. A core competence can be identified by applying three simple tests: is it a significant source of competitive differentiation? Does it provide a unique signature to the organisation, such as miniaturisation for Sony or user-friendliness at Apple? Second, does it transcend a single business? Does it cover a range of businesses, both current and new? Finally, is it hard for competitors to imitate? Is it hard for someone to visit Matsushita or Sony and come back and outline why they are good at manufacturing or miniaturisation respectively.

Following the resource-based strategic management perspective the success of Honda is the result of their decision to define their core competence as engines. With engines as starting point the current portfolio of motorbikes, cars, lawnmowers, etc. can be understood. Technology plays a significant role in the development of this core competence and technology will be in the centre of the company's strategy.

DYNAMIC CAPABILITIES

Following the footsteps of the resource perspective the perspective of dynamic capabilities states that strategy is about developing skills in order to adapt and respond. As a

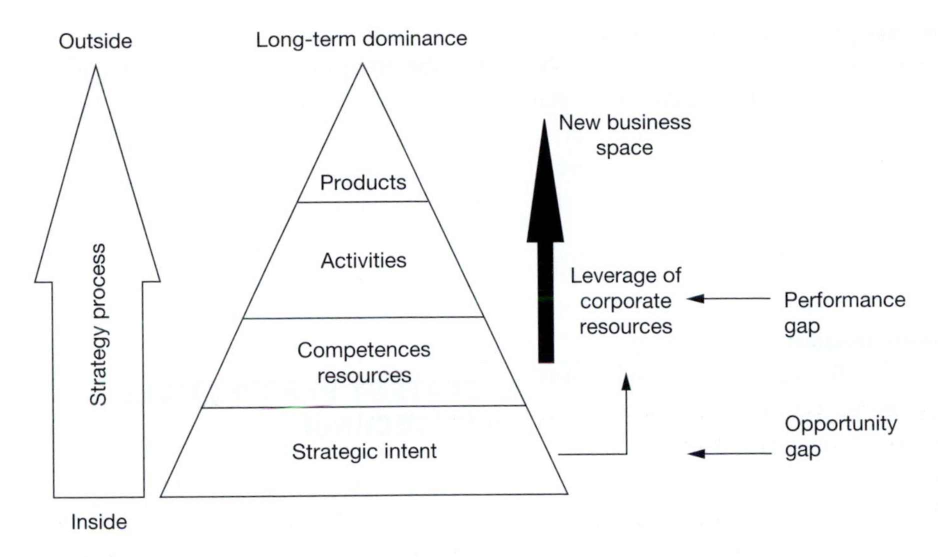

Figure 11.10 *Strategy process of resource-based strategic management*

result the organisation itself is the main body of analysis and is defined as the main source of competitive advantage. Contrary to external or market power perspectives, dynamic capabilities analyses how firm-specific capabilities can be sources of competitive advantage and how combinations of competences and resources can be developed, deployed and protected. It is clear that both internal perspectives focus on firm-specific assets, competences or capabilities. Second, if control of resources is the source of economic profits, then the skills of control, organisation and learning become issues of strategic importance.

Dynamic capabilities focus on the dynamics or renewal of competences such as time-to-market, innovative responses, technological change, while 'capabilities' imply adapting, reconfiguring skills, resources and competences in order to match the requirements of a changing environment. Given the choices and strength of a firm, specific paths or trajectories can be recognised. These paths are not only relevant in order to understand the company today but are also relevant in order to analyse the potentials for tomorrow. According to Teece, Pisano and Shuen (1997):

> The dynamic capabilities approach seeks to provide a coherent framework which can both integrate existing conceptual and empirical knowledge, and facilitate prescription. The development of firm-specific capabilities and how they renew competencies are intimately tied to the firm's business processes, market positions and expansion paths.
>
> (p. 515)

The main difference between competences and dynamic capabilities is the dynamics in the approach of the latter. While competences are related to specified knowledge, such as engines, that forms the company's specific competitive differentiation, dynamic capabilities are related to managerial knowledge in order to learn and so to reduce time to market, improve existing engines, etc.

The dynamic capabilities perspective handles six interconnected layers. A combination of elements of a single layer may create an element in the next layer. From this perspective these layers have a hierarchical structure. These six layers are:

- *Factors of production*. These are defined as undifferentiated and fugitive inputs (land, unskilled labour and capital).
- *Resources*. Resources are defined as firm-specific assets that are difficult to imitate (trade secrets, specific production facilities).
- *Competences*. Competences are defined as organisational routines or firm-specific assets related to a specific set of activities within the firm. Competences can be quality, system integration, etc.
- *Core competences*. Core competences are defined as competences that define the core business of a firm.
- *Dynamic capabilities*. The firm's ability to integrate, build, reconfigure internal as well as external competences in order to adapt to rapidly changing environments.
- *Products*. The final goods and services produced by a firm utilising the firm's competences and, over time, utilising both competences and capabilities.

Dynamic capabilities are the capabilities the company has in order to develop core competences. In this respect the dynamic capabilities follow the resource perspective, but change this perspective from a static point of view towards a dynamic point of view. The dynamic approach implies, in relation to the firm's competences and capabilities, the existence of evolutionary paths or trajectories. In this perspective learning, coordination and reconfiguration are essential aspects in order to ensure a continuous process to gain competitive advantage.

Within the dynamic capabilities perspective the dynamic capability of Honda is related towards the knowledge on translating engineering know-how from one product towards other products and markets. It is not the competence related to engines that is the central point of strategy, but the knowledge to learn and adapt forms the cornerstone of the dynamic capability of Honda.

STRATEGY PERSPECTIVES AND TECHNOLOGY

Both the industry-based strategic management perspective and the perspective that enhances game theory define technology as an exogenous aspect that has to be incorporated within the firm in order to gain competitive advantage through the deployment of one of the three generic strategies. The choice of which technology to choose from is primarily related to the choice of industry the company wants to be active in. Technology might have its impact on the cost function of the firm or on the perceived uniqueness or differentiation of the firm. As a result, technology is, as it influences the firm's activities, functional but does not belong to the core of the strategy process itself. Technology may generate an impetus on both the development of an industry-based vision and on the competitive analysis. In both types of analyses technology will be examined in order to create a better understanding of whether competitive advantage can, indeed, be created, and how competitive advantage might be sustained through the use of the defined technology. The choice of which generic strategy to deploy will influence which technology to choose and not vice versa. The link between type of technology and the industry-based vision as well as competitive analysis is indirect. The exogenous approach of technology implies mainly a one-sided orientation, namely from the development of an industry-based vision and competitive analysis towards the choice of technology and not vice versa. As the transformation from technology to value is insecure, firms should only focus on those technologies that create sustainable competitive advantage themselves, enhance a generic strategy, lead to first-mover advantages or improve the overall industry structure. From this perspective the connotation that high technology is related to high profits is questioned by the followers of the industry perspective and changed in 'valuable technologies are those technologies that can be altered in competitive value chains'.

Within the resource perspective and dynamic capabilities approach both competences and dynamic capabilities are defined as aspects that induce the development of the strategy of the firm. Within these approaches the ambition of the firm is merely based on the experience of the firm through the use of the existing competences and capabilities. Technology plays an important role within the recognition and definition of competences and capabilities; however, these are absolutely not similar. While technology itself is technical both competences and capabilities are related to the functional advantage of what and how to alter technology, and other aspects, into business. The link between the type of technology and resource perspective and the dynamic capabilities approach is existent and direct. The translation of technology into business is important and, within both approaches, recognised as crucial within the strategy process. From this perspective the connotation that high technology is related to high profits is also questioned by the followers of the resource perspective and changed in 'valuable technologies are those technologies that can be altered into firm specific competences'.

Application of the four presented perspectives within Honda, show that the strategy of a single company within the same time frame will result in different outcomes when different perspectives are used. In Figure 11.11 several of the different strategy issues that are related to the use of these four perspectives are presented. This approach can also be related to the use of specific instruments of analysis.

In Table 11.1 different instruments are related to these four perspectives regarding five subjects of study that combine strategy, innovation and knowledge productivity; a step into the future; existentialism; the value of knowledge; organisation of R&D and success.

The resource perspective, like dynamic capabilities, uses the organisation as its point of reference, and therefore mainly focuses on quality improvement, quality development, flexibility and cost reduction as the main sources of company success. The essential of these two perspectives is that companies are defined as bundles of resources that need to be managed through competences and

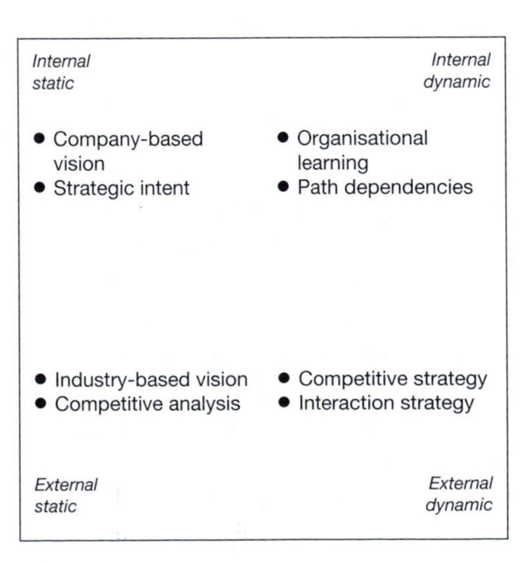

Figure 11.11 *Perspective and subject of analyses*

Table 11.1 *Perspectives and related instruments*

	Industry-based strategic management	*Game theory*	*Resource-based strategic management*	*Dynamic capabilities*
A step into the future	Industry structure	Coopetition	Competing for the future	Firm-specific trajectories
Instruments	Industry analysis Futures analysis Technology analysis	Value net analysis cooperative game analysis non cooperative game analysis	Technology analysis Competing analysis Futures analysis	Capability analysis Technology analysis Futures analysis
Existentialism	Value chain Generic strategies	Organisational learning Right moves	Strategic intent Core competences	Organisational learning Dynamic capabilities
Instruments	5-forces framework analysis Value chain analysis SWOT analysis	Prisoner's dilemma analysis Interactive scenario analysis	Strategic intent analysis Strategic architecture analysis	Path dependency analysis System of innovation analysis
The value of knowledge	Leverage impact of generic strategies		Deploys strategic intent Limits the performance gap Widens the opportunity gap	Strengthens capabilities Increases inimitability
Instruments	Knowledge productivity Intellectual capital analysis Intangible asset analysis	Knowledge productivity Intangible asset analysis	Knowledge productivity Intellectual capital analysis Intangible asset analysis	Knowledge productivity Intangible asset analysis
Organisation of R&D	Position within the value chain Linkages within the value chain	Position within the value net	Strategic architecture	Organisational capabilities
Instruments			System of innovation analysis	System of innovation analysis
Success	Profits	Asset accumulation	Long-term dominance	Short-term gain
Instruments	Generic cash flow analysis	Asset to cash flow analysis	Longitudinal cash flow analysis	Short-term cash flow analysis

dynamic capabilities respectively. Many of these models are reducible to the basic principle of the oligopolistic industrial organisation theory from Chicago. The industry perspective and game theory use an external point of reference such as the position of competitors, suppliers, customers and others. The external focus is reducible to the industrial organisation theory from Harvard. Strategy models that designate time as a reference point, such as game theory and dynamic capabilities, emphasise the impact of change and learning. These strategy models can be defined as dynamic versions of the industry perspective and the resource perspective, respectively. Each of the perspectives has its own research schools and methods of analysis, as well as critique on the other perspectives, and question the assumptions or research findings related to the other schools. This sometimes results in 'who is right and who is wrong' (Ansoff, 1991; Mintzberg, 1991; Goold, 1992).

Technology is a relevant aspect of corporate change and corporate success, however technology itself has no value; the context of its application may generate value and competitive advantage. As presented in this chapter the context wherein technology can be positioned not only differs per company but also per strategy perspective. As a result both analyses on the application of technology as well as the impact of technology on businesses performance need a perspective. Four of these strategy perspectives have been presented here. It is not evident that corporate success is the single result of internal criteria. Neither is it expected that industry structure is sufficiently stable in order to focus solely on external references. The four perspectives are, indeed, only perspectives and therefore do not fully explain the complex and dynamic reality around us. When using one of these perspectives one should always ask the validity of these assumptions and the limitations of conclusions of analysis based on its use. From this point on the four perspectives are valuable instruments in analysing the strategy of a company and the role of technology related to strategy.

REFERENCES

Ansoff, H.I. (1991). Critique of Henry Mintzberg's the design school: reconsidering the basic premises of strategic management, *Strategic Management Journal*, 12: 449–461.

Axelrod, R. (1984). *The Evolution of Cooperation.* New York: Basic Books Inc.

Baaij, J.S. (1996). *Evolutionary Strategic Management: Firm and Environment, Performance over Time.* Nijenrode: Nijenrode University Press.

Bain, J.S. (1968). *Industrial Organization.* New York: John Wiley.

Brandenburger, A.M. and Nalebuff, B.J. (1995). The right game: use game theory to shape strategy, *Harvard Business Review*, July–August, 57–71.

Burgelman, R.A. (1983). A model of the interaction of strategic behaviour, corporate context and the concept of strategy, *Administrative Science Quarterly*, 28: 223–244.

Chaffee, E.E. (1985). Three models of strategy, *Academy of Management Review*, 10 (1): 89–98.

Collis, D.J. and Montgomery, C.A. (1995). Competing on resource strategy in the 1990s, *Harvard Business Review*, July–August, 118–128.

Cool, K. and Schendel, D. (1988). Performance differences among strategic group members, *Strategic Management Journal*, 9: 207–233.

Cyert, R.M. and March, J. (1963). *A Behavioral Theory of the Firm*. New York: Prentice Hall.

Fiegenbaum, A., Hart, S. and Schendel, D. (1996). Strategic reference point theory, *Strategic Management Journal*, 17: 219–235.

Foss, N.J. and Knudsen, J. (eds) (1996). *Towards a Competence Theory of the Firm*. London: Routledge.

Goold, M. (1992). Design, learning and planning: a further observation on the design school debate, *Strategic Management Journal*, 13: 169–170.

Hamel, G. and Prahalad, C.K. (1993). Strategy as stretch and leverage, *Harvard Business Review*, March–April, 75–86.

Hamel, G. and Prahalad, C.K (1994). *Competing for the Future*. Boston, MA: Harvard Business School Press.

Mintzberg, H. (1978). Patterns in strategy formation, *Management Science*, 24: 934–948.

Mintzberg, H. (1991). Learning 1, planning 0: reply to Igor Ansoff, *Strategic Management Journal*, 12: 463–466.

Nelson, R.R. and Winter, S.G. (1982). *An Evolutionary Theory of Economic Change*. Cambridge: Cambridge University Press.

Neumann, J. von and Morgenstern, O. (1944). *Theory of Games and Economic Behavior*. 1953 edition, Princeton, NJ: Princeton University Press.

Penrose, T. (1959). *The Theory of Growth of the Firm*. New York: John Wiley.

Porter, M.E. (1980). *Competitive Strategy: Techniques for Analyzing Industries and Competitors*. New York: The Free Press.

Porter, M.E. (1985). *Competitive Advantage: Creating and Sustaining Superior Performance*. New York: Collier Macmillan.

Prahalad, C.K. (1991). The role of core competencies in the corporation, *Research, Technology Management*, 36 (6): 40–48.

Romme, A.G.L. (1992). *A self-organization perspective on strategy formation*, Faculty of Economics and Business, Administration, University of Limburg, Dissertation 92–4.

Stalk, G., Evans, P. and Schuman, L.E. (1992). Competing on capabilities: the new rules of corporate strategy, *Harvard Business Review*, March–April, 57–69.

Stoelhorst, J.W. (1997). *In Search of a Dynamic Theory of Firm*. Twente: Twente University Press.

Teece, D.J., Pisano, G. and Shuen, A. (1997). Dynamic capabilities and strategic management, *Strategic Management Journal*, 18 (7): 509–533.

Williamson, O.E. (1975). *Markets and Hierarchies: Analysis and Antitrust Implications*. New York: The Free Press.

Zegveld, M.A. (2000). *Competing with Dual Innovation Strategies; a Framework to Analyse the Balance Between Operational Value Creation and the Development of Resources*. The Hague (Netherlands): Werk-Veld.

Chapter 12

Innovation in context

From R&D management to innovation networks

Patrick Van der Duin, J. Roland Ortt, Dap Hartmann and Guus Berkhout

OVERVIEW

Innovation processes have changed significantly over the last five decades. This change is evolutionary; new principles of innovation are developed to overcome the disadvantages of former ones and are the result of the adaptation to the internal (managerial and organizational) and external (e.g., economical, governmental) environment of a company. Four generations of innovation processes can be distinguished, each showing differences in the notion of innovation, in the underlying (technology) philosophy, in the role of R&D, and in the structure of the innovation process itself. In short, the development of innovation processes shows a transition from technologically oriented R&D to dynamic innovation processes that take place within networks. A contemporary (i.e., fourth-generation) model of an innovation process, the Cyclic Innovation Model, will be described. The Cyclic Innovation Model links changes in scientific insights, in technological capabilities, in product design, and in market demand.

INTRODUCTION

Innovating is managing the creation, development, and application of new knowledge in companies. Innovations are developed by innovators from an idea or patent into a new product, service, process or any other type of innovation. Since the 1950s, four generations of innovation management have been distinguished. In short, this development shows a transition from technologically oriented R&D to dynamic innovation processes that take place within networks. Therefore, we consider R&D management as a part of innovation management. The historical development of innovation processes has not occurred in a vacuum, but has been influenced by economical, social, and managerial developments.

This chapter starts with a description of six elements that are part of the concept of innovation. Then, we describe four generations of innovation management. We address not only the basic principles of these generations, but we take into account the internal and external conditions that have influenced the development of the generations as well. It shows that the development of these generations can be characterized as evolutionary. Each (new) generation attempts to overcome the disadvantages of the previous one, and each generation emerges as the result of adaptation to a changing environment. Both "mechanisms" explain the increasingly complex nature of innovation processes. Also, the concept of innovation changes with each new generation of innovation management.

The Cyclic Innovation Model, a contemporary fourth-generation innovation model, is discussed in the third section. A case will be described to illustrate its general principles. The chapter closes with a discussion of the relevant trends.

INNOVATION

There are many definitions of innovation (Cumming, 1999; Garcia and Calantone, 2002). Here, we address six distinguishing elements that, in our view, are relevant to the concept of innovation.

Newness and change

Innovation is strongly related to "something" that is new; a product, service, or process that was not introduced into a market before. Note that it is not necessarily new to the company that develops or implements the innovation. Rogers (1995) emphasizes this in his definition: "An innovation is an idea, practice, or object that is perceived as *new* by an individual or other unit of adoption." Thus, newness must be viewed from the perspective of the user and not from the actor that develops and produces the innovation: "If the idea seems new to the individual, it is an innovation" (Rogers, 1995). The concept of newness is closely linked to the concept of *change*, because the newness of an innovation will often have a specific (new) impact. These changes can occur at the demand-side of the market (users may be attracted by the new product or service, and change their spending pattern towards the innovation), at competitors (they may change their marketing strategy or start innovating themselves), or at the government (the use of the innovation may have negative social consequences that demand legislation).

Tidd, Bessant and Pavitt (1997) describe the change in innovation in two ways: first, in terms of the type of innovation and, second, in terms of the extent to which innovations change existing (market) situations.

Products, processes, and services:

- *Product innovation*: a new product, such as a car or a mobile phone.
- *Process innovation*: a new way to produce or distribute, such as the use of robots in the automotive industry or call centers in the insurance industry.
- *Service innovation*: a (new) combination of the above two. For example, a new way of offering new travel destinations. The new travel destination is the product innovation, and the new way of offering is the service innovation.

Transformational, radical, and incremental:

- *Transformational*, such as the use of steam power during the Industrial Revolution, and the introduction of microchips in the Information Revolution.
- *Radical*, such as the use of mobile phones.
- *Incremental*, such as the introduction of airbags in cars.

Both categories can be combined. Product, process, and service innovations can change market situations in a transformational, radical, or incremental way.

Newness and (subsequently) change do not only refer to the nature of innovation, but can also be used as a way of categorizing different types of innovation. In this respect, Johannessen, Olsen and Lumpkin (2001), on the basis of empirical research into different scales of newness, conclude that the *degree of radicalness* is the most distinguishing factor in determining the newness of innovation.

A broad view on innovation

Traditionally, an innovation entailed some (visible) *technological* change, and therefore only product innovations were called innovations. Nowadays, innovations of a more intangible and non-technical nature, such as service innovations, organizational innovations, or new ways of supply are also considered innovations. This means that the concept of innovation has become much broader than before. Moreover, many innovations are combinations of technical changes and non-technical changes. For instance, for developing prepaid mobile telephony, changes were needed not only in hardware (e.g., the mobile phones), but in the business model as well (i.e., a new way of paying for mobile telephony services). A consequence is that developing innovations has become much more complex, since often different changes (both technical and non-technical) are necessary.

Process

The term "innovation" does not only refer to a new product or service, but is also used to describe the *process* by which an idea or an invention is generated, and subsequently transformed into a product innovation or a service innovation which is introduced into a market. Jonash and Sommerlatte (1999) refer to Schumpeter (1934) in their definition: "Innovation encompasses the entire process that starts with an idea and continues along through all the steps from initial development to a marketable product or service that changes the economy." Chiesa (2001) defines innovation as invention plus exploitation, and views both elements as processes in which new ideas are created and put into work (the invention process), and in which the commercial development, application, and transfer is taking place (the exploitation process). Trott (1998) also views innovation as a process, and draws an analogy with education "where qualifications are the formal outputs of the education process. Like education, innovation cannot be viewed as a single event".

Implementation

Innovation must be clearly distinguished from an *invention* or an *idea*. Buderi (2000) quotes the director of PARC (the research center of Xerox) who states that innovation is "invention implemented". Dunphy *et al.* (1996) view an innovation as a "commercially feasible version of the invention", while Tidd *et al.* (1997) add: "Definitions of innovation may vary in their wording, but they all stress the need to complete the development and exploitation of new knowledge, not just its invention." Rosegger (1986) holds the opinion that *ideas* do not belong to the domain of innovation but to the domain of *invention*, because ideas do not necessarily lead to technological or economic change. All this does not mean that inventions or ideas are not important. On the contrary, every innovation starts with an idea or an invention. Nevertheless, the (potential) value of an invention or an idea for a market or society can only be established and/or assessed after its transformation into an innovation.

Interconnectedness of innovations

It is difficult to view innovations independent from each other. Dunphy *et al.* (1996) state that innovations do not come alone, but in groups or clusters. Rogers (1995) refers to an encompassing *system* in which innovations are developed. Also, innovations often do not receive input from just one source. Smits (2002) regards innovation as "a successful combination of hardware, software and orgware, viewed from a societal and/or economic point of view". In this definition, the hardware is the apparatus, the software is the idea, and "orgware" is the embedding of the innovation into the market and society. Other terms that refer to interconnectedness are "solution innovation" (Shepherd and Ahmed, 2000) and "technology fusion" (Tidd *et al.*, 1997). Tidd *et al.* (2001) illustrate the interconnectedness of innovations with an example of three product generations (or standards) of mobile (cellular) telephony (NMT-450, NMT-900, GSM). They show that each new product generation uses more technologies or innovations than its predecessor. Whereas NMT-450 used five technologies, NMT-900 used ten technologies, and GSM needed 14 technologies (and more than 100 patents).

A product innovation can be the output of one company, and a process-innovation for another company who buys it (Chiesa, 2001), and vice versa. Korbijn (1999) even states that product innovations and process innovations can have contradictory growth curves: the growth of one type of innovation at the expense of the growth of another type of innovation. Just as an innovation can impact on a company or society, it is also influenced by, and relates to, other innovations and their subsequent developments.

Uncertainty and creativity

While innovation is, by definition, related to newness and change, it is also strongly related to concepts such as uncertainty and creativity. During an innovation process, there are many factors that can influence the development of an innovation, but their contribution is difficult to determine beforehand. Radical innovations have a high market and technological uncertainty, while incremental innovations have a low market and technological uncertainty (McDermott and O'Connor, 2002). Trott (1998) refers to Pearson's uncertainty map which divides ways of managing innovation by distinguishing between uncertainty about the end of an innovation process and uncertainty about the means by which this end can be achieved.

Creativity is related to innovation because coming up with new ideas and inventions, and developing them into new products, processes, or services, is not a rational activity that can be managed from a to z. The ability to think newly and differently, and to develop a new view on current problems and opportunities, is an important asset for the innovating company. By doing things differently, the chances for an innovation with a high degree of newness are enhanced. In this way, a company can distinguish itself from its competitors.

THE EVOLUTION OF INNOVATION PROCESSES: IMPROVEMENTS AND ADAPTATIONS

Innovations do not fall from the sky, but are often the result of a process in which many actors are involved, such as innovators, scientists, researchers, general management, account management, and potential (future) customers of the innovation. Although innovation is an uncertain, complex, and risky process, it is too important for companies to leave it to luck alone:

> But real success lies in being able to repeat the trick – to manage the process consistently so that success, whilst never guaranteed, is more likely. And this depends on understanding and managing the process so that little gets left to chance.
>
> (Tidd *et al.*, 1997, p. 13)

The way in which companies have developed new products and services has changed significantly over the last decade. This development can be subdivided into four generations.

Although the first generation of innovation management emerged in the 1950s, the description of R&D management (as innovation in those days can be called) in companies can be traced back to the late nineteenth century. By the end of the 1870s, the first industrial research laboratories were organized in Germany by synthetic dye manufacturers who realized that science could create patentable inventions which, in turn, would yield new and improved products. In the first decade of the twentieth century, industrial research laboratories were organized in the US by, for instance, General Electric, Du Pont, and Bell Labs, and in Europe by, for instance, Philips NatLab. The gap between the late nineteenth century and the 1950s might be explained by the fact that from the 1950s the crucial importance of innovation for companies was widely recognized, and that more often smaller companies were involved in innovation activities as well.

Niosi (1999) provides the following concise description of the successive generations:

> The first generation brought the corporate R&D laboratory. The second generation adapted project management methods to R&D. The third brought internal collaboration between different functions in the firm. The fourth adds routines designed to make more flexible the conduct of R&D function through the incorporation of the knowledge of users and competitors.

The idea of successive generations is a generalization. "Fourth-generation R&D existed as an organized activity, though marginally, since the very beginnings of industrial research. . . . What is new in the late 1980s and 1990s is that these cooperative forms have now become widespread across the industrial spectrum" (Niosi, 1999, p. 112). In other words, these generations refer to mainstream models of best practice in innovation management. Also, the different generations can overlap as illustrated in Figure 12.1.

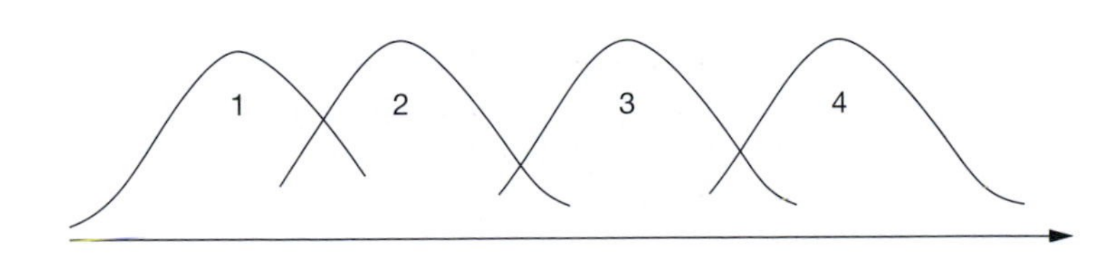

Figure 12.1 *The overlapping of the four generations of innovation management*

Below, we describe the four generations of innovation management. These generations have not developed in a vacuum, but were heavily influenced by changes in the external (e.g. social-economic developments) and internal (strategy and organizational culture) environments of companies. For example, government policies may impact on R&D management practices. With regard to the internal environment, the strategy of a company, for example, determines whether an organization is an imitator, a follower, or a leader, and that, in turn, determines the importance of innovation, and how R&D is organized.

Generation 1

Internal and external environment

During the first generation of innovation management, society generally had a favorable attitude towards scientific advance and industrial innovation. Technological developments were primarily driven by scientific advances and these developments were believed to solve society's main problems. Governments stimulated R&D for multiple reasons. First, technological innovation was needed for military purposes. The cold war demanded technological leadership. Second, technological innovation formed the heart of new and renewed industries. Along with technological progress, economic conditions flourished. The strategy and structure of organizations mirrored these developments in society. Companies were technology oriented and focused on innovation and growth. It was believed that the specialized knowledge to innovate required a functional structure with departments consisting of specialists.

Basic principles

The first generation of innovation management considers scientific discovery as the starting point of innovation processes. Universities are considered as the primary source of scientific discovery. Therefore, R&D organizations (the main actors in this first generation) are structured like universities. Departments within these institutes are essentially mono-disciplinary. The structure of innovation processes is linear sequential and of a technology-push nature. In this process, different departments subsequently contribute to the innovation.

Disadvantages

This first generation of innovation management has significant disadvantages. The final responsibility for an innovation is not always clear when it moves from department to department, because a project management approach is not yet adopted. The lack of a project approach also implies that little attention is paid to the overall transformation process from idea to innovation. Scientific freedom of professionals seems more important than relevance (in terms of commercial results) for the company because the amount of knowledge is considered to be an obstacle to (successful) innovation. Innovation does not always have strategic goals, and there is hardly a relationship between researchers and general management. Market needs and commercial aspects are incorporated late in the process. As a result, failures due to a lack of market need are discovered quite late. During the first generation, a lot of effort was wasted on unsuccessful innovation processes.

> Examples of first-generation innovations: steam-engine, video-telephony, Videotext, color TV, electric light, jet-engine.

Generation 2

Internal and external environment

The second generation marked a period of relative prosperity. However, the economic growth of the previous period slowed down. Demand roughly equaled supply. Competition intensified because the growth targets of companies could no longer be fulfilled solely by the growth of the market. Companies tried to acquire a larger market share and, as a result, the concentration (i.e., the combined market share of the largest four companies in a market) in many markets increased considerably. Attention in society focused on the demand-side of the market. Government policies, for example, tended to emphasize demand-side factors. Along with these changes in society, company strategies focused on growth and diversification to attain economies of scale and reduce financial risks. Corporations developed divisional structures to meet the requirements of their diversified companies. As a result, innovation management became more demand-driven. Also, the interface between R&D and marketing became more important. Divisional structures stimulated R&D (and other innovation) efforts that were more closely related to the business of divisions. To cope with these changes, the mono-disciplinary teams were replaced by multi-disciplinary project teams.

Basic principles

The second generation considers market aspects early on in the process. Even more so, the market is now regarded as the main source of new ideas. Consumer research is the basis for new product ideas (Fornell and Menko, 1981). Innovation processes are managed as multi-disciplinary projects in which R&D personnel and marketing personnel collaborate. In large companies, R&D projects are often performed for internal company clients. A project leader, rather than subsequent managers of departments, has the final responsibility for the overall transformation process. Along with multidisciplinary projects, many research institutes adopt a matrix organizational structure. The structure of the innovation process remains essentially linear sequential, but of a market-pull nature.

Disadvantages

The approach of the second generation leads to new disadvantages, such as the primary focus on small improvements of existing products. Potential consumers can hardly express their needs beyond those solved by familiar products (Tauber, 1974). After the second generation, the influence of marketing in product development is severely criticized because it neglects the long-term need for knowledge aimed at developing the future technological assets of a company (Bennet and Cooper, 1982). Another drawback is that innovation projects are treated separately. As each project serves the goals of different internal company clients, strategic relationships among the innovation projects are not established. Relationships between the projects and the strategic goals of company are not established either.

Examples of second-generation innovations: fuel-efficient cars, healthy drinks.

Generation 3

Internal and external environment

The third-generation management marked a period of decline, due to two oil crises, inflation, and demand saturation. Supply exceeded demand, unemployment figures rose, and resource constraints (especially regarding oil and the products derived from it) played an important role in the market. Companies started to focus on cost control and reductions rather than growth. Fat, hierarchically organized companies were transformed into flat and more flexible companies. One way to achieve this was to split up a corporation into relatively independent and more flexible business units, instead of maintaining the concept of large, hierarchically governed, companies. As a result, innovation-activities were increasingly regarded as investments that should deliver relevant results for the business units' strategies. Therefore, innovation-projects were organized in larger programs, to form a balanced portfolio of projects for separate clients. R&D programs were directly linked to the strategy of a corporation or business unit although because of ethical issues the ties between universities and companies were not always very strong.

Basic principles

The third generation combines the market-pull and technology-push approaches. Projects are combined into programs that are directly related to strategic company goals. The structure of innovation processes remains essentially linear, but with feedback loops and constant interaction with the market and technological developments. Companies approach partners with essential technological and market knowledge. Because of this interaction, communication networks are formed with these partners.

Disadvantages

The main disadvantage of the third generation is its focus on product and process innovations (Miller, 2001), the successful exploitation of which also requires organizational and market innovations. Suppose a manufacturer produces only a limited number of product variants, and suddenly decides to let customers order semi custom-built products. In addition to changing his products and production processes, he must also change the interaction with his customers and the planning of production processes. The necessary changes in work processes will inevitably bear on the organizational structure. Traditionally, R&D labs have no experience with organizational and market renewal, which were the domains of top managers and marketeers, respectively. Involvement of top managers with R&D and close links between innovation processes and strategic company goals, facilitate the transfer of innovations from the R&D labs to the parent company. However, the transfer of innovations from the company to the market is hampered by insufficient experience of the R&D departments with market and organizational renewal. The third-generation R&D management focuses on initiating innovations rather than exploiting them.

> Examples of third-generation innovations: mountain-bikes, ATMs, fast-delivery food, Dell's business model.

Generation 4

Internal and external environment

The fourth generation emerged in a period of economic recovery. Globalization forced companies to focus on their core competences. Alliances were needed to gather the necessary and diverse competences for innovation. Technological developments, especially in communication and information technology, influenced the organization and management of design, manufacturing, distribution, and marketing processes. Management practices of Japan inspired western companies to experiment with team-based organizations and alliances of companies. As a result, innovation management increasingly involved the management of alliances to develop the required technological assets for a company. Experiences in Japan also inspired companies to innovate faster, and to strive for parallel and integrated processes of innovation. Developments in communication and information technology facilitated intra- and inter-organizational cooperation in innovation and R&D.

Basic principles

In the fourth generation, innovation projects are no longer carried out in the isolation of R&D and other innovation departments, but take place in large networks with internal partners (other company departments) and with external partners (universities, suppliers, customers, etc.). The degree of integration of innovating companies with their suppliers and customers increases. New products are developed more quickly and more frequently due to parallel development processes. The fourth generation pays more attention to the market and the organizational innovations required to successfully introduce product innovations. The term innovation broadens out from product innovation to process-, organization-, and market-innovation (Trott, 2002). The traditional R&D department gradually incorporates the new business development department of a company.

Disadvantages

A disadvantage of the fourth generation is the complexity of R&D in general, and the innovation processes in particular. To cope with this complexity, more flexible organizations and the application of information technology are proposed. Although some authors describe a fifth generation, from the early 1990s on (Rothwell, 1994), we believe that it is merely a variation within the fourth generation. In the words of Rothwell (ibid.): "The development of 5G is essentially a development of the 4G (parallel, integrated) process."

> Examples of fourth-generation innovations: UMTS-mobile phones, PC.

Table 12.1 shows that innovation management has changed significantly over five decades. Each new generation tried to overcome the disadvantages of the previous generations, but inevitably contained new disadvantages. In the successive generations, new activities and new links between new actors are added to (rather than removed from) the innovation process. The net result of this evolution is that innovation processes are becoming increasingly complex. The increasing complexity of innovation processes can be summarized as follows:

- Migration from a process in which an innovation is handed over from department to department, to a multidisciplinary project. Eventually, innovation projects are organized in programs directly related to the company strategy.
- "Technology-push" and "market-pull" approaches are combined, because market and technological aspects are considered to be important

throughout the innovation process. Technology and market aspects are sometimes depicted as two levels along a sequential process of innovation.

- Feedback loops are introduced in innovation processes. Like many complex and multi-phase processes, new findings during an innovation process sometimes imply that previous steps must be re-evaluated.
- Activities in innovation processes are organized more in parallel to increase the speed of development.

Generations of innovation management and the concept of innovation

The elements of innovation, described in the former section, have played different roles in the four generations of innovation processes. In the earlier generations (and especially in the first), innovation was mainly considered a technological change. Later, innovation included also non-technical changes, such as the way in which a new service was offered. Also, the earlier generations did not address the interconnectedness of different innovations. The different generations do not only describe the evolution of how innovation took place, but also describe the evolving concept of innovation itself. Table 12.2 shows the concept of innovation in the four generations of innovation processes.

THE CYCLIC INNOVATION MODEL

The former section has shown that developments in many industries generated a new commercial environment with innovation and business processes that cross traditional company boundaries to create combinations across other industrial sectors. Innovation is developing in a new direction, requiring new concepts. These concepts belong to the fourth generation of innovation models (Niosi, 1999). An example of such a model is the Cyclic Innovation Model (Berkhout, 2000), which we will describe in this section.

The Cyclic Innovation Model (CIM) was developed at the end of the 1990s, to describe and analyze the continuous reform that is at the base of ongoing change in public and private organizations. The model describes the generic innovation processes by a "circle of change". It links changes in scientific insights, changes in technological capabilities, changes in product design and manufacturing, and changes in market demand. The model replaces the traditional linear chain concept with a cycle containing four "nodes of change", which are connected through four interacting "cycles of change". They represent the foundations of the complex, boundary-crossing processes that occur in present-day innovations. Figure 12.2 shows CIM at its highest conceptual level. Each node comprises a group of different organizations, and each cycle comprises a network between two complementary groups.

In the *hard sciences cycle* (upper-left part of the model), interaction processes occur that relate to the developing of new technology. Input from different scientific disciplines is required to provide specialistic knowledge in fields such as mechanics, physics, chemistry, biology, and informatics.

In the *systems engineering cycle* (upper-right part of the model), interaction processes take place that relate to the development of new products. Input from different technological areas is required to provide smart methods and tools for new designs and new ways of manufacturing. These two cycles of change border on each other, while sharing

Table 12.1 *Overview of four generations of R&D management*

1st generation (1950s–mid-1960s)	
Conditions in the external environment	New technologies lead to the rise of new industries, the re-generation of existing industries, and the application of technology in traditional industries like agriculture. Economic growth leads to growing (profits for) companies, employment creation, prosperity, and rising consumer demand. The consumer demand significantly exceeds the supply of goods. Society has a generally favorable attitude towards scientific advance and industrial innovation. Government policies stimulate R&D in universities and companies (sometimes for military purposes).
Innovation processes	Linear, sequential process from department to department, starting with scientific discovery.
Strategy	Technology-oriented, and focus on innovation and growth.
Structure	Functionally organized.
2nd generation (mid-1960s–early 1970s)	
Conditions in the external environment	A period of relative prosperity. Manufacturing still grows, but employment is static. Demand more or less equals supply. Many markets show an increase in concentration and competition. Government policies tend to emphasize demand-side factors.
Innovation processes	Linear, sequential process in a project, starting with market need.
Strategy	Focus on growth (organic or acquired) and diversification, to attain economies of scale. Technological change is rationalized; marketing and market need are considered more important than scientific and technological progress.
Structure	Multi-divisional structure. Special targets like innovation are generally organized in multi-disciplinary projects.
3rd generation (early 1970s–mid-1980s)	
Conditions in the external environment	Two oil crises and stagflation (inflation + demand saturation) characterize this period. Supply exceeds demand, and unemployment figures rise significantly. Because of resource constraints, there is a need to investigate product innovation to increase success rate.
Innovation processes	Essentially a sequential process with feedback loops and interaction with market needs and state of the art technology at each stage.
Strategy	Focus on cost control and reduction.
Structure	More flexible and less hierarchical (i.e., more flatly organized). Responsibilities are delegated downwards.
4th generation (mid-1980s–)	
Conditions in the external environment	This is a period of economic recovery. In many sectors globalization is important. Organizations are more aware of the strategic importance of (evolving generic) technologies. The emergence of IT-based manufacturing equipment leads to a new focus on manufacturing strategy.
Innovation processes	Coordinated process of innovation in a network of partners. Coordination is often attained by system integration (with key suppliers and customers) and parallel development (of components or modules of the innovation).
Strategy	Focus on core business and core technologies. Manufacturing strategies, strategic alliances, and external networking activities become more important. Time-based strategies become more important because of short product life cycles.
Structure	More team-based and project-based structures. Structures and procedures are adapted to facilitate alliances.

Partly based on Liyanage *et al.* (1999), Miller (2001), Niosi (1999), Rothwell (1994), and Roussel *et al.* (1991)

Table 12.2 *The concept of innovation in different generations of innovation processes*

Innovation elements	*First*	*Second*	*Third*	*Fourth*
Newness and change	Perceived from a technical standpoint. Relatively large amount of radical innovations.	Mainly perceived from a marketing standpoint. Technical change is instrumental. Many incremental innovations, less radical ones.	Technical and marketing perspectives are integrated in innovation processes.	Newness and change touch every aspect of the innovation.
Broad view	Predominantly technical.	Not only technical, but marketing elements as well.	View on innovation is broadening.	Non-technical aspects are often considered more important than technical ones.
Process	Linear, carried out just by R&D labs.	Linear process, more departments involved.	Parallel processes with feedback loops, but still mainly linear.	Processes are becoming circular (cyclic); processes are taking place in networks or webs.
Implementation	Receives very little attention. It is believed that if the innovation is technically solid, market acceptance will follow automatically.	Heavily influenced by market needs. If these are assessed properly, market acceptance follows automatically.	Aim at shortening the time from idea to implementation.	Shortening time-to-market remains very important.
Interconnectedness	Not much. Innovations are viewed mainly independently.	Increasing interconnection.	More and more interconnectedness.	Majority of innovations is interconnected.
Uncertainty and creativity	Uncertainty mainly on the technical field. Creativity is not emphasized.	Uncertainty from market and society. Increasing importance of creativity.	Uncertainty is addressed in the strategy of the company. Creativity should be in line with this strategy.	Increasing uncertainty because of the increasing systemic character of innovation. Creativity is considered an important input to the innovation process.

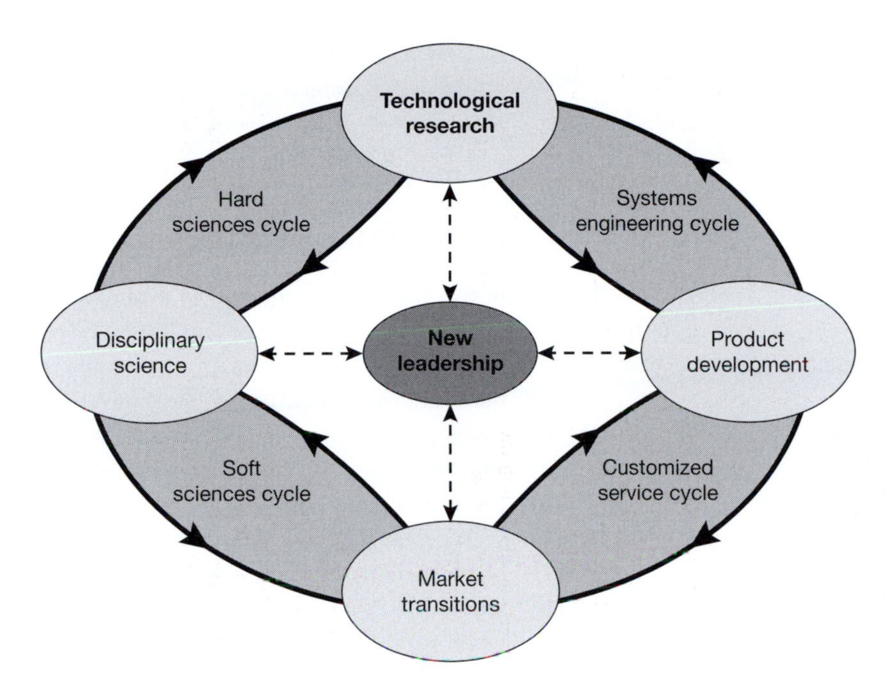

Figure 12.2 *The Cyclic Innovation Model visualizes the "circle of change". It links, in a cyclic manner, the changes in science (left), business (right), technology (top), and markets (bottom)*

the technological research node. Nowadays, the engineering cycle is not just directed towards the development of material products in the traditional fabrication and process industries. In modern engineering, the focus is also on biotechnical products, information products, financial products, logistic products, etc. In addition, products may be the blueprint of a market-oriented process, such as service provision. Products should be regarded as socio-technical functions that a company offers to its customers. Note that material and non-material products may represent components in a complex system, such as a telecom infrastructure. Such an infrastructure may itself be seen as one "product" that functions as a component in the total infrastructural system of a nation.

In the *customized service cycle* (lower-right part of the model) interaction processes occur that relate to the development of (new) markets. Input from a range of products is required to fulfill the (potential) societal demands. Products are increasingly accompanied by services, and advanced service provision is increasingly facilitated by technical products. Innovation increasingly deals with new product–service combinations. The engineering and service cycles share the product development node.

In the *soft sciences cycle* (lower-left part of the model), new scientific insight is gained into the needs and concerns of society. Input from different scientific disciplines is required to provide specialistic knowledge in fields such as economics, sociology, anthropology, psychology, and law. It particularly applies to today's complex commercial processes which take advantage of new product–service combinations. Because today's markets are a melting pot for technical, economic, social, and cultural change, mono-disciplinary models are not appropriate.

Autonomous social transitions manifest themselves as changes in the need for products and services ("demand"), which will stimulate innovations. Autonomous technological developments generate new products and services ("supply"), which will change society. The cyclic interaction of these two innovation drivers will create maximal value to the market.

A fundamental characteristic of CIM is that it describes a full cycle of interdependent processes, and not merely a linear chain of events. Science is no longer at the beginning of a chain and the market is not at the end. Both nodes are part of a perpetual (learning) process along a dynamic path that has no fixed starting or finishing point. The result is an endless building up of value creation that is realized by the reinforcing cycles of the full circle. In CIM, new technologies (developed from new scientific discoveries, for example) and changes in the market (new life styles, for example) continually influence each other in a cyclic manner. This dual nature of innovation will shape the future.

Innovations may create small and large changes in society, often referred to as "incremental innovations" and "radical innovations". The Cyclic Innovation Model allows a more accurate classification. Innovations of class-1 are based on changes in one node only, for instance, the introduction of a new marketing concept in an existing product-service combination. Innovations of class-2 are based on changes in two nodes, while innovations of class-3 require changes in three nodes. Innovations of class-4 are the most radical, as they require changes in all four nodes. In these "four-star innovations" science will have a large impact on society. For instance, the life sciences and nanosciences will drastically change our lives in the near future.

CIM describes a system of coupled networks which interact in a cyclic manner. Management of these networks requires a new type of leadership.

CONCLUDING REMARKS

In this chapter, we adopted an evolutionary perspective on innovation. Innovation refers to the innovation activities (processes) and to the result of these activities (the innovation, a new product, for example). The definition and character of both references have evolved over the past decades.

We described how during the second half of the twentieth century the latter type of innovation (the result) has broadened out from product innovation to service and organizational innovations and, finally, to system innovations. For example, the jet-engine is a science-based technological innovation. At first, it was applied in military aircraft because of its superior performance. For civil use, the cost of buying, using, and maintaining jet-engines was relatively high in comparison with traditional engines. However, jet-engines enabled a higher speed. Over larger distances, the high price of these engines was compensated by the shorter travel times. A further reduction of the price per passenger was achieved by increasing the number of passengers. These changes affected various complementary services before, during, and after flying. Airplane maintenance before and after flying changed with the use of jet-powered planes. The service during a flight changed because passengers on intercontinental flights may want to sleep. The entire organization of flying has changed gradually. Often, small airplanes are used to travel to a major airport (hub), where large airplanes depart for other continents. The jet-engine – a technical innovation – stimulated the use of different types of

CASE EXAMPLE: LUCIO, A MOBILE DATA SERVICE

Lucio is a mobile data service that was introduced into the Dutch telecommunication market at the end of 2002. It enables employees of companies to access their business information (such as e-mail, agenda, address book, and Internet) when they are away from their desk. It is offered as a package of different product-service components implemented by a certified system integrator. The mobile device is a PDA (Personal Digital Assistant). Other product components that are needed to deliver and to make use of the service are a mobile infrastructure (e.g. GPRS), and a VPN-gateway (access to Virtual Private Network) at the premises of the customer. All components are based on a Microsoft Exchange server on a LAN (Local Access Network) and a firewall. Lucio was developed by KPN Mobile, a Dutch mobile operator and service provider, in cooperation with Hewlett-Packard and Microsoft. Lucio was marketed by KPN Mobile as a service that is a "guaranteed total solution", a "reliable service", and "easy to use".

It is essential to view Lucio as an innovation by combination (Schumpeter, 1934; Van den Ende, 2003, p. 1505): innovation is not at a component level but at a systems level. Lucio represents a new combination of several components: infrastructure (GPRS), device (mobile phone or PDA), software (e.g. information manager), hardware (e.g. intranet, mail servers), and the actual services (e.g. e-mail, agenda). Integrated systems such as Lucio cannot be developed by a single company. Until today, no single company possesses the knowledge and experience to develop such a product-service combination.

Figure 12.3 shows schematically how Lucio is connected to the company network and how it uses different components such as a PDA, network elements (VPN gateway, LAN), and software elements (Microsoft Exchange).

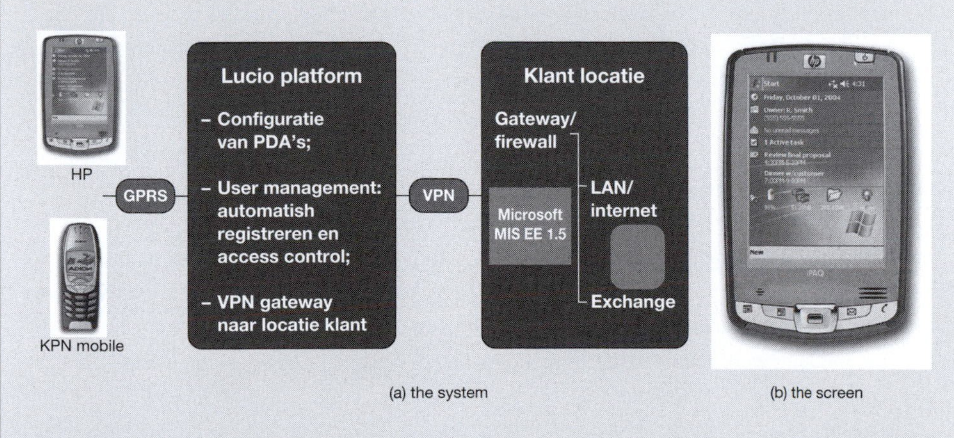

Figure 12.3 *Lucio, a mobile system with a PDA and a mobile phone connected to the company's VPN by using GPRS*

As already stated, Lucio is a cross-company innovation. Below we list the principal business partners and their role in the development of Lucio:

- KPN Mobile: supplier of GPRS-based services, providing secured connection with the customer's local network.
- HP: provider of mobile devices (i.e. the iPAQ 3870), and the ProLiant Server.
- Microsoft: provider of software for the PDA-device (Pocket PC operating system), and the Management Information System.
- Certified system integrators (such as The Vision Web, Flex IT and CSS): providing support to suppliers and customers, and connecting the company network to the mobile network.

When we reconstruct the innovation processes involved in Lucio in terms of CIM, we notice that neither the science nor the technology nodes play any direct role. New science and technology components were not required to develop Lucio. On the other hand, the market transition node of CIM played a vital role. Great attention was needed to assess the emerging business requirements for access to in-house information and applications at any time and any place. Results of this study were translated to an estimate of the market potential at a very early stage of the project. This was an important starting point for the development of Lucio. An internal paper at KPN Mobile referred to a particular study carried out by IDC (2001) which concluded that almost 30 percent of the business customers were interested in using a broadband mobile data service that would give them access to their intranet as well as to the Internet. More than 20 percent was very interested in using this new service. Because this study also predicted a significant market growth, a major market transition towards the "mobile business age" was foreseen by KPN Mobile. To realize the predicted market transition, the added value of Lucio was presented to the business community as a product–service combination that offers "fast and easy mobile access to in-house business applications". This yielded the functional specifications for the product development node of CIM: a suitable PDA, interface to the GPRS-infrastructure, and connection to the LAN of the client. These hardware and software specifications led to requirements for the technological capabilities of partners needed in the engineering cycle: Microsoft for the software, HP for the PDA, and KPN Mobile for the telecom network. In a next step in the project, the constructed image of the future business in the market transition node could be backcasted via the service cycle (lower right-hand side of CIM) and via the product development node towards the engineering cycle to establish the partners needed (upper right-hand side of CIM). In the engineering cycle of CIM, the design of Lucio was a joint venture of KPN, HP, and Microsoft, using existing technology. Only minor technical adjustments of existing modules were required (optimization). The development of a specific Lucio-gateway was undertaken by KPN Mobile.

In terms of CIM, Lucio is a multi-sector, class-2 innovation (see Figure 12.4). This means that Lucio had a low technical risk (using existing science and technology), a medium marketing risk (using known market segments), and a high cultural risk (using sector-crossing partners).

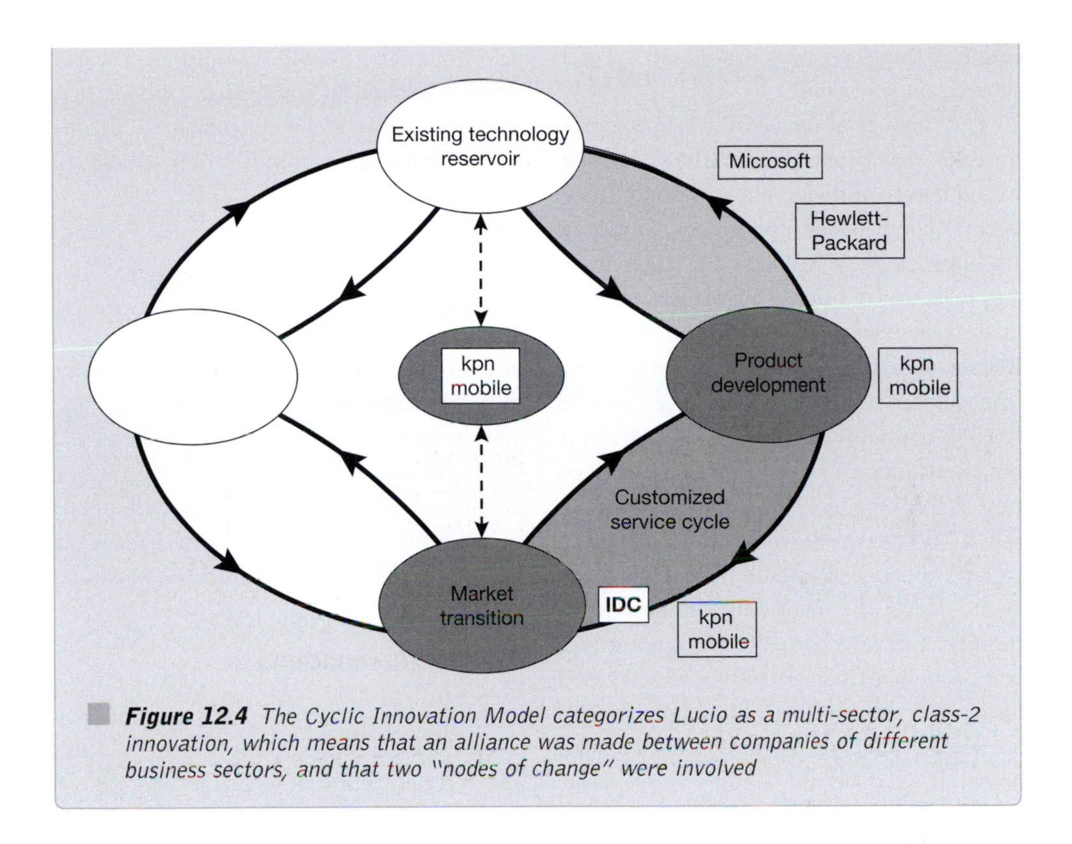

Figure 12.4 *The Cyclic Innovation Model categorizes Lucio as a multi-sector, class-2 innovation, which means that an alliance was made between companies of different business sectors, and that two "nodes of change" were involved*

airplanes (product innovations) and also initiated new services (e.g. the possibility to sleep on board) and organizational innovations (new airports). All these innovations, together, form a coordinated system – a system innovation.

We also described the evolution in thinking about the best way to structure innovation processes. Initially, linear models were used in which innovations were handed over from department to department. At that time, the innovation process was regarded as a science-based process in which technology was transformed into marketable products. Later on, innovation processes were also seen as market-led processes in which consumer demands initiated the development of new technologies, products, and services. Next, innovation processes were organized in projects (i.e. project members from different departments joined the innovation process rather than handing over the innovation from department to department) with stage-gates ("go"/"no go" decisions after each phase of the project), and projects were combined in programs to attain specific business goals. Currently, innovation processes are more iterative, and organized in coalitions of organizations.

The evolution of innovation processes was strongly influenced by macro-economical and business developments. Companies are capable of influencing this development by applying the aspects of the models of innovation processes that best fitted their company. Each new generation of innovation processes was influenced by the previous one, as it tried to overcome its disadvantages. The result was an increasing complexity of each subsequent generation.

The historical account of innovation processes has been criticized. First, the historical nature of the generations can be questioned. For instance, the first generation is set between the 1950s and the mid-1960s, but even today, many companies are still using principles from that first generation. Recent books and articles on innovation management still present the linear process (technology-push or market-pull) as a way for companies to organize their innovation (e.g. Douthwaite *et al.*, 2001; Dundon, 2002; Yu, 2003). Moreover, the linear innovation process was already put into practice at the end of the nineteenth century, when the first company-owned R&D laboratories were founded (Bassala, 2001). Niosi (1999) even states that fourth-generation innovation has existed since the beginning of industrial research. Second, it is not always clear whether these different models or generations of innovation processes *prescribe* (how innovation should take place) or *describe* (the actual practices and principles of innovation within companies). Third, different authors place different generations in different time intervals, which makes the different historical accounts together rather confusing. Liyanage *et al.* (1999) state that doing research in projects with milestones, project accountability, project evaluation, etc., began in the second generation, while Rothwell (1994) places that development in the third generation.

Despite these criticisms, the historical view on different innovation processes provides a clear and logical structure to the many different ways and principles of innovation management. Many scientists in the field of innovation processes and management accept it as a good way to look at its development. This development is also strongly related to some of the trends that were described in the introduction of this book. Five of these trends will be discussed in more detail.

Specialization (of knowledge workers)

In many industries, new products integrate more technologies, while separate technologies become more advanced. Tidd *et al.* (2001) illustrate both trends for the case of mobile telephony. Keeping up with these trends requires specialization, or a focus on core competences on behalf of companies.

Bundling/unbundling

In some industries, companies are not specialized but incorporate entire value chains. The telecom industry in the second half of the twentieth century in Europe serves as an example. National telecommunication companies used to be state-owned monopolies that governed the entire chain of telecommunication from network building up to telecommunication services. Liberalization in the telecom industry is one of the causes of increased competition in this industry. Another source of increased competition is the merger of information technology and communication technology. To remain competitive, the former monopolies have to focus on their core activities. Furthermore, these companies were forced by law to unbundle the chain. Nowadays, different companies build networks, maintain networks, sell network capacity to service providers, sell phones to customers, and so on. The specialization, or the focus on specific activities, in the value chain of an industry is referred to as "unbundling".

Coalition forming

The example of the jet-engine illustrates that innovation in one specific component (the jet-engine) may require changes in a large system (the entire air traffic industry). The necessary specialization that was required to remain competitive has the adverse effect that governing changes in the entire system of air traffic becomes more difficult. Tushman and Rosenkopf (1992) show that these types of changes require lengthy negotiations among relevant organizations. To increase the coordination across the value chain, companies form coalitions.

Societal responsibilities

Societal responsibilities refer to different aspects. Two of these aspects, liability of producers regarding the misuse or abuse of their products, and the effect of product use and disposal on the natural environment, had a particularly large effect on product development processes. Liability affects testing procedures during the development process, while environmentalism requires that products are designed in such a way that they can be easily taken apart and re-used after disposal.

Client orientation

The emergence of the second generation of (market-led) innovation processes reflects the increased attention for clients. The positive effect of a client orientation in innovation is disputed. On the one hand, making technological innovations that do not fulfill a need seems a waste of resources. From this perspective, the client is considered to improve the business impact of innovation activities. On the other hand, merely doing what consumers need leads only to minor innovations (or small improvements) in existing products. A client orientation should therefore focus on the (long-term) interests of potential consumers. It does not necessarily have to do so by relying entirely on what consumers *say* they need.

FURTHER READING

Tidd, J., Bessant, J., and Pavitt, K. (2001). *Managing Innovation. Integrating Technological, Market and Organizational Change.* Chichester: John Wiley & Sons (second edition). This book provides a rather complete overview of the field of innovation management. It is well grounded in the recent literature.

Trott, P. (1998). *Innovation Management and New Product Development.* Harlow: Pearson Education. This book provides a clear overview of innovation management that differs in many respects from Tidd *et al.*

Liyanage, S., Greenfield, P.F., and Don, R. (1999). Towards a fourth-generation R&D management model–research networks in knowledge management. *International Journal of Technology Management.* 18(3/4), 372–394. This article provides a well-written overview of the generations of R&D management.

Garcia, R. and Calantone, R. (2002). A critical look at technological innovation typology and innovativeness terminology: a literature review. *The Journal of Product Innovation Management.* 19, 110–132. This article discusses and categorizes the different terms that are used in the scientific literature to describe the degree of newness of innovations.

REFERENCES

Bassala, G. (2001). *The Evolution of Technology.* Cambridge: Cambridge University Press.

Bennett, R.C. and Cooper, R.G. (1982). The misuse of marketing: an american tragedy. *Business Horizons.* 25(2), 51–61.

Berkhout, A.J. (2000). *The Dynamic Role of Knowledge in Innovation. An Integrated Framework of Cyclic Networks for the Assessment of Technological Change and Sustainable Growth.* Delft: Delft University Press.

Buderi, R. (2000). *Engines of Tomorrow. How the World's Best Companies are Using their Research Labs to Win the Future.* New York: Simon & Schuster.

Chiesa, V. (2001). *R&D Strategy and Organisation. Managing Technical Change in Dynamic Contexts.* London: Imperial College Press.

Cumming, B.S. (1999). Innovation overview and future challenges. *European Journal of Innovation Management.* 1(1), 21–29.

Douthwaite, B., Keatinge, J.D.H., and Park, J.R. (2001). Why promising technologies fail: the neglected role of user innovation during adoption. *Research Policy.* 30, 819–836.

Dundon, E. (2002). *The Seeds of Innovation. Cultivating the Synergy that Fosters New Ideas.* New York: AMACOM.

Dunphy, S.M., Herbig, P.R., and Howes, M.E. (1996). The Innovation Funnel. *Technological Forecasting & Social Change.* 53, 279–292.

Fornell, C. and Menko, R.D. (1981). Problem analysis: a consumer-based methodology for the discovery of new product ideas. *European Journal of Marketing.* 15(5), 61–72.

Garcia, R. and Calantone, R. (2002). A critical look at technological innovation typology and innovativeness terminology: a literature review. *The Journal of Product Innovation Management.* 19, 110–132.

Johannessen, J.-A., Olsen. B., and Lumpkin, G.T. (2001). Innovation as newness: what is new, how new, and new to whom? *European Journal of Innovation Management.* 4(1), 20–31.

Jonash, R.S. and Sommerlatte, T. (1999). *The Innovation Premium. How Next Generation Companies are Achieving Peak Performance and Profitability.* Reading: Perseus Books.

Korbijn, A. (ed.) 1999. *Vernieuwing in productontwikkeling. Strategie voor de toekomst.* Den Haag: Stichting Toekomstbeeld der Techniek.

Liyanage, S., Greenfield, P.F., and Don, R. (1999). Towards a fourth-generation R&D management model-research networks in knowledge management. *International Journal of Technology Management.* 18(3/4), 372–394.

McDermott, C.M. and O'Connor, G.C. (2002). Managing radical innovation: an overview of emergent strategy issues. *The Journal of Product Innovation Management.* 19, 424–438.

Miller, W.L. (2001). Innovation for business growth. *Research-Technology Management.* September–October, 26–41.

Niosi, J. (1999). Fourth-generation R&D: from linear models to flexible innovation. *Journal of Business Research.* 45, 111–117.

Rogers, E.M. (1995). *Diffusion of Innovations.* New York: The Free Press (fourth edition).

Rosegger, G. (1986). *The Economics of Production and Innovation. An Industrial Perspective.* Oxford: Pergamon Press.

Rothwell, R. (1994). Towards the fifth-generation innovation process. *International Marketing Review.* 11(1), 7–31.

Roussel, P.A., Saad, K.N., and Erickson, T.J. (1991). *Third Generation R&D. Managing the Link to Corporate Strategy.* Boston, MA: Harvard Business School Press.

Schumpeter, J. (1934). *Theory of Economic Development: An Inquiry into Profits, Capital, Credit, Interest and the Business Cycle.* Boston, MA: Harvard University Press.

Shepherd, C. and Ahmed, P.K. (2000). From product innovation to solutions innovation: a new paradigm for competitive advantage. *European Journal of Innovation Management.* 3(2), 100–106.

Smits, R. (2002). Innovation studies in the 21st century: questions from a user's perspective. *Technological Forecasting & Social Change.* 69, 861–883.

Tauber, E.M. (1974). How market research discourages major innovation. *Business Horizons.* 17(3), 22–26.

Tidd, J., Bessant, J., and Pavitt, K. (1997). *Managing Innovation. Integrating Technological, Market and Organizational Change.* Chichester: John Wiley & Sons (first edition).

Tidd, J., Bessant, J., and Pavitt, K. (2001). *Managing Innovation. Integrating Technological, Market and Organizational Change.* Chichester: John Wiley & Sons (second edition).

Trott, P. (1998). *Innovation Management and New Product Development.* Harlow: Pearson Education.

Tushman, M.L. and Rosenkopf, L. (1992). Organizational determinants of technological change: towards a sociology of technological evolution. *Research in Organization Behavior.* 14, 311–347.

Van den Ende, J. (2003). Modes of governance of new service development for mobile networks. A life cycle perspective. *Research Policy.* 32, 1501–1518.

Yu, T.F.L. (2003). Innovation and coordination: a Schutzian perspective. *Economics of Innovation and New Technology.* 12(5), 397–412.

Chapter 13

Operation management with System Dynamics

Zofia Verwater-Lukszo

OVERVIEW

The process industry has to cope with a rigorous competition caused by more short-term dynamics in supply, more unpredictable and turbulent demand patterns, stronger requirements on product variety, delivery lead-time and quality of product. This forces companies to spend efforts on improving their competitiveness and productivity. Appropriate strategies or actions in the area of production planning and scheduling as well as inventory management can contribute to survival in these conditions. This chapter describes an effective performance measurement system for supporting operational decision making in batch-wise plants contributing to the improvement of a company's competitiveness and productivity. To assess possible improvement options with respect to the selected evaluation parameters a generic System Dynamics model of an internal supply chain in a batch-wise company is developed. The application of the model in a specific industrial plant in the food sector results in recommendations to reduce the inventory level and to increase the effective utilization of capacity in the plant.

INTRODUCTION

Rapid changes in the global economy and in the local markets increase the market dynamics and challenge industrial companies to respond faster. Agility – an important trend in contemporary business – refers to the ability of a company to respond quickly to unpredictable and turbulent demand patterns and to produce short series of products with a growing number of grades. In addition to accelerating the pace of product innovation, companies also improve their competitive position by more efficient and effective production. The latter is not only a matter of technology but also involves the organization and management of production activities. The interactions between the technical and organizational aspects of manufacturing operations need to be recognized in an effective efficiency-improvement strategy. Agility, a strategy maintaining a flexible production system, could be an important weapon to increase a company's competitiveness and to secure business (Stevenson, 2005). The flexible batch-wise mode of operation is recommended for situations, in which companies have to adapt their production by manufacturing new grades or completely new products using new raw materials and new procedures.

In principle, in the industry two main production systems can be distinguished: production of goods such as cars, called assemblage or discrete industry, and production of products such as polymers via (bio-) chemical or physical transformation of inputs (feedstocks), called process industry. The process industry converts raw materials into intermediary and end products – for other industries and consumers. There are two ways of processing materials: continuous and batch-wise. A *continuous process* is fed by a constant flow of feedstocks, successively runs through different process steps, and yields a constant product flow. Typically, the continuous production mode is a steady state, which means that at a given point in the system there is no change in the process conditions over time. Of course, during the start-up, shutdown and changeover, the production mode is different. Examples of continuous production are the large-scale production of fuels in the petrochemical industry and the production of solvents in the chemical industry. A *batch-wise process* is fed at the beginning by feedstocks, which then undergo a sequence of processing activities over a finite period of time. Finite quantities of material are produced using one or more pieces of equipment. Here the production mode is dynamic: to make different products the composition of the batch equipment changes continuously. Each batch operation runs through several phases, e.g. in the case of a chemical batch reaction: process initialization and charging of ingredients, heating (and possibly pressurizing) to establish the prescribed reaction conditions, chemical transformation, cooling of product mix, discharging, and reactor cleaning. The frequent changing of products in a batch plant, the variability in product recipes, the sequencing problems and/or the necessity to clean the installation between batches, the dynamic character of each batch process step, etc. all make batch processes particularly difficult to manage.

The choice between batch and continuous processing is an economic one. Investments in continuous processes are high. The process is optimized for one product, which results in a very efficient but inflexible process. For these reasons, continuous processing is used for product categories with a small product range, low product differentiation, low added value and a long production life span. Batch processing is chosen where more flexibility and less efficiency are required. Several operations may be carried out with the same equipment, and the same operation may be performed with different types of equipment. One-to-one mapping between equipment and operations does not exist.

It should be stressed once more, that the attractiveness of batch processing plants, despite their complexity, lies in the flexibility they offer to produce different (types of) products with the same equipment and to use the same pieces of equipment for different processing operations. This feature makes batch plants eminently suitable for producing a large number of product grades, short series of (tailor made) products, or new pilot products. As a consequence, market-driven batch manufacturing of higher added-value specialties has been a fast growing segment in the process industry in most industrialized countries.

This chapter describes an effective performance measurement system for supporting operational decision-making in batch-wise plants, and especially planning and scheduling as well as inventory decision-making, contributing to the improvement of company's competitiveness and productivity. It should be stressed that most of the concepts presented here are also applicable for continuous or discrete production plants.

OPERATION MANAGEMENT IN CURRENT INDUSTRIAL PRACTICE

Operation management, one of the basic functional areas in business management, is primarily responsible for production of products in an effective (meeting the company's objectives) and efficient (using resources well) way. The operation functions include many interrelated activities such as forecasting, planning and scheduling, inventory management, quality management, training, and motivating employees. Nowadays, more and more companies are recognizing that manufacturing operations cannot be isolated from the external environment of customers and suppliers. Supply chain management, linking the production process with other actors that are involved in producing and delivering, enlarges the scope of operation management. Keeping inventory levels low requires a well-organized coordination with suppliers, and planning production levels smoothly requires accurate forecasting and a clear understanding of customer requirements (Davis and Heineke, 2005). To support the fast and accurate communication along the supply chain, companies use advanced technology, such as Internet and Electronic Data Interchange. Changes in supply and demand can be monitored on-line then, and the company can adjust the operation accordingly. As mentioned before, for batch processes the complexity of the management of manufacturing operations, and especially planning and scheduling, increases considerably. This complexity stems from the non-steady state behavior of each batch operation in itself, and from the need to align the various operations and production runs as efficiently as possible. Moreover, designing an effective inventory policy in a batch-wise production plant turns out to be a hard task, too (Verwater-Lukszo and Christina, 2005). Production of a large number of product grades and short series of products means that inventory management, which has a great influence on the economic performance of a batch plant, is very complicated as well.

Production planning concerns long- or mid-term decisions about production activities, whereas production scheduling concerns short-term aspects, such as the production sequence and the actual start- and end-data of the production activities. Usually, research into planning and scheduling involves developing algorithms for optimal planning or scheduling in a specific situation. In some planning algorithms the uncertainty about the future demand is incorporated (Heijnen and Grievink, 2004). Production planning and scheduling optimization looks especially effective in multi-product and multi-purpose situations where the production run sequence determines the frequency of equipment cleaning operations and, hence, production capacity utilization. However, initial schedule optimization would be of limited value when disturbances within the process (e.g. equipment failure, off-spec product) or outside the process (e.g. rush orders, delayed feedstock supply, incorrect stock level data) necessitate almost constant schedule correction. In these situations, any improvement strategy must include a strategy for dealing with disturbances (Schumacher, 2003). Moreover, when looking at various improvement options, it is generally very difficult to determine the added value of a change in planning and scheduling policy even when compared to a current bottom line. This is an additional reason why advanced scheduling algorithms are not widely adopted in industrial practice. It is difficult to compare their added value with other possible policy changes in planning and scheduling aimed at performance improvement. The same can be said about improvement of complex inventory management in a batch plant. From the above it is clear that industrial decision makers – plant managers and the planning personnel – would be supported in their decisions by a method estimating the improvement potential of various improvement options (Roeterink *et al.*, 2003a). Such a method should then support them in deciding whether to invest in hardware i.e. production and storage capacity, or software i.e. batch scheduling and sequencing, or to focus on information management. For this, the actual planning and scheduling situation in relation to inventory management should be characterized properly and next, the systematic exploration of improvement options and their benefits should be performed in a repeatable and well-structured way.

DECISION SUPPORT: PERFORMANCE MEASUREMENT SYSTEM

The developed system assists decision makers by providing a systematic structure to arrive at potential improvement options related to planning, scheduling, and inventory management when compared to the performance of the actual operational situation. The latter is defined by the given internal supply chain of a batch production system, including the given inventory management system and the given planning and scheduling activities.

The decision-support method consists of four main phases as presented in Figure 13.1. The first phase is identification of the improvement options and the evaluation parameters that measure the influence of potential improvements in production planning and scheduling or inventory management on the achievement of the company's objectives. The internal or external researcher in cooperation with process people identifies potential improvement options. This is a creative process supported by a causal relationship diagram representing relations between variables in the internal supply chain with regard to customer orders, materials, and production resources.

Mostly, the following evaluation parameters are chosen: work in progress (inventory of partially finished products), backlog (stockout), lead-time, capacity utilization and production capacity, bottleneck resource utilization, throughput, material availability, end-product inventory level, shipment rate, rework rates, delivery lead-time and lead-time variability, order fulfillment ratio, and sales.

In the second phase the assessment of evaluation parameters is carried out. For this step a System Dynamics simulation model is developed. The idea is to develop a general model, which can be easily adapted to specific industrial situations. The model is capable of simulating physical and informational aspects of the planning and scheduling as well as inventory management situations. The developed general System Dynamics model represents interrelated subsystems: raw material, production, end product inventory, and customer service subsystem (see also Verwater-Lukszo and Roeterink, 2004). The System Dynamics model is the cornerstone of the proposed decision-support system. It should be mentioned further, that experimenting with this model

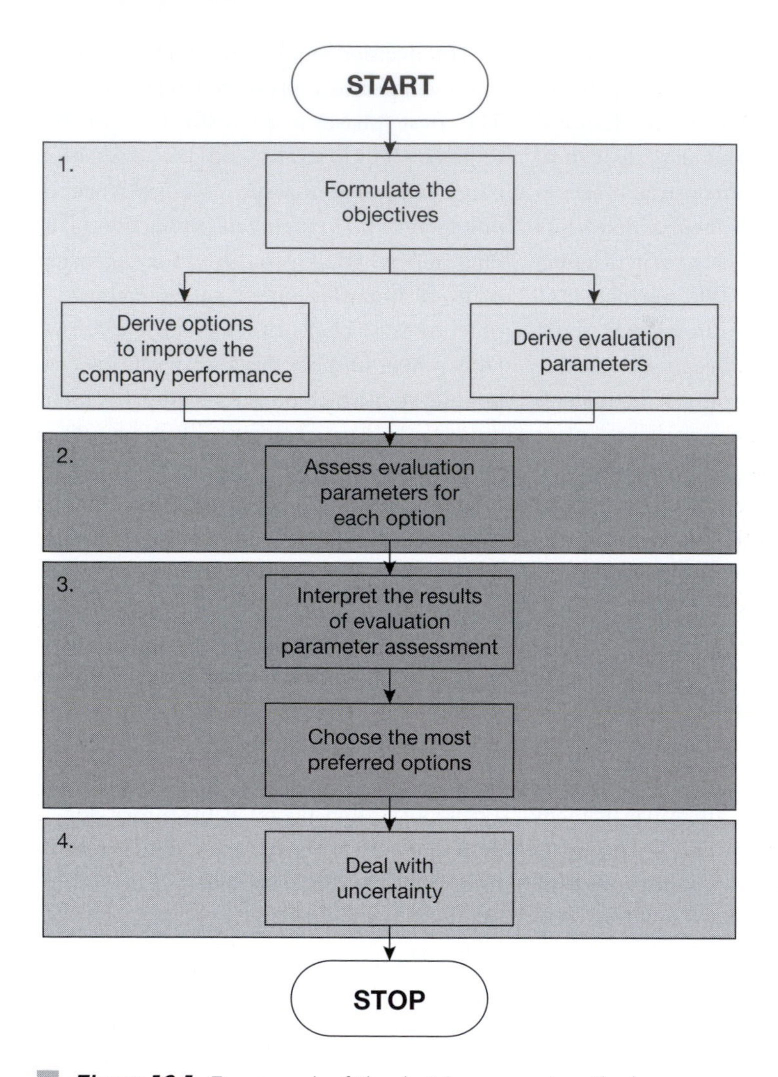

Figure 13.1 *Framework of the decision-support method*

in a specific plant would result in a so-called impact table. This table shows the impact of the intended improvement options on the evaluation parameters.

In the third phase, information of the previous phases is used to identify the most preferred choice with regard to the identified options for improvement.

The last stage in the developed method deals with uncertainty. Decisions are always based on assumptions about future conditions when uncertainties exist. Treatment of uncertainty means, for example, that the most robust options are identified.

System Dynamics model for internal supply chain

System Dynamics is a modeling approach to study complexity: it deals with understanding how complex systems change over time, whereby internal feedback loops within the system influence the entire system behavior. It was originally developed at the

Massachusetts Institute of Technology by Jay Forrester in 1961 (Forrester, 1961) to support (industrial) managers and policy makers in solving business and organizational problems. To present the feedback structure of the system, System Dynamics approach uses the so-called Causal Loop Diagrams (CLD). A causal diagram consists of variables connected by arrows denoting the causal influence among variables (Sterman, 2000). The overall causal diagram for an industrial plant is pictured in Figure 13.2.

The causal diagrams are used to start formal model development of the system. For this, the stocks – another central concept of System Dynamics – should be identified first. Stocks can be seen as accumulations: they describe the state of the system and generate information upon which decisions can be taken. The basic building blocks of System Dynamics are based on integration methods (see Figure 13.3). X(0) means the initial state of the stock value. The stock level X(t) accumulates by integrating the flow rate (dX). Auxiliary variables (Y) control or convert other entities (g(X(t))).

CASE STUDY IN A FOOD COMPANY

The performance measurement system is successfully applied in a batch-wise production plant producing more than 30 different potato starch derivatives (see Roeterink, *et al.*, 2003b; Verwater-Lukszo and Roeterink, 2004; Verwater-Lukszo and Christina, 2005 for other applications). The starch material in suspension is processed in four main reaction routes that produce more than 15 work in progress. After getting through the packaging stage, the end products are kept in the warehouse to be shipped to customers. Figure 13.4 shows the summary of this transformation which, due to its diverging character, is much more complex than suggested by this figure. The question is: how to identify the most promising operational improvement options?

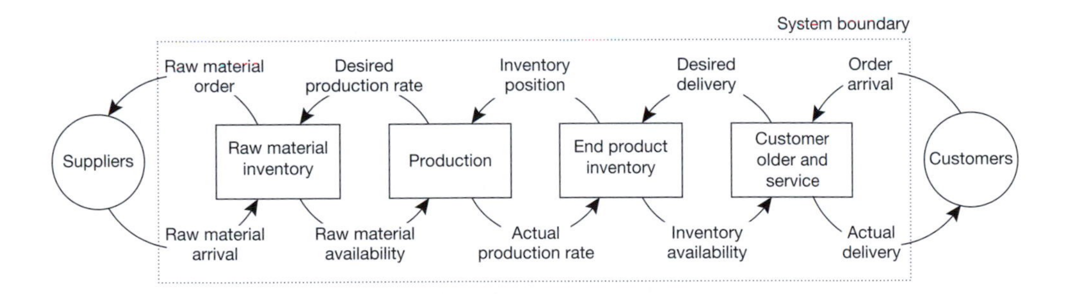

Figure 13.2 *Causal Loop Diagram overview of the internal supply chain model*

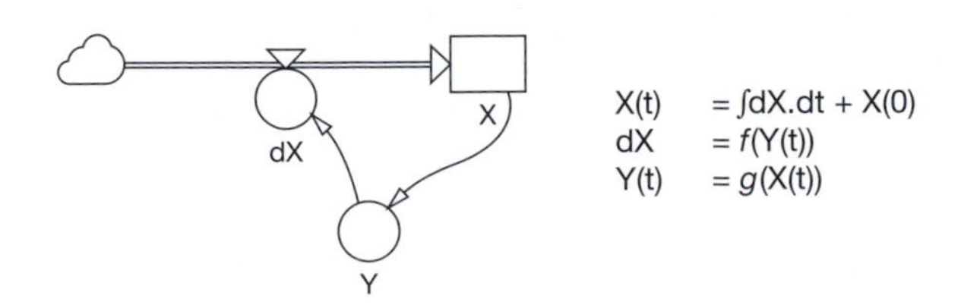

Figure 13.3 *Basic building block of System Dynamics*

Starch-suspension + Main reactants → Work in progress → End-products

Figure 13.4 *Summary of the transformation of raw material to end product as modeled in Figure 13.2*

Identification of improvement options and evaluation parameters

The process of deciding which improvement options and evaluation parameters will be investigated is valuable due to the fact that it forces the management to be very explicit about the performance priorities and the relationships between them (Neely, 2000). Therefore, the first step of the analysis begins with discussing the enterprise objectives. To structure these objectives in a hierarchical way the objective-tree technique can be used (Verwater-Lukszo and Heijnen, 2001). First, an overall objective will be determined that defines the breadth of concern. Next, other objectives that contribute to this main objective are specified. Every objective is defined in terms of a factor and a certain direction in which the factor is desired to change. The specification procedure continues until all main objectives (at the strategic level) are translated into operational objectives. An objective is called operational when the factor defined in the objective is operational, i.e. when this factor can be measured in practice. The subset from these measurable operational objectives will be chosen as a set of evaluation parameters. The general framework for the objective tree is the same for all industrial plants. Figure 13.5 represents a graph with (a part of) an objective tree. The continuity of the company is seen as the company's main objective, and translated into economic, ecological, and social sustainability, which are each divided into more and more specified sub-objectives at the lowest level. By refining the main objective into several sub-objectives, the objective tree is used to scrutinize evaluation parameters that can be used to measure the achievement of objectives. Good evaluation parameters both define precisely what the associated objective means and serve as a scale to describe the consequences of improvement options. This case study focuses on objectives related to economic sustainability such as competitiveness, productivity and profit, and results in the following evaluation parameters: capacity utilization, manufacturing/delivery lead-time, throughput, end inventory; rework/order fulfillment ratio, sales and inventory costs.

Generation of improvement options is carried out in three steps: first, a large number of possible improvement options are generated; second, these options are screened to eliminate the least promising alternatives and finally, strategies are designed by using the promising options. The use of causal diagrams, as presented in Figure 13.6 for the production subsystem helps the decision maker(s) to identify potential improvement options. A connection between variables in the causal diagram is called a causal link and is presented as an arrow. These causal links illustrate how one variable affects the other one. Take, for example, the causal link from processing time to manufacturing lead-time. A plus (+) sign on the causal link indicates that if processing time increases (decreases) then manufacturing lead time will also increase (decrease) under the condition that all else remains equal. A minus (–) sign as seen in the causal link between processing time and production capacity means that if processing time increases (decreases) then production capacity will decrease (increase) with all else equal. Based on this causal relation diagram several improvement options can be identified, such as: reduce product variety; employ undertime or overtime; find the optimal product sequence; add new capacity; reduce or increase batch size; and invest in new hardware.

Causal diagrams are also modeled for other subsystems in the internal supply chain

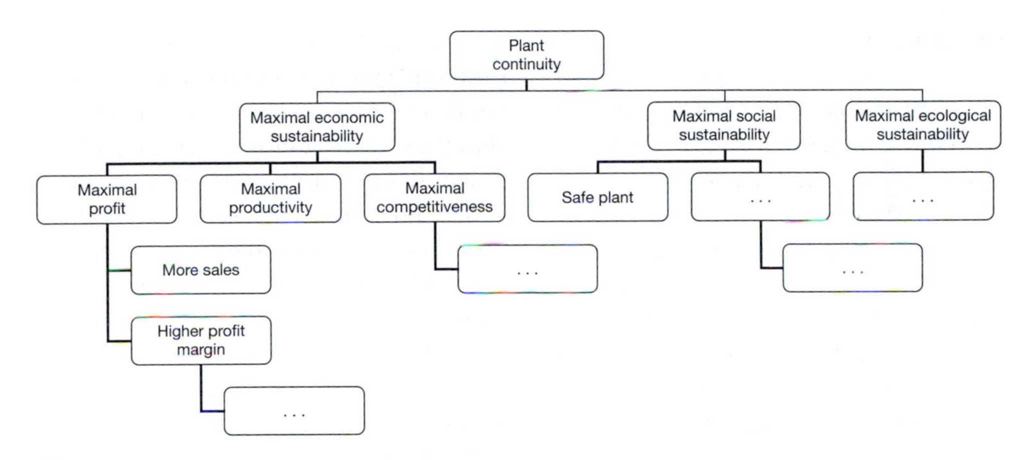

Figure 13.5 *Part of the objective tree*

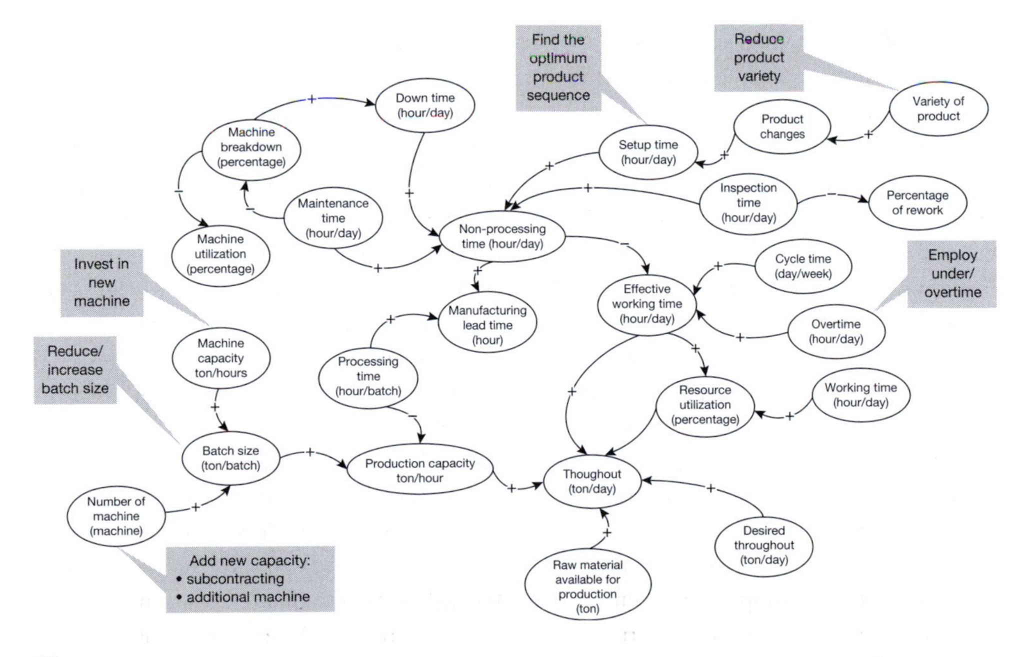

Figure 13.6 *Causal diagram for production subsystem used for the generation of improvement options*

to identify other improvement options. A next step is to screen these options to work further only with the most promising ones. Screening of options is necessary to narrow the range of alternatives that have to be examined in more detail (Walker, 2000). The following criteria can be taken into account:

- Feasibility deals with technical, economic, or administrative/ organizational constraints, e.g. "increasing storage capacity" could be an infeasible option due to the limited area the company has.
- Management acceptability, such as "reducing product variety" can be unacceptable due to strategic reasons.
- Dominance means that from two improvement options with almost the same purpose and benefit the worst one can be skipped.

With these three criteria, the number of options can be reduced. Next, combinations of options into so-called strategies should be examined, too.

Assessment of potential improvement options

As mentioned before, to represent the internal supply chain in a company all activities, from the receiving of the raw materials to the shipment of the products, are incorporated into a model. A System Dynamics approach is used to capture the dynamic relationships and feedback structures. In the developed model the internal supply chain is divided here into four sectors: raw material, production and rework, inventory, and customer service. As an example, the model of the production subsystem developed for the considered batch-wise plant in the food sector is presented in Figure 13.7. This subsystem is modeled on an aggregate level in which the raw material (production start rate) is transformed into an end product. This transformation takes as long as the manufacturing lead-time that represents the average transit time for all aggregated items. As a consequence, the production start rate will be delayed with the manufacturing lead-time to become the end product. The manufacturing lead-time is composed of the processing time in the reactor, dryer and packing, the disturbances, changeover time, and inspection time.

The production start rate is determined by the control mechanism of inventory, work in progress, and rework. The planner makes corrective actions when the inventory and work in progress are lower or higher than the desired level. The planner also makes corrective actions when there are expected rework occurrences. Based on these corrective actions and customer orders, the planner can determine the desired production order and then the desired production start rate. The work in progress shows the raw material being produced on the production floor. It shows the amount of raw material being transformed, which is influenced by the raw material availability as shown with the feasible production start rate, the actual production capacity, and also the minimum production rate that the company wishes to maintain. The adjustment for work in progress (WIP) modifies production starts to keep the WIP inventory in line with the desired level. The desired WIP is set to provide a level of work in process that is sufficient to yield the desired rate of production (given the current manufacturing cycle time).

As shown in Figure 13.7, the production planning mechanism controls how much material should be acquired and how many

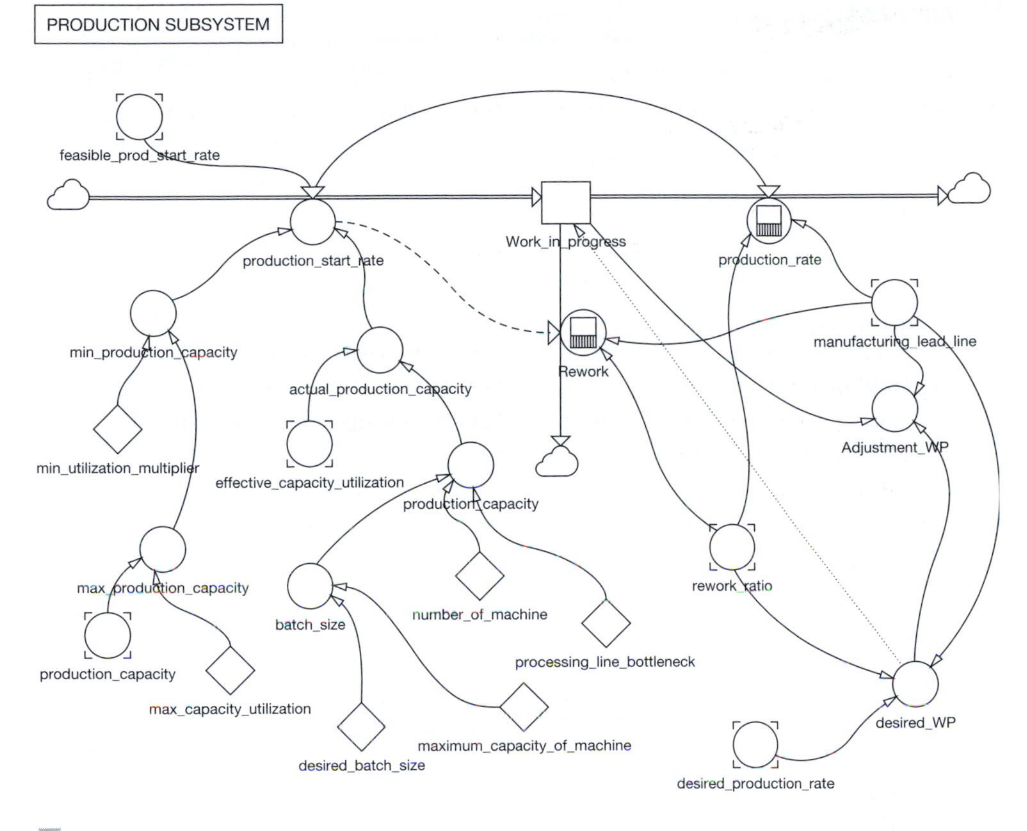

Figure 13.7 *Representation of the production subsystem*

products should be produced. The production planning is based on the customer orders. When the company cannot fulfill the order, the order will be put into backlog so that it can be fulfilled later. This illustrates interactions between the two subsystems, i.e. production and customer services.

Model simulations show how the improvement options influence the selected evaluation parameters. This information is organized in the form of the Improvement-Potential-Assessment (IPA) matrix presented in Table 13.1, which is further used as the basis for choosing the most preferred options.

Choosing the most preferred options

Information that is collected from the previous phases, is used to decide on operational situations that potentially lead to the most promising improvements in the company's productivity and competitiveness. First, from the IPA matrix it is concluded that options related to production cycle, product changeover and changeover time are influencing the performance of capacity utilization and rework ratio. Furthermore, the case study indicates that the company has high levels of end product inventory to cover its daily demand and the company produces almost at the maximum available capacity.

Table 13.1 *Improvement-Potential-Assessment matrix*

	Productivity			*Competitiveness*		*Profit*	*Cost*
Improvement option	*Capacity utilization (%)*	*Manufact-uring lead-time*	*Rework ratio*	*Delivery lead-time*	*Order fulfilment ratio (%)*	*Sales (euro)*	*Inventory cost (euro/day)*
1 Lengthen the production cycle by 50%	2.48	−0.01	36.5	0	0	0.45	4.84
2 Shorten the production cycle by 50%	−7.45	0.04	133.5	0.29	−0.02	−1.89	−32.01
3 Reduce product changeover	0.78	−0.01	11.92	0.00	0	−0.15	1.90
4 Reduce changeover time	0.64	0	−1.27	0	0	0.02	1.28
5							
6							
7 Reduce target delivery lead-time	0	−0.01	0	−14.29	0	−0.01	−2.93
8 Reduce safety stock coverage by 10%	0	−0.01	0	0	0	0	−11.45
9 Reduce safety stock by 50%	0	−0.01	0	0	0	0	−59.93
10							
11 Strategy I toward "Lean Manufacturing"	0	−15.35	0	0	0	0	−54.20

As the company's primary preference is profit, sales and inventory costs are chosen as the most important criteria. Multi-criteria decision analysis results in the following five most preferred options:

1 reduce safety stock by 50 percent (option 9);
2 reduce safety stock coverage by 10 percent (option 8);
3 lengthen the production cycle by 50 percent (option 1);
4 reduce target delivery lead-time (option 7);
5 reduce product changeover (option 4).

For the time being, option 11 aimed at introducing "Lean Manufacturing" is not selected here because of the very high implementation costs.

Treatment of uncertainty

Decision-making does not end with choosing the most preferred option. The framework to treat uncertainty in Figure 13.8 gives a systematic way to structure the uncertainty

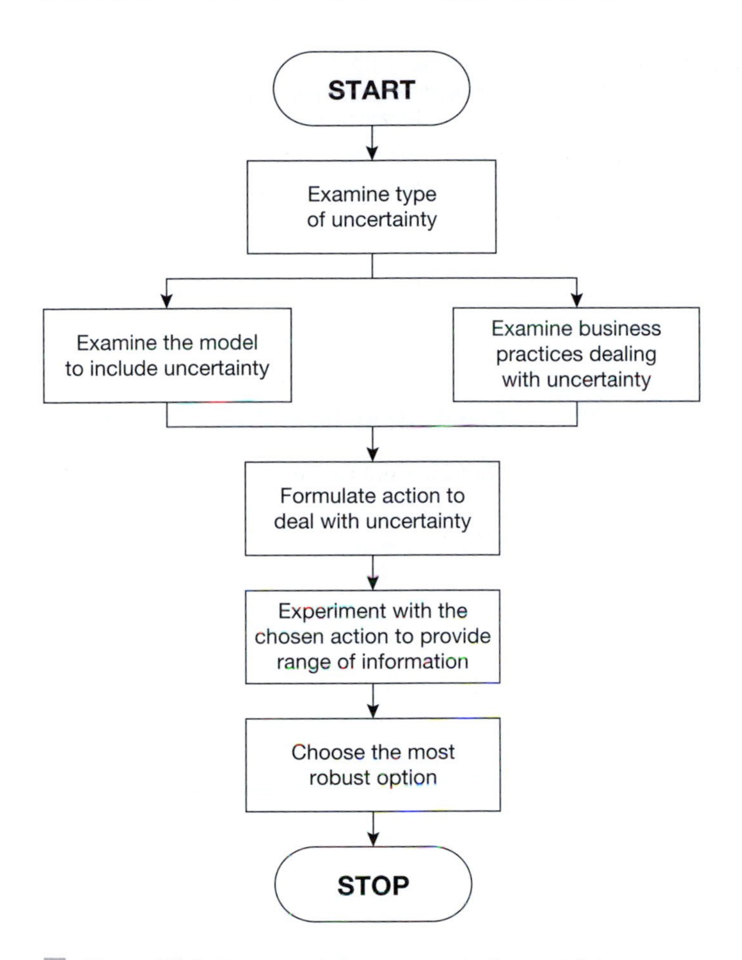

Figure 13.8 *Framework for treatment of uncertainty*

in this study. It starts with examining how the decision maker(s) will respond to the existence of uncertainty. Afterwards, it is necessary to identify the types of uncertainty that exist in and around the internal supply chain. From all the gathered information, actions to deal with uncertainty can be formulated supporting the decision maker in choosing the most robust options.

Three types of uncertainty can be identified: external uncertainty, system response uncertainty, and value uncertainty (Geenhuizen and Thissen, 2002). External uncertainties refer to inputs that are beyond the control of the decision maker. In this case they concern market and the raw material characteristics. Furthermore, the uncertainty in the surroundings of the internal supply chain should also be taken into account, such as the government regulation, i.e., environmental and food safety regulation.

System response uncertainty is related to the system responses to the external uncertainties. In this category three subtypes of uncertainty are distinguished: parametric uncertainty, system uncertainty, and stochastic uncertainty (Walker, 2000).

Finally, value uncertainty is related to the uncertainty about the valuation of system outcomes. Values and standards used to evaluate future outcomes of options can change with regards to time, place, and people, e.g. due to changes in decision makers' preferences or changes in the company's environment such as competition and industrial structure.

As an illustration, we would like to discuss four scenarios that are developed to deal with external uncertainty, see Table 13.2.

For example, Scenario 4 is based on two assumptions: the existence of strong forces driving performance and high rush orders in the production requirement. In this scenario, it is assumed that:

- Strong forces driving resource performance are represented by significant changes in the effective working time. For example a reduction of 25 percent of wastes (non-processing time).
- High rush orders are represented by high increase in the customer order by pattern break.

Meanwhile the predetermined factors are:

- Market growth for one year in the future is represented by change in the customer order. For example an increase of 3 percent in customer order.
- Raw material supply is always available.

For the defined scenarios four new impact tables are calculated so that the decision maker can determine the most robust options, i.e. the options that perform well in different scenarios. We conclude that without any effort from the company to provide reserve capacity, only two options will be likely to survive in the future: "reducing safety stock by 10 percent" and "lengthening production cycle by 50 percent."

Table 13.2 *Matrix with scenarios to deal with external uncertainty*

	Forces driving resource performance	
	Weak	Strong
No rush order	Scenario 1	Scenario 2
High rush order	Scenario 3	Scenario 4

Conclusions from the case study

The case study shows that the company has a high level of end product inventory to cover its daily demand and the company produces almost at the maximum available capacity. As a consequence, options to reduce inventory level (reducing safety stock coverage) and to increase effective capacity utilization (lengthen production cycle by 50 percent) perform better. Moreover, these robust options will probably survive in the future according to the market analysis.

FINAL REMARKS

Based on the applications of the presented decision-support system, it can be concluded that the developed system is appropriate to predict the consequences of potential improvements options with respect to the selected evaluation parameters. It assists the decision makers in the identification of improvement options in an operational situation. Another effect of the method appearing during industrial application is that it helps to break down the information and communication barriers between different departments within a company, such as those often found between operations, planning, and sales.

To sum up, we would like to stress that like all aspects of operations management, planning and scheduling as well as inventory management are affected by larger trends in the global economy. Shortening of product life cycles and increased client orientation challenge enterprises to respond quickly to changes on the market. Agility is a *conditio sine qua non* to secure business. In short, fast and effective operational decision-making will remain a necessary part of the operation of an enterprise for the foreseeable future, and with a presented model-based approach it will remain a less difficult task.

ACKNOWLEDGMENT

Work reported here was carried out at the Delft University of Technology in cooperation with five batch processing companies and TNO TPD in the framework of the multi-disciplinary project "Batch processes – cleaner and more efficient." The Dutch Ministry of Economic Affairs and the Ministry of Education, Culture and Science financially supported the project in the framework of the EET program (Ecology, Economy and Technology). Many students and researchers were involved in the project. Jason Lim, Elisa Anggraeni, Susi Christina, and Huub Roeterink are kindly acknowledged. For more information on the research program that generated this result, visit http://www.batchcentre.tudelft.nl/.

REFERENCES

Davis, M.M. and Heineke, J. (2005). *Operations Management: Integrating Manufacturing and Services*. Boston, MA: McGraw-Hill Irwin.

Forrester, J.W. (1961). *Industrial Dynamics*. Cambridge, MA: MIT Press.

Geenhuizen, M. and Thissen W. (2002). Uncertainty and Intelligent Transport System: Implication for Policy. *International Journal of Technology, Policy and Management*. 2(1), 5–19.

Heijnen, P.W. and Grievink, J. (2004). *Dealing with Uncertainty in Mid-Term Planning*. IEEE International Conference SMC 2004. The Hague, October 2004.

Neely, A. (2000). Performance Measurement System Design: Developing and Testing a Process-based Approach. *International Journal of Operations and Production Management*. 20(10), 1119–1145.

Roeterink, H.J.H., Lim, J., Verwater-Lukszo, Z., and Weijnen, M.P.C. (2003a). *A Method for Improvement Potential Assessment in a Batch Planning and Scheduling Situation*. FOCAPO Conference 2003. Florida, January 2003.

Roeterink, H.J.H., Verwater-Lukszo, Z., Weijnen, M.P.C., and van Daalen C.E. (2003b). *Improving the Logistic Performance in a Food Company: A Case Study*. International System Dynamics Conference 2003. New York.

Schumacher J. (2003). *A disturbance management approach to improving the performance of batch process operations*. PhD Thesis. Delft University of Technology.

Sterman, J.D. (2000). *Business Dynamics: Systems Thinking and Modeling for a Complex World*. Boston, MA: McGraw-Hill Irwin.

Stevenson, W.J. (2005). *Operations Management*. Boston, MA: McGraw-Hill Irwin.

Verwater-Lukszo, Z. and Christina, T.S. (2005). *System-Dynamics Modelling to Improve Complex Inventory Management in a Batch-wise Plant*. ESCAPE-15 Conference. Barcelona, May 2005.

Verwater-Lukszo, Z. and Heijnen, P.W. (2001). *Integrated Plant Management for Better Economic and Ecological Business Performance.* Conference ENTREE2001 on Integrated Green Policies: Progress for Progress. Florence, Italy. November 2001.

Verwater-Lukszo, Z. and Roeterink, H.J.H. (2004). *Decision Support System for Planning and Scheduling in Batch-wise Plants.* IEEE International Conference SMC 2004. The Hague, October 2004.

Walker, W.E. (2000). A Systematic Approach to Supporting Policy Making in the Public Sector. *Journal of Multicriteria Decision Analysis.* 9(1), 11–27.

Chapter 14

Managing knowledge processes

J.H. Erik Andriessen

OVERVIEW

Knowledge management is the art of systematically organizing and managing knowledge processes, such as identifying knowledge gaps, acquiring and developing knowledge, storing, distributing and sharing knowledge, and applying knowledge. A systematic and cohesive approach, covering all of these aspects, is increasingly needed because the primary processes in many organizations, profit or non-profit, service or industrial, are either knowledge processes or are strongly supported by knowledge processes. Moreover, rapid market changes, turnover of employees and geographically dispersed work require that these processes are managed systematically and explicitly. Knowledge is defined in this context as tacit insights and skills and not just as information held in documents. Knowledge management strategies can roughly be divided into "codification strategies," where the role of explicating knowledge and storing it in ICT-based repositories is emphasized, and "personalization strategies," where the communication flow between people, such as in so-called communities of practice is emphasized. Many forms and structures used to support the knowledge processes are discussed in this chapter. The chapter ends with the identification of a number of dilemmas faced in this field, and a call is made for solutions to these dilemmas to be found.

INTRODUCING THE PROBLEM

Imagine that you are a top manager of a large multinational firm producing audio and video equipment. One product line is DVD systems. Your company has established factories in several countries all over the world and has recently acquired several other companies producing elements for DVD systems. Innovation in production processes is taking place in many of these factories, but the transfer of novel ideas from one factory to another appears to be limited. You decide that an information system should be set up containing all the documents and basic information concerning the production of DVD systems. Existing production methods, and also reports and memos concerning innovative approaches must be stored systematically in this database. Using a system of keywords this information should then be accessible for all managers and experts at all the DVD-related factories in the company.

No sooner said than done. The company spends considerable energy and money on collecting all the relevant documents and making a systematic database. However, it is not just the formal documents that are required by the manager, he also wants documentation on innovative ideas, solutions to problems in the set up of new production lines, and the results of experiments, to be entered in the system. This knowledge is often not (yet) formalized and registered, it exists only in the heads of the experts in the separate factories. Therefore, about twenty of these experts, from all the relevant factories, are brought together for a week-long workshop. They gather at a nice seaside resort and have a very intensive week of presentations and exchanging experiences and insights. They also chat often at the bar and take time to swim in the sea. Facilitators from the company note down all this exchanged knowledge and systematize it in a report. Soon afterwards the repository is ready and the people involved in DVD production in the various factories are informed that they can find answers to all their questions in this database. It is accessible through the company intranet, but only via a very strict password system, because it also contains information that would be extremely valuable to competitors. A central corporate department is made responsible for keeping the system up to date, in collaboration with the experts.

After a year the new system is evaluated. The responsible top manager is quite disappointed to learn that relatively few people make use of the system. The reasons are diverse, including the fact that, while it is updated, it always lags some months behind current development; moreover, the experts do not like to enter their knowledge and ideas into the system, because it takes so much time and they do not expect the system to give really useful answers to what they consider to be their highly specific questions and problems. The top manager is, however, pleasantly surprised to learn that the group of twenty experts has kept in contact and has even organized a second meeting. It appears that when the experts have problems or innovative ideas they much prefer to consult each other rather than use the system. The top manager now decides to manage the exchange and development of knowledge by facilitating and supporting these groups and other similar "communities of practice." The idea of repositories containing useful documents and information is not discarded, but these are now monitored by community members.

Many of the issues related to the subject of "knowledge management" are touched upon in the example given above. The example covers the role of people who exchange knowledge, to the role of computer systems as repositories of documents. But it also points to a number of basic questions such as: What is knowledge? How does this differ from information? What is knowledge management? Is it possible to manage knowledge? To what extent are knowledge processes related to learning processes? Can an organization learn? These questions will be dealt with extensively in a later section. Here it is sufficient to say that the term knowledge management is in fact shorthand for "knowledge process management." The central issue is to manage processes such as the development, sharing, and storage of knowledge.

In this chapter a few traditional aspects of knowledge management will not be discussed, i.e. recruitment of people and training courses. These subjects are of utmost importance and are covered in Chapter 3 on Human Resource Management for advanced technology. In this chapter the focus is on the development and exchange of practical experiences and insights of professionals. These issues are becoming more and more relevant in modern organizations where routine production processes are replaced by non-routine services.

The term "knowledge management" in itself is a suspect term for some people. It is sometimes associated with past conceptualizations of management control or hijacked by the IT community for the marketing of new IT systems. Nevertheless, in this chapter I will use the term, because it is very much accepted in the field of management of modern organizations; and, added to this, in the past years many authors have used this terminology in their treatment of the field (e.g. Senge, 1990; Jashapara, 2004; Awad and Ghaziri, 2004; Davenport and Lawrence, 1997).

In this chapter I will first explain why knowledge management is important. I will then discuss various central concepts and theoretical notions related to knowledge and knowledge processes. This will be followed by a section in which the focus will be on the strategies and structures used by companies in this area of management. These strategies can be roughly divided into "codification strategies," emphasizing the role of explicating knowledge and storing it in repositories, and "personalization strategies," emphasizing the communication flow between people, e.g. in communities of practice. The two strategies are discussed in separate sections. This is followed by a short excursion into the realm of the technical tools and systems that are available to support knowledge processes. The chapter ends with a reflection on the pitfalls that may be met and potentials shown in the field of knowledge management.

WHY KNOWLEDGE MANAGEMENT?

Why is knowledge management important? Managing knowledge processes is of course not a goal in itself. It is a means to improve an organization's capacity to achieve business goals and to innovate, and it can also be a means to develop the capacity of the employees to master their work and develop their abilities. Nowadays, the primary processes in many companies, and in non-profit organizations such as hospitals and governments, either are themselves knowledge processes or are strongly supported by knowledge processes. Indeed, in consultancies and banks, in insurance companies and schools or universities, developing, applying and/or transferring knowledge is the core business. This is also the case in high-tech companies such as Shell or Unilever, IBM or Philips, where the primary production processes also depend heavily on knowledge work such as research and development, documentation, human resource management, or sales and marketing. So the first reason for systematic knowledge management is that organizing the development, exchange, and storage of the primary means of an organization's work, i.e. knowledge, is the key to the success of such organizations.

Second, in response to rapid market changes and the short life cycle of projects, the knowledge held in many companies needs to be regularly updated and renewed. Keeping up with the competition, or better, achieving competitive advantage, requires constant alertness to new external developments, requires that employees share experiences with solving problems, and requires systematic innovation processes. This applies particularly to technology-oriented companies. Example: one of the world's largest producers of mobile phones develops its various production processes, located in different countries, using the same system of integrated design and design steps. However, it appeared that the comparable specialists at the different production lines hardly communicate, so that there was no transfer of new ideas and methods. This problem is found in many companies, i.e. the problem of organizing and motivating horizontal contact of specialists in different parts of an organization.

Third, knowledge management is becoming increasingly important due to the rapid turnover of employees. The mobility of workers, internally within their organization and to other organizations or to retirement, is often very high. This makes it difficult to support continuity of skills and knowledge. It is therefore very important for companies, and countries, to find ways to share the knowledge of employees moving in a company or leaving the company. Example: a subsidiary of a large multinational company manufactures specialized sonar equipment. These sonar systems can be found in submarines, large offshore plants, and at other places all over the world. The systems have a very long life cycle. Some are over thirty years old. Considerable repairs have been made and the equipment has been updated. Although this is always precisely documented in manuals and reports, it is very difficult for new maintenance personnel to find their way through this jungle of documentation, particularly when a repair is urgent, which is almost always the case. The company has a large number of highly experienced maintenance men, who have been repairing many of these systems over the years, so they are very knowledgeable and have little need for documentation to determine how to diagnose and repair a breakdown. However, in the coming five years a large proportion of these repairmen will retire and the company is now desperately seeking a solution for this future problem when they will lose valuable tacit knowledge. Although this is a very special and urgent case, it is not unusual in many organizations where key employees change jobs every few years and take their tacit knowledge with them.

Sharing and cooperation is not only required inside organizations. Companies increasingly cooperate with other organizations, such as governments or research agencies and even with competitors, to develop new ideas or technologies (which become more and more expensive), to find solutions to common problems or to explore new market areas. Such collaboration processes need to be systematically organized to function effectively. Example: Habiforum is the expert network for multiple space use in the Netherlands. It was established to initiate and stimulate innovations in this area. In this network, representatives of building companies, local authorities, real estate developers, and researchers cooperate to exchange experiences and to develop solutions to problems concerning multiple space use. A website for sharing documents and finding information on all aspects of the topic (technical, policy, legal, financial, etc.) has been designed to make systematic discussion and innovation possible.

A final trend that necessitates systematic attention to knowledge processes is the increasing decentralization and geographical dispersion of organizations. Globalization of commerce and production implies that information and knowledge has to be distributed over wide distances and has to be accessible in many places at the same time. Decentralization of organizational structures and the growth of networking organizations ask for constant alertness to exchanging experiences and the development of common concepts, practices, and systems. Example: A German company providing training and consultation in the area of computers and software development is, in fact, a network of about 60 independent consultants and trainers. The members cooperate in teams to work on projects. They all have their own office, mostly at home, and they coordinate much of their work through email and telephone. On top of these means of communication they use a groupware system that supports a common document database, email, a bulletin board system, and the coordination of the work of project teams.

Summarizing, the pressure for constant innovation, the short life cycle of project teams, the geographic distribution of interaction and expertise, and rapid employee turnover necessitate systematic attention to knowledge creation and exchange. According to a national Canadian Survey, the threat of losing key personnel ranked as the primary trigger to implement more knowledge management practices (mentioned by three-quarters of the firms). Losing market share was placed second, followed by "difficulties in capturing workers' undocumented knowledge."

So managers do, and have to, ask themselves increasingly: What knowledge do my employees need and how do they acquire this knowledge? What knowledge is available in the organization, i.e. who has knowledge about problem X or client Y? How can we store the knowledge in such way that it is easily available to employees? Are we re-inventing the wheel regularly? And, what should we do to stimulate people to share their knowledge with their colleagues?

Yet it is not only organizations that have an interest in systematically caring for their knowledge processes, this is also relevant for teams and for individual employees. Teams and groups are the best mechanisms to use in an organization for knowledge exchange and creation, and to enable individual learning. A major reason for this is that interaction in groups supports the development of shared mental models, i.e. common ways of viewing problems and solutions. Although much sharing and learning will happen without any explicit structuring and managing, teams may find it beneficial to organize and facilitate these processes by introducing meetings, and to use repositories.

Individuals also have to pay deliberate attention to activities that enable them to accomplish a task. These activities may be acts of interpersonal behavior, e.g. asking a co-worker, passing on information to another person, as well as acts that are performed in individual settings, e.g. consulting a knowledge database, searching for knowledge. Workers face questions such as, how do I keep up to date in my field? How can I learn from the experience of others? Where do I find the knowledge I need but how can I prevent information overload, i.e. how do I organize all the information and knowledge I have and receive?

Sonnentag (2004) distinguishes *intensity* and *selectivity* in personal knowledge management. She has shown that some individuals are motivated to search and distribute as much information as possible while others have a much more selective strategy. This last strategy is important to prevent information overload.

CONCEPTS AND THEORIES

Basic terms

Thus far I have used the word "knowledge" many times; but what does this term actually mean? What is knowledge and is it different from information? What is its relation to learning? Usually a distinction is made between data, information, and knowledge. The European Committee for Standardization (CEN, 2004), gives the following definitions for data and information:

- *Data* are discrete, objective facts (numbers, symbols, figures) without context and interpretation. Example: 1209, 1600, 04.
- *Information* is based on analyzing and interpreting data, and adds value by understanding the organization of data. The data above become information when one understands that it means: On 12 September at 4.00 o'clock in the afternoon the outside temperature was 4 degrees centigrade. *Information Management* covers the processes of selecting, capturing, categorizing, indexing, and storing information.
- *Knowledge* can now be defined as information in personalized context; it is information that is experienced, interpreted, and processed by a person in a particular situation and in that way developed into insights and skills. It implies insight in how to deal with information in a certain situation, based on positive or negative experiences.

So knowledge is personalized and situation-bound. The information presented above, concerning temperature at a certain moment, is related to knowledge for the janitor in an Italian apartment building, who knows that with this extreme cold he has to turn the central heating on, which is normally only done on October 1 of each year. Weggeman (1997) defines knowledge as: information + experience + skills + attitudes. Others specify knowledge as knowing which information is needed (know what), how information must be processed (know how), why information is needed (know why), where information can be found to achieve a specific result (know where), and when which information is needed (know when).

Knowledge management is, then, according to CEN (2004), planned and ongoing management of activities and processes for leveraging knowledge to enhance competitiveness through better use and creation of individual and collective knowledge resources. Later I will present a slightly different definition, based on the model for knowledge management developed in this chapter.

In certain philosophical discourses knowledge is regarded as information you have in your mind and of which you are certain that it is correct. Others, however, argue that "knowledge is fluid, social and evolving" (Krogh *et al.*, 2001). This implies that there is no true knowledge, in contrast to information for which one can often establish whether it is true or not. All one has is the best available knowledge so far, which can, and will always, be superseded by new insights.

Some scholars argue that, strictly speaking, personalized knowledge cannot be stored in manuals or databases. The moment it is distinct from its previous owner, it has lost its personalized character. It has become information that becomes knowledge by being read and interpreted and used by others, who then have developed their own version of this piece of knowledge/information. In the literature and also in the rest of this chapter this "mistake" is often made, when the line between information and knowledge is not explicitly mentioned. However, this line is not always as sharp and clear as it may seem. People can write extensively about their insights in such a way that others can sometimes very rapidly internalize this knowledge. Expert systems also embed certain expert knowledge in such a way that personalized and local knowledge can be translated almost directly to other situations. So it is better to distinguish various kinds of knowledge such as the following (see CEN, 2004):

- *Explicit Knowledge* is knowledge that has been codified, typically in objects, words, and numbers, in the form of graphics, drawings, specifications or manuals, which can therefore easily be shared and understood.
- *Tacit Knowledge*, sometimes also called *implicit knowledge*, consists of mental models, skills, insights, and perspectives, largely based on experience. This knowledge is difficult to transfer, but KM techniques such as learning by doing or collaboration in communities can help people to share this knowledge. Cooperating employees develop a shared repertoire of routines, vocabulary, stories, symbols, artifacts, and heroes that embody the accumulated knowledge of the community. This shared repertoire serves as a foundation for future learning.

Finally, some people distinguish a third form, i.e. *Embedded Knowledge*, that exists in the constellation of an object, a machine, or construct. The pyramid of Cheops can be used as an example. All the knowledge of the Egyptians about building pyramids is embedded in it. This includes not only how these pyramids were built but also, e.g. why they were built the way they were. Organizational structures and processes also have a lot of embedded knowledge, which is often taken for granted.

Codification and personalization

Knowledge management was defined above as the planned management of activities and processes for leveraging knowledge. This is still very general and abstract, so some years ago a large research project was initiated to try to find out what types of knowledge management were actually used by large companies (Huysman and Wit, 2002). Knowledge management was already quite popular and many promotional stories and experiments could be found, many of which boasted that they had found the ultimate solution. The objective of the study, however, was to collect examples from operational practice in order to learn about success and failure factors. The search for cases of knowledge management beyond the conceptual stage started and the first examples were found in companies, such as Unilever, Cap Gemini, ING Barings, IBM, the Dutch National Railways, and the insurance company Aegon. Many approaches were encountered, from company-wide systems with best practices and project data, to initiatives concerning knowledge sharing meetings and communities. Cap Gemini, for instance, an IT consultancy that operates worldwide and has about 40,000 staff, uses a kind of intranet, called CapCom. This system provides information on a wide variety of subjects, and includes a database with project data, with best practices and with data on consultants ("Yellow Pages"). The rule is that after a project is finished, the consultants have to add their experiences to the system.

Some experiences with the approaches and systems encountered in the study were positive, but many were negative. Knowledge management appeared all too often to be far from achieving its goals. Usage of the databases or intranets was often very limited and keeping these systems up to date appeared to be quite difficult, and because of the tendency for these systems to expand continuously, the users felt themselves to be overloaded with information. In some cases knowledge management was viewed as a control mechanism, since it required employees to systematically record all their project activities and contacts. There was much registering required, with little actual knowledge *sharing*.

Various knowledge management traps were identified:

- The *IT-systems trap*: knowledge appeared to be viewed as an object, a commodity, that could be extracted, stored, and distributed wherever needed.
- The *Opportunity trap*: knowledge management systems were sometimes not developed to solve specific company problems, but because IT managers had discovered an interesting software application on the market and had persuaded management to buy it.
- The *Management trap*: managerial motives dominated the development of the approaches, such as efficiency, having an overview, having control, instead of usefulness for the work of the knowledge workers.

Of course the experiences were not only negative. Several companies appeared to benefit strongly from their strategy, and employees were quite satisfied. In these companies, the focus was not (only) on explication and storage of knowledge, but (also) on personal relations and communication. Knowledge processes in these organizations were much more dependent on interpersonal knowledge exchange and mutual learning through master–apprentice relations, and through certain functionaries, sometimes called knowledge brokers or intermediaries, who brought information to people and brought experts in contact with each other. An important role in this context appeared to be played by "communities of practice," loose networks of people who were interested in the same professional issue and exchanged their experiences.

The research project reflected the existence of, roughly speaking, two knowledge management strategies (Hansen *et al.*, 1999), (1) *a codification strategy* and, (2) *a personalization strategy*. The great advances made in the 1980s in information and communication technology persuaded management to develop a *codification strategy*. Knowledge processes were considered to benefit enormously from investing millions of dollars in "knowledge technologies." Procedures to elicit knowledge from employees, to convert it into a systematized form, and store it in company-wide repositories are the core activities of this strategy. Indeed, information storage, retrieval, and exchange can benefit strongly from digital information systems. As explained above, however, knowledge is often very implicit and tacit, it is built upon personal experiences and reflected in personal skills.

The codification approach with regards to *knowledge* has seen many failures. Unfortunately, many organizations believe that implementing a technical solution for capturing and transferring corporate knowledge is the only step required to make employees participate and use the system. However, there is often psychological resistance against using such a knowledge system. Of course, basic standard documents have to be available, preferably via the company intranet. When, however, people want to find solutions for specific problems they encounter in their work, they often have to search far and wide. Even then it is quite uncertain whether that piece of information is applicable to their situation. The "not invented here" notion often prevents people from searching for solutions in, for instance, repositories of best practices. Furthermore, certain knowledge, such as skills and insights, is often difficult to explicate. How do you write down the way you have painstakingly succeeded in solving a failure in a client's intranet, and the way you persuaded that client to use another classification system? You cannot and you do not want to do it, because you are already overburdened with the next project.

Finally, there may be psychological resistance against making this knowledge *impersonal* and removing it from its context. In such a case personal rewards are rarely provided and others may use your experiences without you knowing it. The "knowledge is power" idea determines your behavior, in this case meaning "knowledge *keeping* is power."

Another way of dealing with sharing, applying, and developing knowledge in organizations is what Hansen *et al.* (1999) called the "personalization strategy." In this strategy the focus is on the exchange of *tacit* knowledge and on interpersonal knowledge sharing through personal contacts, master–apprenticeship relations, meetings, and communities of practice. For the transfer of skills, competences, and insights, it is important to have social networks with other people. These networks generally will consist of people from different and geographically distributed units of the organization but often, also, from outside the organization. The "knowledge is power" idea here means "knowledge sharing is power," because it enhances your status and position when others consult you for your expertise.

The personalization strategy does not exclude the role of ICT. However, the emphasis lies not on databases and repositories of documents, but on applications that support the interaction between people. That means, providing "yellow paper" applications, i.e. databases concerning people and their experiences, their expertise, and e.g. the projects they have been working on. Such a system helps a person to find the experts they are looking for. Other relevant tools are the communication tools; email, instant messaging, videoconferences, and shared applications.

Situational learning and the learning organization

The theoretical basis for the personalization strategy can be found in the notions of the two related disciplines: that of social, or situational, learning, and that of the learning organization. Gaining of competences and certain skills can, of course, take place in schools and through systematic training courses. Developing real knowledge and insights, however, requires what is called "situational and social learning." Situational in the sense that it is learning from personal experiences in concrete situations, social in the sense that it occurs often in interaction with other, more experienced, people. People learn from each other, and particularly the new apprentices from the old experts, through looking over their shoulders. This type of learning is often supported by discussion of problems and solutions, through storytelling and actually showing each other how to solve certain problems.

An organization that views its future competitive advantage as based on continuous learning and use of knowledge and an ability to adapt its behavior to changing circumstances, is called a learning organization (CEN, 2004).

According to Nonaka and Takeuchi (1995) knowledge and information can be created and transferred in four ways or, better, in a spiral of four steps, which also reflect the various ways of learning (see Figure 14.1):

- *Socialization* of knowledge: "tacit" knowledge, such as skills and insights, is learned by others through direct interaction and imitation (= from tacit to tacit).
- *Externalization* of knowledge: tacit information is articulated and thereby made explicit, i.e. documentation. Examples of this type are the writing down of insights and storing them in a database, or extracting knowledge from experts and building an expert system (= from tacit to explicit).
- *Combination* of knowledge: combining pieces of knowledge into larger systems of integrated knowledge, such as a new method, theory, or model (= from explicit to explicit).
- *Internalization* of knowledge: a process of internalizing explicit information to personal insights and skills; this can be done through e.g. reading instructions or following courses and then learning by doing (= from explicit to tacit).

A smart combination of these systems gives the optimal way to transfer knowledge and learning. In this way organizations can learn from their past. But *can* organizations learn, like individuals? Do organizations have a memory, like individuals have? Lehner (2004) warns:

> the term "organizational memory" should in no way be considered analogous to a "brain" to which organizations have access. The term is simply meant to imply that the organization's employees, its written records, or data "contain" knowledge, that is readily accessible.
>
> (p. 54)

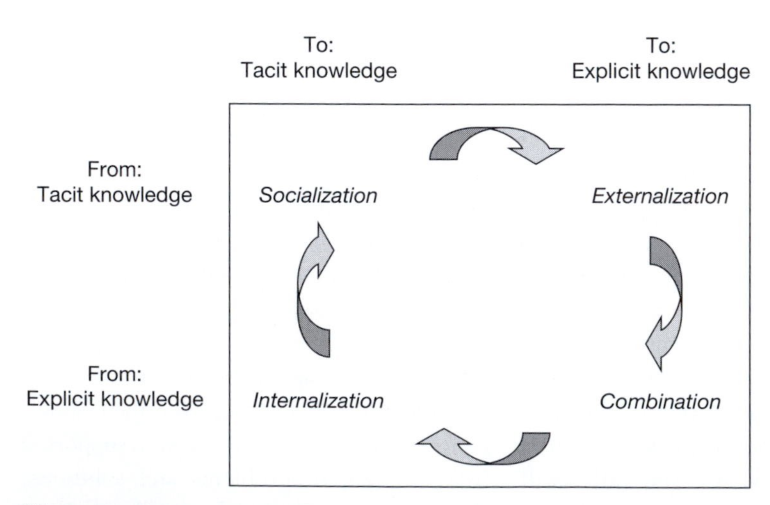

Figure 14.1 *Four ways of knowledge creation and transfer*
After Nonaka and Takeuchi (1995)

I would say that knowledge is also embedded in the way organizations work, in their structure or even workplace layout. However, the idea of organizational memory is only effective when it is considered in the same way as an individual memory. Human remembering is not a simple picking of rigid data from a fixed repository, but an active process whereby memory elements are reprocessed in relation to an actual setting. For instance, the aroma of coffee and croissants takes you back to Paris, which reminds you of a French colleague that you have to call. In the same way organizational memory only works effectively when it can be contextualized and adapted to the present setting (Bannon and Kuutti, 1996).

Organizational learning means that the solutions individuals have found for their problems are somehow shared and transferred to the organization. The question is how to transfer this individual learning. This may best be viewed as a circular process in which individuals, groups, and the organization are involved (see Figure 14.2). Individuals with certain skills are recruited by the organization. Internal employees learn from external sources such as clients and conferences (short incoming arrows at the top left of Figure 14.2) and from inside sources, i.e. from organizational knowledge systems and from personal activities. Individuals share their knowledge with others, which may result in shared (group)

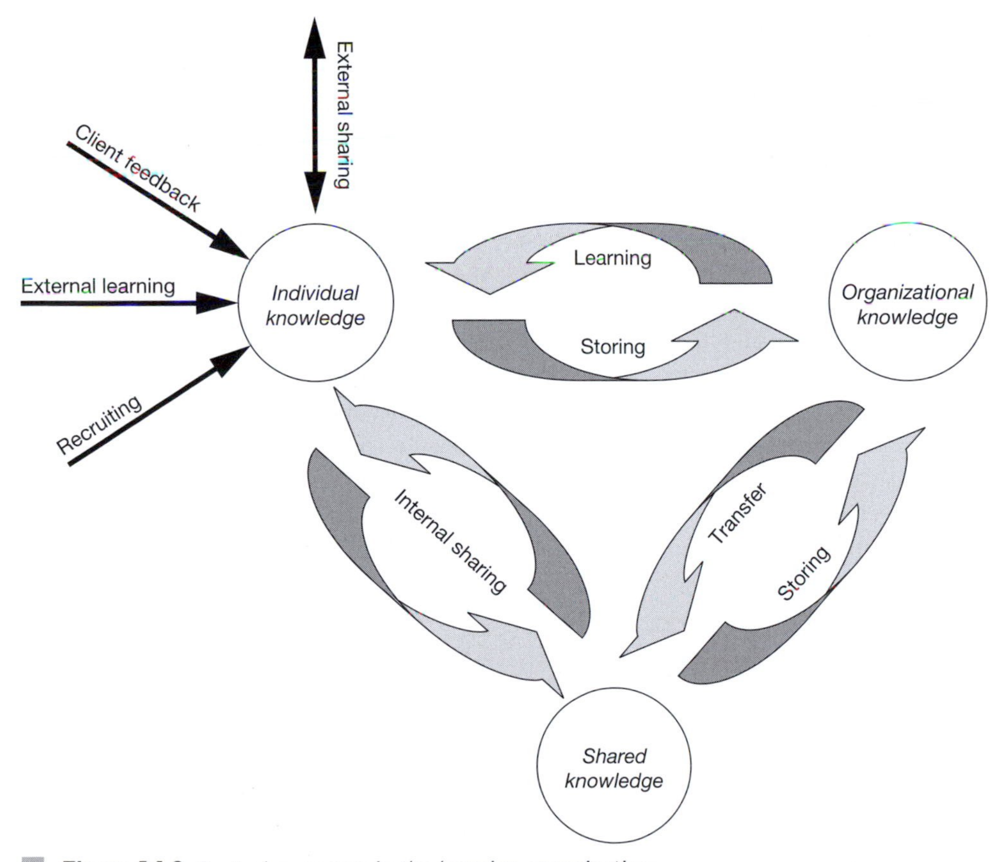

Figure 14.2 *Central processes in the learning organization*
From Andriessen *et al.* (2004)

knowledge. This can be explicit knowledge, exchanged through e.g. presentations and discussions in communities, or implicit, tacit knowledge exchanged through working together ("socialization").

The problem with communities and interpersonal sharing is that the transfer of what is learned remains limited to the few people involved. Elsewhere in the organization people cannot benefit from this knowledge, since the local knowledge is not "translated" into new organizational procedures and ways of working. When shared knowledge is accepted by the organization it becomes organizational knowledge, which is then available to be embedded in organizational practices and to be distributed again to individuals or groups. But there has to be a special agency to ensure that the experience becomes embedded in the organization (see also the learning theory of Argyris and Schön 1978).

These ideas suggest that communities and knowledge networks may have a double function, that of facilitating the interaction and learning of individual members, and that of bridging the gap between experience-sharing individuals and the organization. And, indeed, some organizations have communities that "translate" their members' knowledge into overviews of best practices. The tomato products community in Unilever even provided a set of requirements when a new tomato products plant was built in Africa.

A SYSTEMATIC STRATEGY

Knowledge management implies systematic attention to and facilitation of all the earlier mentioned processes and their supporting structures (see also Figure 14.2). This requires activities on three levels:

- developing a knowledge philosophy and strategy;
- managing the operational knowledge processes;
- organizing the structures for the relevant knowledge processes, and the conditions necessary for this purpose: functions, departments, procedures, a supporting culture, the technology, and the right human and social capital

Knowledge philosophy and strategy

Developing a knowledge management strategy is not a matter of hit and run, of picking a few measures haphazardly. It implies identifying the knowledge processes that have to be taken care of on the basis of a vision concerning the company's mission and then systematically developing a strategy to optimize the knowledge processes. In a previous section the distinction was made between a codification strategy and a personalization strategy. Hansen *et al.* (1999) suggest that an organization should not pursue a combination strategy, but should choose one strategy on the basis of the type of work processes and business strategy they have. If a company focuses on a rather standard type of work processes (they give Andersen Consulting as an example) it should opt for a codification strategy. In the case of customized solutions (e.g. McKinsey) the company should opt for a personalization strategy. Others (De Bruijn and De Neree tot Babberich, 2000; see also Chapter 17) accept that these two approaches imply competing values. However, they conclude that most large modern organizations are complex organizations that contain both standardized and customized processes. These companies therefore have to strike a balance between the two strategies and should be able to manage the tension between the two types of values involved. This can be done by giving attention to the various elements of a knowledge management strategy as discussed below

Operational knowledge processes

Several authors have proposed models that classify the knowledge processes into a limited number of categories, integrated in what may be called a "knowledge processes cycle" (CEN, 2004; Weggeman, 1997). Here I present the elements of the CEN model, in a slightly adapted version.

Identifying knowledge gaps

Discovering this gap implies comparing the knowledge that is required for the work to be done with the available knowledge. This sounds simple, but it is of course quite a challenge to find out what knowledge is needed, given the chosen business strategy, as much as it is a challenge to find out what knowledge is available, and where in the organization. This concerns not only knowledge about how to provide the products or services adequately, but also knowledge about markets and clients, knowledge about managing production processes and managing project teams.

Example: At Pink Roccade, a Dutch IT company, the HR department provided every new employee with a mentor. Twice a year the mentor and employee had a discussion in which the present knowledge of the employee was compared with the skills the company and the employee needed. Based on this comparison a training program for the employee was set up.

Acquiring and creating knowledge

Having determined where the knowledge gaps are, the required knowledge has to be acquired. This can be done in several ways. One is to bring it in from outside via consulting the clients, by recruiting people with new skills, through cooperating with research institutes, or by visiting fairs, meetings, and conferences. Creating new knowledge inside an organization may imply systematic problem solving, brainstorming, setting up expert groups, or doing research and development in specific departments; but new knowledge is also created in less formalized ways, i.e. during daily problem solving or during discussions in meetings or in communities of practice (see below). New knowledge may also be acquired through analyzing large databases. High capacity computers have made this possible. "Data mining" of customer data has allowed companies to discover buying patterns, or subgroups of customers for which specific campaigns may be developed.

Storing knowledge

Considerable knowledge is stored in people's brains. This can be very explicit but is often quite tacit. Such knowledge is stored in the form of skills, insights, and professional expertise. It is also stored in working routines, such as in the ways people have learned to do their tasks, or in machines and layouts of physical production departments. Other knowledge can be stored in information systems, although the problems with this approach have been discussed in earlier sections. It may therefore be better to store (also) easily accessible information about who is an expert in what areas and topics. These "Yellow Pages" provide co-workers with the possibility to contact other people and to ask them for advise in solving certain problems or in finding documents. TNO, a large independent Dutch research organization, has developed a system that provides this information. It is quite sophisticated in the sense that it systematically connects information on people (demographics) and their expertise and competences, with information about projects, organizational units, and documents. It is thereby possible to search for a certain person and then identify what department they are in, what expertise they have, what projects they have done, and what documents they have produced at both the group and individual level. It is, however, also possible to start with an area of expertise and discover in which departments this expertise is available, and in what projects this is being, or has been, applied. The system allows for any other "walk" through the data and the linking of all the elements. At the Philips Center for Production Technology, employees interview their colleagues to collect their knowledge, after which it is documented, structured, and entered in a knowledge base that is accessible for all employees.

Distributing/sharing knowledge

Knowledge can be shared by sending documents or digital memos around, in the hope that the readers will, indeed, share its content. Practical knowledge, as distinguished from information, can, however, best be shared through interpersonal contacts, because two-way communication is needed to share context and insights. Four types of mechanisms for this purpose can be found: (1) *meetings*; seminars, workshops and the like; (2) *master–apprentice* models, in which a newcomer accompanies an old-timer and learns from him or her the tricks of the trade; (3) *communities of practice*; contacts between people who are interested in sharing knowledge concerning a particular topic (see below); and (4) *knowledge intermediaries*; specialists whose task it is to collect knowledge, bring it to the attention of people who may need it, and also bring people with a problem into contact with others who might help them solve the problem using their expertise. Technical devices such as digital Yellow Pages or communication tools can be used to support these interpersonal interactions.

Applying knowledge

The aim of all the above is to use knowledge for value creation. When applying knowledge for work processes, it will sometimes become clear that there is a lack of some knowledge and that it cannot be found in the organization. Once such a knowledge gap is identified the cycle begins again.

THE CONDITIONS: STRUCTURE, CULTURE, PEOPLE, AND TECHNOLOGY

What is needed to make knowledge management a success? What conditions will support sharing and storing, developing, and using knowledge in such a way that better products and services, innovation, and client satisfaction will follow? The results of a European survey (Heisig and Vorbeck, 2001) point to the importance of the following conditions. Of the companies surveyed 47 percent mentioned corporate culture as one of the main enablers for good knowledge processes, 30 percent mentioned structures and processes, 28 percent cited information technology, 28 percent skills and motivation, and 27 percent management support. Some of these conditions have been discussed briefly in previous sections. The main groups of conditions will be presented below, but first a comprehensive model of all the factors involved will be given in Figure 14.3.

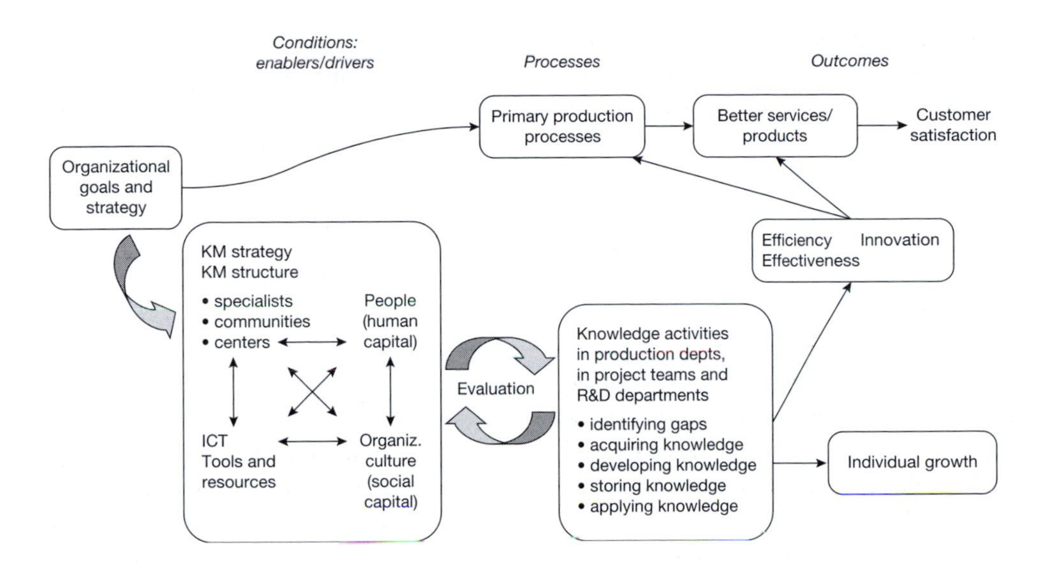

Figure 14.3 *A comprehensive knowledge management model*

Figure 14.3 represents the idea that the primary production processes and (product and process) innovation are supported by the activities in the knowledge process cycle (the box in the middle of the figure). The quality of these knowledge processes is determined by four clusters of conditions, i.e. the knowledge management strategy and structures, the norms, values, i.e. the culture with regard to knowledge sharing, the capacities, networks, and attitudes of the people involved, and the technological support provided by the organization.

The left box contains *enablers* – such as specialists, tools, etc. – and *drivers*, that is, people's goals, motivations and interests, and organizational goals. The arrows in this box indicate that it is the *interaction* of these factors that determines the knowledge activities. This interaction refers, for instance, to the question what, in a certain context, the optimal division is between human and technical support of certain processes. By evaluating the knowledge processes, an organization can learn where and how to adapt the conditions.

Knowledge management can now be defined more precisely as follows: Knowledge management is the planning, organizing, and managing of the knowledge processes and its individual, structural, cultural, and technological conditions in such a way that realization of the organization's objectives and strategy is advanced.

The various conditions will be discussed briefly below.

Structures and procedures

Corporate structures and procedures can be found, first, in the existence of certain functionaries and groups or centers, such as a Corporate Knowledge Officer, knowledge intermediaries, expertise centers, and communities of practice (see below). Second, procedures such as reward systems play an important role. As far as reward systems are concerned, it is strongly debated whether, and to what extent, people can be motivated to share knowledge by providing extrinsic incentives such as the possibility to attend conferences. *Siemens ICN* is known for its ShareNet initiative, a global collaboration and knowledge-sharing network for its sales force. Contributions to ShareNet are rewarded with ShareNet "Shares." Peer ratings assess the quality and (re)usability of the contributions people make. Siemens not only rewards the contributors, but also re-users of ShareNet content. The "Shares," which can be compared to air miles, can be exchanged for real (Siemens) products. Besides this, top ShareNet contributors are rewarded with an invitation to attend the Siemens ShareNet global knowledge-sharing conference.

It is argued that monetary rewards only have a temporary effect on knowledge sharing (APQC, 1999). This means that these incentives have to be given repeatedly and that the effect disappears when the rewards are no longer given. Tangible rewards may only provide temporary compliance, and may inhibit organizational learning. This type of reward may also stimulate unwanted behavior, i.e. distributing information of inferior quality, only with the intention of receiving a reward. So, most organizations expect knowledge sharing to be stimulated mainly by the general culture and by integrating it in the normal work process, i.e. by making knowledge sharing part of an annual performance evaluation.

Corporate culture

The organizational culture of companies in general, and of knowledge sharing in particular, can, perhaps, be found in official documents, but can be deduced better from existing practices. Nonaka and Takeuchi (1995) illustrate the difference between knowledge-related cultures by comparing those at General Electric and Honda (see Table 14.1). In a Honda type of culture, employees are much more willing to share, use and create knowledge than in a General Electric type of culture.

Human and social capital

"Human capital" refers to the knowledge and skills, attitudes, and motivation of individual employees. These human characteristics are relevant for knowledge sharing in the sense that both willingness and ability to share are a prerequisite for adequate knowledge

Table 14.1 *Honda versus General Electric*

General Electric	*Honda*
• People live to compete	• People live to create
• Winning means everything	• Life should be a joyful experience
• People are difficult to change	• The potential of people is limitless
• Knowledge production is based on comparison: are we better than the others?	• Everything is based on finding our purpose: Why do I exist? What is the core concept of our work? Why do we want to do this?
• External motivation of employees (extrinsic rewards)	• Internal motivation of employees (intrinsic rewards)
• Goal: to be number 1	• Goal: to serve the customer
• Relation to knowledge: exploitation, i.e. using existing knowledge	• Relation to knowledge: exploration, i.e. building constantly new knowledge

sharing. "Social capital" refers to the relations between the employees. It is reflected in three primary dimensions (Nahapiet and Ghoshal, 1998), i.e. in the number and intensity of existence of connections and networks among the employees; second, in the degree of mutual trust between employees; and, third, in the development of a common language of knowledge and norms for the employees involved. Companies are assumed to thrive better and to be more effective and innovative when they have sufficient social capital (Prusak and Cohen, 2001).

Technology support

Guide, a Swedish IT consultant agency, was founded in 1988. By 1994 it had 133 employees and two offices, and five years later it had 800 employees and 10 offices, in several countries. This enormous growth required the development of basic administrative and personnel information systems, and of overviews of project management data, of existing expertises, and of media for communication and cooperation between dispersed employees. The company developed several knowledge management systems in the course of the years. In 1995 it started with a database that listed the type of training and competences of the consultants. In the next year it developed a system that provided management with an overview of the availability of consultants for projects. In 1998 the company got an intranet and this gave general accessibility to a database with personal information about the consultants.

Then the office in Gothenburg (Sweden) developed a more advanced system for systematically assigning jobs to the consultants on the basis of their availability and competences. This system was, however, not accessible in the other offices and job assignment was done largely manually by project managers. The Gothenburg system was then introduced in a few other offices, on a Lotus Notes platform. The problem arose that this platform was not compatible with the other platforms used by Guide, and this took quite a while to amend. Later this system was augmented with visualization tools. Finally, a community oriented website was added, providing articles, discussion lists, co-writing tools, and working spaces for communities of practice.

This example illustrates the development path many organizations have to follow, and the various types and generations of information and knowledge tools that they will use on the way. It also hints at the complexity and difficulties involved in the development and introduction of such systems in a large and dispersed organization. Technical tools to support knowledge processes can be divided into four categories (Jashapara, 2004):

- *Knowledge capturing tools*, search engines: tools to make tacit knowledge explicit such as cognitive mapping tools, to visualize the associations around a concept, and expert systems, which contain advise in a particular domain based on the experience of practitioners, e.g. concerning medical illnesses, or paint making.
- *Knowledge storing tools*: all kinds of repositories.
- *Knowledge analyzing tools*: tools for datamining, online analytical reasoning.
- *Knowledge sharing tools*: Internet and intranets, skills directories (expertise yellow pages) and groupware.

The term "groupware" refers to those ICT applications that support communication, coordination, cooperation, learning, and social encounters in dispersed groups of people. *Communication* tools make the interaction between geographically distributed people possible. Both *asynchronous message systems* (email, computer conferencing) and *synchronous communication systems* (telephone, video, sms, shared applications) serve this function. The characteristics that make asynchronous message systems attractive are the time they allow for considering a reply, the distribution list facility and the systematic storage and processing facilities. *Collaboration tools* improve teamwork by providing document-sharing or co-authoring facilities. A subgroup consists of *Group decision support systems*, to support brainstorming, evaluating ideas, and decision making. *Coordination tools* support the coordination of distributed teamwork. Group calendars and workflow management systems are the best-known tools of this type.

Communities of practice

Knowledge sharing and developing can take place between separate individuals, but also in groups. Specific groups, aimed exclusively at knowledge sharing and developing, are often called *Communities of Practice* (CoPs). CoPs form a framework, by which explicit and tacit knowledge can be shared, for the purpose of solving individual or common problems, or for developing new knowledge. Of course, communities are not the only vehicles used for the transfer of knowledge. Ad hoc meetings and encounters, master–apprentice relations or knowledge intermediaries may also realize this objective. But in the last few years many modern organizations, and networks of organizations, have been experimenting more or less formally with CoPs. These communities are different from project teams and can be defined as follows: CoPs are networks within or between organizations (not Internet communities), whose members work organizationally and geographically dispersed, are focused on exchange and development of knowledge (not on a planned product) in a specific domain ("practice"), having a certain level of group identity. Such communities can be found in many large companies, but their forms and functions are very diverse. Examples are the CoPs in AtosOrigin, an international company that provides ICT services including consultancy for implementation and system integration (Andriessen *et al.*, 2004). Two types of communities can be found within AtosOrigin, known in the company as (regional) "Expertise Groups" and (national) "Networks of Performance."

Expertise Groups (EGs) are initiatives within several regional sections of the AtosOrigin enterprise. They consist of consultants working in this section, who exchange experiences concerning a work-related topic. Examples of EGs are those focusing on Oracle databases, on Microsoft software, or on Java programming, but also those concerning project management. The members come together face to face about once a month. During the meetings people give presentations and talk about projects, sharing literature and feedback on training sessions and courses attended. They have a formally appointed coordinator. Most groups also have social events to support the process of building group identity and trust. The use of ICT is limited since the main interactions take place during the monthly meetings.

Networks of Professionals (NoPs) are bottom-up networks of company employees, distributed nationwide. The goal of NoPs is to exchange existing knowledge and to create new knowledge. The topics of the NoPs are to some extent parallel to those of the expertise groups. The members interact over the company intranet, but some have face-to-face meetings once in a while. NoPs have to have at least five members to be granted official status and be given company backing and support. There are no rules relating to membership or functioning of the community. Individuals may be members of more than one such group.

It appears that CoPs can have several functions. CoPs can be useful for the organization, because they may develop best practices, result in more efficient work methods or develop innovative ideas. Individual members can find solutions for their work problems, can increase their general knowledge, and can find useful contacts. But CoPs can also have a social function. Many organizations, in their search for a viable knowledge management strategy, and inspired by concepts of organizational learning, now opt for a "personalization strategy." They expect CoPs to accomplish what knowledge technology has failed to do, i.e. sharing tacit insights, skills, and knowledge and "translating these in new solutions and initiatives." Many types of CoPs can be found, differing in purpose, size, formality, and cohesion, and the precise success conditions for each of these types have yet to be established. Nevertheless, the success of this type of knowledge sharing and knowledge creation is widespread and generally acknowledged.

KNOWLEDGE MANAGEMENT AND INNOVATION

To what extent are the various knowledge management practices, discussed in the previous sections, to be found in organizations and what is their influence on innovation? Few systematic and generalizable studies have been done. Some large surveys may shed light on these questions. One is a national Canadian survey (Earl, 2003) on the use of knowledge management practices by firms in a variety of sectors. Knowledge management was presented in terms of 23 practices grouped under six headings: Policies and Strategies; Leadership; Incentives; Knowledge Capture and Acquisition; Training and Mentoring; and Communications.

On average, firms in all five sectors used about 11 knowledge management practices, just under half of the practices listed. The average number of practices used increased with firm size. Almost every firm (94 percent) ascribed the responsibility for their knowledge management practices to managers or executives and this most frequently used practice shows the importance of leadership for knowledge management. Capturing and using knowledge obtained from other industry sources such as industrial associations, competitors, clients, and suppliers ranked second (92 percent). The third and fourth most popular practices, both fell under training and mentoring. Four-fifths of knowledge management practitioners encouraged experienced workers to transfer their knowledge to new or less experienced workers and provided informal training related to knowledge management. These results give a reasonably positive picture of the knowledge management activities in these companies, although one has to take into account the fact that the respondent in each company could have been inclined to present the situation in its most favorable light.

According to a recent working paper based on the French Third Community Innovation Survey (Kremp and Mairesse, 2004) knowledge management practices are especially found in large firms and in the high and medium-high tech industries, such as the pharmaceutical industry, aeronautic and space construction, or electronic component manufacturing. Knowledge management policies appeared to be more frequent in firms that implemented new management methods, such as project-based management, in firms making high R&D investments, and in firms that use the Internet and ICT to acquire and share information. This last result suggests that knowledge management is part of a general strategy towards making use of modern and innovative practices.

Concerning the relation between knowledge management and innovation the same paper reported that there is a correlation between certain general indications of knowledge management intensity and the innovation in firms, in terms of number of innovative products or patents. The data are, however, quite vague and superficial and generalizable conclusions concerning the impact of knowledge management practices on the innovative capacity of companies are not well supported.

The term "innovation" may refer to widely diverse issues, such as the development of innovative ideas, or the number of patents, the renewal of internal production processes, or the diffusion of new products in the market. All these processes may have different determinants and are determined by many other factors than knowledge management practices. The development of new products, particularly in large high tech firms, may depend basically on the investments made in, and creativity of, the firms' R&D departments. In other cases, such as consulting firms, innovation is indeed the fruit of the work of many employees, dispersed throughout the organization. In this latter case a general knowledge management strategy aimed at creating and sharing new ideas may be very suitable for innovation.

Knowledge management is not only focused on innovation, because its objective may also be to support the better use and exchange of existing knowledge and to preserve existing knowledge in an organization, despite high employee turnover. This may be particularly important for reducing cycle times, achieving shorter time-to-market and cost reductions.

DILEMMAS AND SOLUTIONS

The exposé above has shown that knowledge management is accepted in many organizations as a crucial part of managing technology and as a condition for achieving a competitive edge. Strategies and structures for knowledge management are being developed, although the art of managing knowledge processes is still in its infancy. The field needs research and experimentation, particularly because a number of central dilemmas still have to be solved.

The first one is the organizational dilemma between the need to systematize knowledge into codified information for wider accessibility and standardization and the fact that learning and sharing situational knowledge can only be done via personal contact. In this chapter the opportunities and problems of the two sides of the knowledge management coin have been discussed repeatedly. The general solution lies in combining the two, which seems to be self-evident, and which implies that the maintenance and use of codified knowledge should be linked to expert support, something that is very difficult to organize in practice.

Another dilemma is the one individuals face when they have to choose between the short-term demands of their daily tasks and the long-term need for developing their knowledge. This dilemma is less for the case of trying to find knowledge in databases or finding an expert with the required knowledge, to solve an immediate work problem. However, in many cases professionals, individually or in project teams, are under such time pressure that they simply do not find the time, either to reflect as a group on how they are doing, or to spend time on knowledge-sharing activities that may, but one can never be certain, in due time improve their work. If management wants to change this situation, it can only be done by a combination of a change in management philosophy and by changing its structures and procedures used to support knowledge sharing.

A third problem results from the globalization of many organizations. This implies that cooperation and knowledge exchange by employees may be hampered by diversity in language proficiency and by cultural differences in perspectives and interaction habits. Although this appears particularly tricky for personalized knowledge sharing and communities, the views on codifying may also diverge.

Finally, a dilemma is found between the need for company-wide, or even inter-company exchange of knowledge versus the requirement for security. There is a cartoon that bears the following text: "We don't pay much attention to information security. We are hoping our competitors will steal our ideas and become as unsuccessful as we are." In many organizations, however, it is crucial to protect critical knowledge from spying eyes. This dilemma is reflected in two separate challenges. One is related to the accessibility of codified information. It is crucial that databases and communication are protected from hackers, viruses, or competitors. At the same time this may imply that certain information or contacts will not be accessible for part of the personnel of the organization, e.g. because they are working at clients' sites. This problem can be solved to a certain degree by technological means.

The other challenge is one of being able to cooperate with partners in an alliance, without giving away company-sensitive knowledge (e.g. Soekijad and Andriessen, 2004). Many organizations cooperate in vertical (e.g. producer–suppliers) or horizontal networks. This situation of so-called co-opetition can be found in networks of competitors that cooperate for the sake of e.g. sharing R&D costs or exchanging knowledge in new areas. Managing successful knowledge sharing under these circumstances is one of the main challenges of the future.

FURTHER READING

Awad, E.M. and Ghaziri, H.M. (2004). *Knowledge Management*. Upper Saddle River, NJ: Pearson Education. A recent textbook on knowledge management with particular emphasis on the technical and codification side.

Hansen, M.T., Nohria, N., and Tierney, T. (1999). What's your strategy for managing knowledge? *Harvard Business Review*, 77(2), 106–116. The classical article on the difference between the codification and the personalization strategies of knowledge management.

Huysman, M. and Wit, D. de (2002). *Knowledge Sharing in Practice*. Dordrecht, the Netherlands: Kluwer. An interesting study of knowledge management programs in ten large Dutch and multinational companies.

Jashapara, A. (2004). *Knowledge Management. An Integrated Approach*. Harlow: Prentice Hall. A recent textbook on knowledge management with particular emphasis on the organizational and personalization side.

REFERENCES

Andriessen, J.H.E., Huis in't Veld, M., and Soekijad, M. (2004). Communities of Practice for Knowledge Sharing. In J.H.E. Andriessen and B. Fahlbruch (Eds), *How to Manage Experience Sharing: From Organizational Surprises to Organizational Knowledge*. Oxford: Elsevier Science Ltd.

APQC (1999). *Creating a Knowledge-Sharing Culture*. Houston, TX: American Productivity and Quality Center.

Argyris, C. and Schön, D.A. (1978). *Organisational Learning: Theory, Method and Practice*. Amsterdam: Addison-Wesley.

Awad, E.M. and Ghaziri, H.M. (2004). *Knowledge Management*. Upper Saddle River, NJ: Pearson Education.

Bannon, L.J. and Kuutti, K. (1996). *Shifting Perspectives on Organizational Memory: From Storage to Active Remembering*. Paper presented at the 29th Annual Hawaii International Conference on System Science, Maui, HI.

Bruijn, H. de and Neree tot Babberich, C. de (2000). *Competing Values in Knowledge Management*. Utrecht: Lemma.

CEN, European Committee for Standardisation (2004). *European Guide to Good Practice in Knowledge Management*.

Davenport, T.H. and Lawrence, P. (1997). *Working Knowledge: How Organizations Manage What They Know*. Boston, MA: Harvard Business School Press.

Earl, L. (2003). *Knowledge Management in Practice in Canada, 2001* (No. 88F0006XIE No. 07).

Hansen, M.T., Nohria, N., and Tierney, T. (1999). What's your strategy for managing knowledge? *Harvard Business Review*, 77(2), 106–116.

Heisig, P. and Vorbeck, J. (2001). Benchmarking Survey Results. In K. Mertins, P. Heisig, and J. Vorbeck (Eds), *Knowledge Management. Best Practice in Europe*. Berlin, Heidelberg: Springer-Verlag.

Huysman, M. and Wit, D. de (2002). *Knowledge Sharing in Practice*. Dordrecht: Kluwer.

Jashapara, A. (2004). *Knowledge Management. An Integrated Approach*. Harlow: Prentice Hall.

Kremp, E. and Mairesse, J. (2004). *Knowledge Management, Innovation and Productivity: A Firm Level Exploration based on French Manufacturing CIS3 Data* (No. Working Paper 10237). Cambridge, MA: National Bureau of Economic Research.

Krogh, G. von, Nonaka, I., and Aben, M. (2001). Making the Most of Your Company's Knowledge: A Strategic Framework. *Long Range Planning*, 3, 421–439.

Lehner, F. (2004). Organisational Memory. In J.H.E. Andriessen and B. Fahlbruch (Eds), *How to Manage Experience Sharing: From Organizational Surprises to Organizational Knowledge*. Oxford: Elsevier Science Ltd.

Nahapiet, J. and Ghoshal, S. (1998). Social Capital, Intellectual Capital, and the Organizational Advantage. *Academy of Management Review*, 23(2), 119–157.

Nonaka, I. and Takeuchi, H. (1995). *The Knowledge Creating Company*. Oxford: Oxford University Press.

Prusak, L. and Cohen, D. (2001). How to Invest in Social Capital. *Harvard Business Review*, 79(3), 86–93.

Senge, P. (1990). *The Fifth Discipline: The Art and Practice of the Learning Organization*. New York: Random House.

Soekijad, M. and Andriessen, J.H.E. (2004). Conditions for Knowledge Sharing Competitive Alliances. *European Management Journal*, 21(5), 578–587.

Sonnentag, S. (2004). Task Orientation Matters: Knowledge Management from an Individual Level Perspective. In J.H. Erik. Andriessen and B. Fahlbruch (Eds), *How to Manage Experience Sharing. From Organizational Surprises to Organizational Knowledge*. Oxford: Elsevier Science Ltd.

Weggeman, M. (1997). Cultuur en managementstijl in kennisintensieve organisaties. *Holland/Belgium Management Review*, 54: 62–72.

Part V

Dilemmas and strategies

INTRODUCTION

The book concludes with a section on dilemmas that managers operating in a technology context are faced with. The idea behind this section is that the manager is confronted with many developments, both inside and outside the organization. Managers seek to respond adequately to all these developments, whether it is globalization, the development of new technologies, new legal requirements affecting financial management or the production line, or demographical changes resulting in the need for different products.

In their response to all kinds of developments, managers experience that different tendencies ask for conflicting responses: an action that is an adequate answer to one development might have an adverse effect on a different part of the organization. An example can illustrate this point. From the viewpoint of the design of technological firms, it might be an excellent idea to make the organization more transparent in order to standardize procedures, and to control the entire organization. With a standardized and quantifiable output measurement system, the whole organization becomes easier to manage. For innovation, however, creativity is needed. Unfortunately, creativity processes do not go well with control and standardized procedures. It is precisely in the unusual that innovations develop. In other words: employees and parts within the organization need some freedom in order to be able to innovate. How to control the uncontrollable? What should CEOs do in this case? Should they exert control and standardize procedures because otherwise the organization will be out of control? Or should they trust the R&D departments to come up with something good in the end, and take the risk that it will be a financial disaster or that R&D develops a product that the market does not ask for?

This last section of the book focuses on the question how managers deal with such dilemmas. The dilemmas have their origin both in developments outside the organization (e.g. globalization and liberalization) and within the organization (e.g. conflicting requirements of different parts within the organization). Managers have to respond to all these developments. How do they weigh the impact of their actions on different parts of the organization?

The first three chapters of this part describe the theory behind three major dilemmas. Chapter 15 describes the dilemmas around innovation (Peer Ederer). In Chapter 16, Michel van Eeten and his colleagues describe how managers can maintain reliability in circumstances that are characterized by a great deal of uncertainty. Chapter 17 by Hans de Bruijn analyzes the dilemmas around the use of performance management.

The final chapter of the book reports on a number of interviews with managers, who deal with the management of technology and innovation on a daily basis. Willemijn Dicke interviewed six senior managers for the particular purpose of the book: a recruitment manager R&D (Unilever), a senior strategist (Gasunie), a global business manager CO_2 (Shell), an innovation manager (Siemens), a manager of the network operating center (KPN Mobile), and an open source developer/CTO of a Software Solutions Company. These managers tell in their own words about the particular dilemmas of their daily work and what their strategies are to tackle these problems and ambiguities of technology and innovation management.

Chapter 15

Making the impossible possible

Controlling innovation

Peer Ederer

OVERVIEW

This chapter describes an organizational solution to the special challenges involved in organizing and controlling innovation processes in a company. The specific challenge arises from the fact that innovative practices inherently require various kinds of freedoms for approaching the unknown, whereas any form of organization inherently tries to restrict freedom in order to create control. The chapter suggests that the management literature shows four fundamental principles of how resources of a company can be steered and controlled towards a desired strategic target. A careful consideration and application of a proper mix of these four principles can help the innovation manager to overcome the paradox of control and freedom. Two case examples are included to illustrate how different mixes have been successfully used in innovation-driven companies.

INTRODUCTION

An excellent research organization is always slightly out of control

Dr Fopke Klok, Head of Philips Research, October 15, 2003 in Amsterdam

In the Oxford Dictionary, technology is defined as "the application of scientific knowledge for practical purposes." Even if not always, still much of that "application" requires or leads to making changes in existing products, processes or methods – or in other words technology typically requires or leads to innovation. Several of the major trends in technology discussed in this book so far, such as technology development becoming ever more complicated and expensive, that companies specialize ever further, or that product life cycles become shorter and shorter, are the effects of innovation through technology. Therefore, technology management and innovation management are usually tightly linked – and hence technology managers typically inherit a special challenge of innovation management: namely, how to control the uncontrollable.

Innovation introduces this challenge because, by definition, innovation is concerned with the new, the changing or the unknown. Since the new is not yet known, it cannot yet be harnessed, it cannot yet be tied down and tamed, and, hence, the innovative processes used for creating the "new" are inherently uncontrollable (Eisenhardt and Brown, 1998). However, to a manager that is not acceptable. Managers are members of an organization, and it is the purpose of the organization to collectively achieve certain targets or perform various functions – otherwise there would not be a point to their existence. Organizations cannot achieve targets, unless they have control over all their resources, including those busy with innovation (Picot *et al.*, 2003). Organizations and managers must therefore also achieve control over their innovative process by linking the abilities of the whole of the company to the utilization of its innovation processes (Iansiti, 1998).

The central dilemma or paradox of this chapter is: the challenge of controlling the uncontrollable, or the paradox of freedom v. control. A paradox is a situation in which two seemingly contradictory factors appear to be true or necessary at the same time (Quinn and Cameron, 1988). A problem that is a paradox has no real solution, as there is no way to logically integrate the two opposite sides into an internally consistent understanding of the problem. Dealing with paradoxes is at the core of strategy management (De Wit and Meyer, 2004). It also applies to the issue of innovation management. This chapter will not, and cannot, provide a recipe for how to dissolve the paradox – it does aim, however, to offer to the manager of technology some of the most prominent armory available in management sciences for how to overcome the paradox of freedom and control inherent to innovation and use it for his competitive advantage. As the quote by the head of Philips R&D, Dr Klok, illustrates, a technology organization cannot be fully under control, but it cannot be out of control either – being "slightly out of control" illustrates the necessity to manage a paradox.

Achieving control is a function of exerting power over a particular organizational resource, (which are typically people), so that the object experiencing this power accomplishes or reaches a predetermined desired target. Achieving control is not about setting these targets, nor is it about recruiting the resources that are supposed to achieve the target. The designer of the control mechanism takes both the target and the resources for granted. The task is to install organizational mechanisms to guide or control the given resources towards the given targets.

In practical terms of technology management, examples for this necessity to install control mechanisms in an organization could be how to organize a group of engineers who are charged with developing a new product, or a process for how to decide the allocation of resources between competing research projects. All too often, the more innovative a process becomes, the more it is assumed that organizations can actually not provide a controlling guiding function towards the resources employed in this function. The stereotypes of the "techies" in the R&D departments, or its "flippy" scientists come to mind, those people who live in an organizational world on their own, and whose creativity may not be disturbed lest their innovative power be reduced.

Other stereotypes stemming from the same helplessness of how to control the innovation process, include the "rebel entrepreneur" who by breaking boundaries of convention can create innovations more effectively, or the "mad genius" loner, who in mysterious ways is cooking up the next technological breakthrough. However, organizational management in innovation has come much further than these stereotypes would imply. Two short case studies about successful innovation-driven companies illustrate two very different organizational control systems for dealing with the paradox of control v. freedom in innovation and technology management. First, the actions by Carlos Ghosn in the turnaround story of Nissan, and second the organization created by Ricardo Semler at Semco.

Throughout this chapter, the term "resources" is used to include the following four main resources available to innovation (and technology) management:

- people with technology skills (researchers, developers, engineers);
- formalized intellectual properties (patents, copyrights);
- machineries (as an already existing application of a technology);
- financial resources (usually necessary for investment, because innovation tends to be an investment activity).

In the sense that the intellectual properties, the machineries, and even the money, must be transacted and handled by people, one could say that there is only one resource that needs to be controlled, and that is the people and their actions. If this distinction matters, the word "people" is used instead of "resources" in the remainder of the text.

FOUR SOURCES OF POWER FOR ACHIEVING CONTROL

It is argued in this chapter that there are four principal sources of power for achieving control. The following will outline in detail what these are, how they function, what their main advantage and disadvantage is, and what measures prominent management thought leaders are suggesting on how to implement them in an organization. The four powers are:

1. *The power of rules*: Achieving control with rules is a matter of defining the right set of rules and regulations. If the resources are subjected to the right rules and regulations, along with proper incentives and punishments for following or not following them, then resources will be controlled towards achieving the aspired target.
2. *The power of leadership*: Achieving control with leadership is to make the resources follow a leader or a leading principle. In both cases, either the leader or the leading principle inspires or guides the resources towards achieving the aspired target.
3. *The power of competition*: Achieving control through competition is a function of creating markets where competition allocates resources towards the achievement of the aspired target.
4. *The power of complexity*: Achieving control through complexity is a function of creating a community of self-organization that can self-regulate itself towards achievement of the aspired target.

To state the major conclusion from this chapter upfront: none of these four powers can be, or even should be, employed on an exclusive basis. On the contrary, it is probably the carefully selected blend of all of them that will create the best results of achieving control – in particular achieving control over innovation processes in organizations. Nonetheless, managers will usually have strongly predisposed preferences to use one source of power over the others. In some cases these predispositions are based on strong belief of superiority in one particular source, in other cases these predispositions rest on more familiarity and practice with one source of power over the other. However, these managerial biases apart, from an empirical point of view it is impossible to ascertain the general superiority of one source of power over the other, or even to correlate the suitability of one in certain situations over the other. All four sources of power are represented by schools of thought with long and deep histories of scholarship as well as long traditions of successful application in practice.

The power of rules

Among the four principal powers for control mechanisms, rules are possibly the most frequently used to achieve control over resources. A hypothetical innovation-oriented company called "Inova" might enact a rule that says: "work at Inova company starts at 8:30 am." Then all the human resources of Inova are due to start work at that time. The employees at Inova have thus lost their freedom to start working whenever they are ready for it in the morning – instead, their work starting time is under control.

The key advantage of rules is that within the parameters set by the rule, they have universal and timeless validity. The example "work at Inova company starts at 8:30 am" is true for all Inova employees, for all Inova locations and for all times, until or unless a new rule is made. With relatively small effort, rules can therefore achieve a very high degree of control. The main disadvantage is that this very same universality of rules can make them very costly or annoying to obey. Imagine the cafeteria worker at Inova Company, who would have to arrive at 8:30 am in the morning, even though his work only really starts at 11:00 am.

The challenge of rule making is to make the rules be obeyed and be as useful as possible. Typically rules are therefore accompanied by a regime of rewards and punishments for following them, because that makes rules more useful for the ruled. A prominent field of study in the social sciences has always been why people follow rules. One of the more recent and most influential explanatory models in practical management application has been developed by Michael C. Jensen from Harvard University. In one of his classic papers "The Nature of Man," (1994, p. 19), Jensen concludes:

> Whether they are politicians, managers, academics, professionals, philanthropists or factory workers: individuals are resourceful, evaluative maximizers (REMs). They respond creatively to the opportunities the environment presents to them, and they work to loosen the constraints that prevent them from doing what they wish to do. They care about not only money, but almost everything – respect, honor, power, love, and the welfare of others. The challenge for our society, and for all organizations in it, is to establish rules of the game and educational procedures that tap and direct the creative energy of REMs in ways that increase the effective use of our scarce resources.

For Jensen, controlling resources towards a certain desired outcome is like solving a mathematical equation, even if one of enormous difficulty. The difficulty stems from the nature of Jensen's "REM" people. According to him, these REMs are far from simple-minded creatures, they are instead "Resourceful." They are so resourceful in fact that it is likely to be a continuous struggle to invent and design the right mix of rules, rewards and sanctions for making people do what is expected of them. People will also find ways to game the system towards their increased personal benefit and to the contrary of the intended outcome.

In addition, REM people are "Evaluative," in contrast to being for instance merely calculative. They barely ever calculate the net present value of rewards and sanctions, but instead make their judgment based on values. These values could be, and often are, numerically expressible economic values, (such as comparing gasoline prices), but they will also relate to much fuzzier value sets like religion, morals, recognition, respect or simply the mood of the day. Moreover, people are unpredictable in which value they are going to use when judging a course of action.

Finally, whichever value people use, they will then want to "Maximize" its outcome. In contrast to what might often be stated by people, in practice they will want to have not only more love, but as much love as possible, not only more money, but as much money as possible, etc. According to Jensen, in practice people do not draw a line where they realize that enough might be enough – instead, even if they have reached such a line, they will then want to have even more.

In summary, rules are an important design ingredient for the innovation manager trying to achieve control over the innovation process, because rules can create a lot of control with only *small effort*. The key criteria to watch out for when making rules is to respect people's ability to be resourceful about rules, to be evaluative about them, and to realize that people will usually want to maximize their benefit.

The power of leadership

Leadership may well be the oldest control principle employed in human organizations, reaching all the way back to the alpha individual of the horde informing the others on how to behave. Sticking to the hypothetical company of "Inova" and how to control the beginning of its working hours by using the leadership principle: "people will start working in the morning once the designated and accepted leader tells them to do so."

A weakness of using leaders to control the resources is that if a leader does not issue a command, then the control is not achieved. Imagine the leader in a company who calls his team every morning and tells them when to come to work. If he does not call, then the workers do not show up. If it turns out that he would have to call them at the same time each morning, then it would be easier to make a rule – and make the people follow the rule, rather than the leader. However, it could also be that the conditions for the starting time of work change every morning, in which case it would be better not to make a standard rule, but rather employ a personal wake-up call by a leader. If conditions change a lot, then creating standard rules would be less effective than using the flexible control that an experienced leader can achieve. In this way flexibility is the main advantage of the leadership principle.

Using leadership as a mechanism of control is not to be confused with leaders using their authority to be a rule maker, which falls under the power of rules. In the management literature the controlling power of leadership in its own right, independent of rulemaking, is only a relatively recent field of study. One of its protagonists is Jay A. Conger who taught this subject in renowned business schools around the world. In a defining commentary on the research in his field in 1999, he recounts the history of the study of "Charismatic and Transformational Leadership in Organizations." How the notion of this kind of leadership is supposed to reach far beyond the rules-based approach typified by Jensen, is captured well by Bennis and Nanus in 1985:

> Management typically consists of a set of contractual exchanges, "you do this job for that reward, . . . a bunch of agreements or contracts." What gets exchanged is not trivial: jobs, security, and money. The result, at best, is compliance; at worst, you get a spiteful obedience. The end result of leadership we have advanced is completely different: it is empowerment. Not just higher profits and wages . . . but an organizational culture that helps employees to generate a sense of meaning in their work and a desire to challenge themselves to experience success.

According to the leadership school of thought, charismatic and transformational leadership would, to an almost transcendental degree, yield results from resources that would otherwise be unimaginable. Hence also the notion, that "exceptional times require exceptional leaders." Various behavioral models have been developed to ground the nature of charismatic leadership in more measurable terms – or at least in descriptive terms. According to Conger (1999, p. 156), all these models include nine converging attributes of leadership, such as:

- vision;
- inspiration;
- role modeling;
- intellectual stimulation;
- meaning-making;
- appeals to higher-order needs;
- empowerment;
- setting of high expectations;
- fostering collective identity.

These attributes need not be exclusively attached to a single person as a leader. They can also be applied to a "leading principle," a "leadership vision" or even a whole "leading community" of people. For the designer of a control mechanism, these nine attributes are a helpful checklist if the leadership mechanism is to be employed. The more the individual leader or the leading principle can represent these nine attributes, the more effective will the leadership be.

In summary, the power of leadership is an important mechanism for controlling resources towards a certain target, due to its *flexibility*. However, this power of leadership is not to be confused with rule making, instead it is based on the transformational, charismatic power of individual persons or ideas to make people do certain things, which otherwise they would not do on their own accord. The nine features by which charismatic leadership can be identified are vision, inspiration, role modeling, intellectual stimulation, meaning-making, appeals to higher-order needs, empowerment, setting of high expectations and fostering of collective identity.

The power of competition

The principle of competition permeates modern industrial societies so deeply that it seems a most natural phenomenon. Yet, it has become established as a mechanism of control only a lot more recently than the power of rules or the power of leadership – having become mainstream only in the course of industrialization in the western world during the nineteenth century, and spreading from there to most other societies. With the downfall of Soviet communism, the principle of competition has become accepted as the major mechanism to allocate (control) resources in national economies and, by extension, also within companies.

Nonetheless, as is true for each of the four control mechanisms, competition is rarely used exclusively even in national economies, let alone within companies. That has mainly to do with its main disadvantage, which is that competition creates a lot of losers. The essence of competition is for one resource to be "superior" over the other. But that implies that the other will be "inferior." Again, taking the hypothetical case of the Inova Company trying to control the starting work time of its employees each morning by using the principle of competition: "Work starts as soon as a qualifying majority of the people have shown up." In this case, the starting work time would be like a bid: whoever comes first, or arrives early enough to be within the first group, has won, and all the others come late and have lost. One of the reasons why organizations will typically restrain the power of competition with various kinds of rules and regulations, is in order to soften the impact on the losers and reduce the potentially disproportionate costs that such losses might inflict on people.

The unbeatable advantage of competition is that it is the most efficient of the four control mechanisms. It matches any given need with the most superior resources available to it. The inferior resources will then need to look for other places of employment. In the case of Inova company, the starting work time at the company is optimally matched with the earliest time when the resources are willing to start, and will thus require the least remuneration for their services. No other control mechanism allocates resources with this degree of precision and efficiency.

But it took a long time until the social sciences discovered and understood the intrinsic efficiency of the principle of competition. Adam Smith's introduction of *Invisible Hand Theory* in his classic "The Wealth of Nations" in 1776, is considered to be the founding moment of any sort of economics, macro or micro. Smith illustrated the "invisible" efficiency maximization power of competitive markets or, stated differently, the optimization of resource distribution in markets. A more recent path breaking protagonist of this thinking was the Nobel Prize winner Friedrich von Hayek whose insights explained the functioning of competitive markets in more detail (translated from 1996, p. 11, and 1976, p. 115): "Markets are institutions of information collection. They enable us to utilize widely dispersed information for developing extra individual patterns, . . . which make it possible to use widely dispersed know how and capabilities for various purposes without further intervention." And: "The pricing system is a mechanism for transmitting information. Its highest importance is its efficiency in utilizing knowledge, i.e. how little the individual participants need to know, in order to do the right thing."

These insights are not only true for national market economies, they are also seen by many managers to be a relevant principle for controlling an organization. A very prevalent use of competition logic in innovation management is the use of return of investment (ROI) calculations for judging whether to go ahead with a project or not.

In summary, the principle of using competition for controlling resources towards a certain target is important due to its *efficiency* in doing so. When installing a competition system in an organization, the designer of this control system needs to understand that it functions by being a pricing system that discovers information. Therefore, central to the functioning of competition, is that there is a stable currency available in which the prices can function (either financial or artificial currencies such as point systems), and that sufficient transparency is provided – otherwise there is no information that can be discovered. Competition cannot work without transparency and a common currency between the competing resources.

The power of complexity

The most recent understanding of how to achieve control in organizations, has been advanced from system theory in the form of *Complex Adaptive Systems*. It proposes that organizations, whether they are companies or national economies, are not just rule- or market-based exchanges of goods – but instead they are self-organizing, emergent communities (also called chaotic systems). Where competition-oriented economic thinking emphasizes the need for equilibrium, for instance, by finding prices that bring demand into line with supply, complex adaptive systems stresses the need to be in constant and dynamic flux.

The reason why complexity is increasingly realized as a source of power in its own right, is that many organizations have become so enormously complex, that achieving control through the other three mechanisms yields less and less results (Gratton, 2004). However, instead of this leading to an absence of control, it turns out that complexity also possesses the power to allocate and control resources, if it is understood and installed appropriately. Applying for a final time the hypothetical "Inova" case and using the power of complexity to control the working starting time might mean the following: "work starts as soon as the relevant group of individuals self-organizes towards doing so."

To be calling on this principle at the Inova Company as a way to achieve control may appear strange at first sight. Yet, this is exactly how companies such as Semco, one of the two cases described later, have organized their working time hours. There is a very important distinction between saying: "work starts as soon as the *relevant individuals in a group decide* to do so," which would be inviting non-controlled anarchy, and saying: "work starts as soon as the *relevant group of individuals self-organizes* to do so," which would be utilizing the dynamics of complexity.

Complexity works through group dynamics to achieve self-organization, and by that explicitly discourages individual dynamics of self-determination. Nonetheless, the main disadvantage of complexity as a source of power is obvious: it remains rather fuzzy. Also, it is sometimes difficult to tell whether something is chaotic (that is self-organized) or anarchic (that is uncontrolled). On the plus side, complexity, if made to work correctly, is probably the most forceful power of the four. Complexity can still achieve control even where all other control mechanisms have long run out of steam.

Eric Beinhocker, a former partner at McKinsey&Co, venture capitalist, entrepreneur and recognized thought leader in the field of strategy and complexity, explained complex adaptive systems like this (1997, p. 30):

> Examples of complex adaptive systems include ant hills, forest ecosystems, the immune system and the internet. All are open systems comprising a number of agents whose dynamic interactions self-organize to create a larger structure. Over the past twenty years, aided by advances in mathematics, physics, chemistry and biology, and by the wide ability of cheap computing power, scientists have begun to find that complex adaptive systems are governed by deep common laws.

Some of these laws are thought to be due to the fact that complex adaptive systems are driven by cognitive behavior of people. People only rarely base their decisions on cold-blooded deductive reasoning assumed by traditional economists. However, people are highly skilled at recognizing patterns and developing instincts. This allows them to make decisions in the face of incomplete information, where computer models would spin error messages instead.

Another law is that webs of relationships are more than a merely intensively connected network. These relationships are characterized by multiple dimensions of interaction and feedback dynamics, which makes them impossible to predict in detail, but reasonably predictable overall. This is also called the soccer game effect: if a premier league team meets an amateur village side, it can be certain that it will win the game. However, even then it is impossible to predict which players will score in what minute and in what combination.

A third component of complex adaptive systems are waves, or recurring rhythms of activity patterns. Economic systems have an eerie quality to be ebbing up and down in regular wave patterns, and virtually never to progress in a linear fashion. These wave dynamics follow their own sets of rules, which can be exploited for benefit.

In summary, the principle of using complexity for controlling resources towards a certain target is important due to the *enormous momentum and force* it can create. Utilizing the power of complexity does not mean anarchy, instead it means arranging the conditions in such a way that self-organization will spontaneously erupt and generate momentum for change. The main characteristics of these conditions are to make use of the complex cognitive and instinctive behavior of agents (people), to utilize the dynamics of webs, and to recognize the dynamics of wave rhythms.

The organizational control matrix

The four powers for achieving control explained so far in this chapter, the powers of rules, leadership, competition and complexity, represent fundamental control principles. By that is meant that their sources of power are independent of each other: the power of rules can exist independent of the power of leadership and vice versa, and the power of competition can exist independent of the power of complexity, etc. But these four powers share common features, by which they can be sorted along two dimensions into a matrix. The two dimensions are:

- how the control is exerted towards the resource (subject of control);
- and how the control is experienced by the resource (object of control).

In addition, power of control can be exerted either relative to or bound by the contextual environment, or it can be independent of it. Likewise, the experience of control can be either relative to and bound by the contextual environment, or it can be independent of it. This creates, then, the matrix of achieving organizational control shown in Figure 15.1.

The power of rules

Rules *exert* their power by being bound to a contextual authority or various parameters of time, place or target group validity. Often rules are also bonded to each other, creating systems and structures of rules. By contrast, on the *experiencing* side, being controlled by a rule is not bound to something or somebody else. A rule is a rule, unrelated and

EXERTION of power by the SUBJECT of control	Independent of context	Bound by context
Bound by context	**Leadership based designs**	**Rules based designs**
Independent of context	**Complexity based designs**	**Competition based designs**

Figure 15.1 *Overview of the control design mechanisms matrix*

independent of other circumstances. Thus, rules are bound as a control subject, and unbound as a control object.

The power of competition

Exertion of competition is bound to the context. It can only be exerted if it defines the playing field where it is active, which resources are inside and which are outside of the bid. But, in contrast to rules, competition can also be *experienced* only relative to other objects of this control mechanism. It is the essence of competition to be compared, to be relative to the others, and thus to be either superior or inferior to them. Thus competition is bound in both instances, as subject and as an object.

The power of complexity

Complexity requires no boundary for *exertion*. On the contrary, its very openness, randomness and chaos are, in fact, what creates the conditions of complexity to begin with, and are the fountain of its power. But that is not true for the objects of complexity *experiencing* the power – they, on the other hand, are, by definition, connected and intertwined, bound to each other in many undifferentiated ways. Thus, complexity is unbound as a control subject, and bound as a control object.

The power of leadership

The principle of leadership is unbound on both sides, on the exerting and on the experiencing side. Even if there is only one object of leadership power, which is unrelated and unbound by anything, except by being led by a leading power, then this power is being *experienced* and fully effective. Likewise, the *exertion* of leadership requires no external authority such as the rules power does, and, unlike the competition power, requires no drawing of boundaries – all it requires is the quality of internal charisma to be effective and exertive. Thus, leadership is unbound in both instances, both as a subject or as an object.

The non-overlapping dimensions of the organizational control matrix prove the argument that these four sources of powers are functioning as principles, independent of each other. What matters for the practitioner is that each of these four principles has important and valid instruments and applications in the field of innovation management. Each of these four has decisive advantages and disadvantages, which are summarized in Figure 15.2. The most important result from this table of advantages and disadvantages is that it suggests a combination of these instruments should always be used when trying to achieve control over an innovation organization. Smart and deliberate combinations of different powers for control can level or reduce their respective disadvantages against each other and make the advantages reinforce each other. How this has been achieved in two very different companies will be demonstrated with the following two case studies.

TWO CASES: CONTROLLING INNOVATION IN PRACTICE

The following two cases, Carlos Ghosn at Nissan, and Ricardo Semler at Semco, illustrate how these two executives have used a variety of different control mechanisms in the companies they were in charge of in order to bring them under control. In both cases, gaining control over technology-driven innovation played a particular role to achieve a superior performance. Shorter versions of these two cases have been written by the same author and published in the strategy textbook *Strategy – Process, Content, Context* (De Wit and Meyer, 2004, pp. 487 and 490, quotes in the two cases were among others from Magee and Semler, see also references for this chapter).

Where Carlos Ghosn and Ricardo Semler have employed the power of one of the four control mechanisms, the text denotes this with parentheses, for instance (*leadership*).

EXERTION of power by the SUBJECT of control	Independent of context	Bound by context
Bound by context	**Leadership** + : flexible – : limited reach	**Rules** + : simple – : costly
Independent of context	**Complexity** + : forceful – : fuzzyness	**Competition** + : efficient – : harsh on losers

Figure 15.2 *Summary of advantages and disadvantages of each control power*

Carlos Ghosn at Nissan in Japan: benevolent autocracy

In 1998 the Japanese carmaker Nissan was flirting with bankruptcy. Its market share in Japan had been sliding for 26 years straight, and while its key domestic rivals Honda and Toyota were reporting record profits, Nissan had not been able to make a profit for seven of the eight previous years. Daimler Chrysler had declined to buy Nissan, even for the symbolic amount of one dollar, while Ford, too, had lost interest. At the urging of Carlos Ghosn, then a senior manager at Renault, it was this unlikely French carmaker who gained a controlling stake in March 1999. Just three years later, Nissan was one of the most profitable automobile manufacturers in the world, even surpassing Toyota, and was set to recapture the no. 2 market share position in Japan. What happened?

Between 1992 and 1998, three different presidents had been behind the wheel at Nissan, but none was able to get the skidding company under control. No fewer than four restructuring plans were announced, but each ran into the sand and gained nothing for the company. So when Renault eventually stepped in and sent the 45-year-old, non-Japanese speaking, Brazil-born French/Lebanese Carlos Ghosn, to take control of Nissan in summer 1999, his task was widely hailed as "Mission Impossible." Later in that year, this assessment was toned down to "Mission Improbable," and in 2002, *Fortune Magazine* named Carlos Ghosn "Asia's businessman of the Year." In 2003 and 2004, Nissan unveiled cars that received accolades around the world for their innovative features (for instance the new March, or the new Z cars), and for 2005, Nissan was well within reach to sell 40 percent more cars than it did before, mostly on the strength of its renewed high-quality engineering and innovative design. What had changed since Carlos Ghosn arrived?

According to one senior executive at Nissan, "Ghosn stresses action, speed and results. He follows up closely. If there are any deviations he goes after them immediately.

He is relentless in following up." In his own words at the time:

> I have one goal, that Nissan will be profitable in 2001 . . . This is not like buying a Persian rug: The guy says he wants 100, but if he gets 50 he will be happy. We want 100, and we are going to get 100. If we do not get it in 2001, that's it, we will resign . . . From now on, financial objectives will entail accountability. (*Power of rules used to make targets stick by not compromising on them – a rule is a rule.*)

Accountability is Ghosn's credo. He sees no value in business relationships that are not characterized by clear and controllable targets. Starting at the top, the number of directors on the board was reduced from 43 to nine. The traditional lifetime employment and seniority-based reward system was completely revamped. Several hundred key managers received stock options instead. Promotion and rewards were linked to performance against an annual set of objectives (*power of rules used for organizational structure and pay systems*). Ghosn created six program directors with worldwide profit responsibility for a range of cars under their management, who were in charge of designing, making and selling the cars. In this way, the program directors could harness the innovations found by the engineers for features that the customers actually wanted (*power of competition used to allocate resources between the competing product lines*).

Externally, by the end of 2002, Nissan's 67 equity investments in *Keiretsu* (group) companies were reduced to 25, while all 1,400 cross shareholdings with other Japanese companies were undone. The 300 global banking relationships were centralized into a single treasury function. The number of suppliers was reduced by half to 600, with each remaining supplier committing to at least 20 percent cost reduction over three years. The way the supplier reduction was achieved was typical of the overall approach: on a first come, first served basis any supplier who stepped up to the new deal at Nissan was awarded the contract. The new deal promised doubled volumes at lower prices, to be jointly achieved by intensive engineering cooperation between Nissan and the supplier (*power of competition used for supplier selection*).

The pressure was equally fierce inside the company, both in terms of cost reduction as well as growth promotion. Headcount was reduced by almost 20 percent, dropping from 148,000 to 127,000 employees, and five manufacturing plants were closed. All the while, Ghosn planned to introduce 28 new car models within three years. Ghosn's advent to Nissan was not all slash and burn. Off-site R&D centers were immediately granted funds to refurbish their facilities and a completely new automotive plant was launched and commissioned in Mississippi in the record time of only six months. Ghosn: "I only take extremely well considered decisions – and I usually decide fast." For Ghosn as an accomplished bridge player, and committed family father of three children, this is not a contradiction.

When Ghosn arrived, he came with a clean slate. In the first few weeks he interviewed the entire company; all functions, all levels (*power of leadership used to build personal credibility and charisma with the organization*). Within five days of his arrival he already instituted his trademark turnaround strategy: installing nine cross-functional teams with up to 10 middle managers and hundreds of subteam members, to work out the entire "Nissan Revival Plan" (the now famous NRP) within only three months (*power of complexity used in cross-functional teams*). The NRP would become the blue print of action for the next three years. Team members were not responsible for implementation, but their recommendations had to be aggressive, specific, backed up by numbers, and not respectful towards current practices. In particular, the actions were supposed to emphasize innovative solutions, preferably technological, instead of just numb cost-cutting or investment delays. Any team that did not live up to these targets was sent straight back to redo the numbers. Moreover, Ghosn arrived only with a handful of senior managers from outside Nissan – almost the entire previous top management remained in place throughout the restructuring and even later. Also, no consulting company was involved in the drafting of the NRP – so virtually all the difference to the existing conditions was made by Ghosn and his actions himself (*power of leadership used by taking personal action, instead of working through consultants*).

Besides the application of his organizational and strategic skills, communication enjoys highest priority for Ghosn. Much to the anger of the business and financial community, virtually nothing emerged from Nissan during the first months of Ghosn's reign. Only once the NRP was finished did he announce it personally at the Tokyo Motor Show. All major decisions at Nissan since then have been taken, announced and defended in the public by Ghosn himself. So far he has never been forced to take back any of his announcements (*power of leadership*).

For Ghosn, the Nissan assignment was his fifth radical restructuring of a business he was put in charge of – and not the last. In May 2002, the NRP targets were achieved one year ahead of schedule. Ghosn then unveiled the new Nissan 180 plan – by 2005, Nissan would increase car sales almost 40 percent, from 2.6 to 3.6 million vehicles, reach 8 percent operating profit on sales (top of the industry), and have reduced net automotive debt to zero. To industry insiders, this sounded like "mission impossible" all over again . . . but sure enough, by the end of 2004 the Nissan 180 plan was about to be achieved ahead of schedule. For Ghosn this was no surprise: "Nissan 180 is an ambitious plan, but we have very detailed analysis backing up its feasibility." Ghosn's next challenge is already outlined: from 2005 he will assume the post of CEO in Renault and Nissan at the same time – a novelty in the modern corporate world – but for Ghosn, only another challenge of innovation.

Ricardo Semler at Semco in Brazil: democracy at work

Ricardo Semler took over his father's pump-making business in 1980, when Semco was a US$ 4 million company, focused on the domestic Brazilian market, and heading for bankruptcy in a severe recession that was to last for most of the decade. By 2003, Semco had expanded beyond pumps to dishwashers, digital scanners, cooling units, mixers, real estate services, environmental consulting and high technology software development, operating as a federation of ten businesses, with revenues totaling US$ 160 million and about 3,000 employees.

Semco has no traditional organizational hierarchy for decision-making and control. Major decisions affecting the entire organization, such as the purchase of a new plant site or an acquisition, are put to a democratic vote, while other decisions are taken consensually by all employees involved (*power of complexity used for making the relevant group decide themselves*). There are no internal audit groups, no controls on travel expenses, and inventory and storage rooms remain unlocked – but all information is made available to everyone, encouraging self-control (*power of complexity*). Already in 1992 the central headquarter building was replaced by a network of office spaces dotted throughout the city of São Paulo. Any employee was free to walk into any office in the morning, occupy space there and make it his place of work for the day. According to Semler: "Freedom is no easy thing. It does not make life carefree – because it introduces difficult choices." (*Power of complexity*).

Communication is seen as the life blood of the company. To stimulate information exchange, the offices have no walls and all memos must be kept to one page, without exception (*power of rules*). Furthermore, everyone is trained to read financial statements, and everybody is expected to know the profit and loss statements of the company and their business unit (*power of competition used to understand the market allocation of capital and creating transparency about it*). In order to avoid any possible suspicion towards the formal reporting, the financial literacy training is conducted by one of Brazil's most aggressive unions – incidentally, union membership is not discouraged at all. In fact, Semco has experienced strikes, walk-outs and lawsuits by its employees.

The alternative organizational configuration of Semco is made up of four concentric circles. The innermost circle consists of six Counselors, who serve as the executive team and take turns as chairperson every six months. Despite being the majority owner of the company, Semler is not even one of these six – he calls himself "gainfully unemployed." Around the Counselors is a circle of Partners, who act as business unit managers. Around them is a circle of Coordinators, who function as first-line supervisors. Everybody else is in the fourth circle, and is called Associate. Very critically, there are Nucleuses of Technology Innovation, which are "no-boss" temporary project teams who are freed from their day-to-day work, in order to focus on some kind of business improvement project, a new product, a cost reduction program, a new business plan, or the like (*power of complexity used for the team structures inside the NTI*). Additional emphasis is placed on keeping small cell structures. No business unit is allowed to grow to more than 200 members or so, or to extend its reach beyond a limited number of core customers or core technologies. If a cell becomes too large, it is expected to split (*power of complexity used for creating relevant groups*).

The managers of Semco decide among themselves what their pay will be, and the target is to extend this practice to all employees. The amount is made transparent to all others by regular participation in salary surveys, thus everybody knows what the pay is of everyone else (*power of competition used by making pay for work transparent*). The top managers will be selected by their future subordinates, not by their future superiors, and all managers must participate in quarterly 360 degree manager ratings (*power of leadership used for appointing managers*). Furthermore, every member is part of the company-wide profit sharing program that pays out 23 percent of a business unit's profits per quarter as a bonus to the employees. The members of a business unit decide among themselves how the bonus is distributed – in fact, the payout ratio of 23 percent has also been decided by the employees. Members of a Nucleus of Technology Innovation receive royalties on the achievements of their projects (*power of competition*).

At any given moment, who belongs to the Semco company and who doesn't, can be rather fuzzy. Semler explains:

> When we walk through our plants, we rarely even know who works for us. Some of the people in the factory are

> full-time employees; some work for us part-time; some work for themselves and supply Semco with components or services; some work under contract to outside companies (even competitors); and some of them work for each other. We could decide to find out which is which and who is who, but . . . we think it is all useless information.

This does not mean that Semco is soft on financial performance targets. If a business unit does not perform, it risks being dissolved quite soon. Nor is Semco a big family. Semler explicitly states that Semco is a business, and that it will not mix up personal concerns of its employees with the company interest. Only under extraordinary circumstances will the company extend loans to its employees for instance, and as a general principle, family circumstances or even education are not taken into account when hiring or promoting employees (*power of competition and rules*).

As for strategy, Semco has no grand design. Semler readily admits that he has no idea what the company will be making in 10 years time: "I think that strategic planning and vision are often barriers to success." Semco's approach is largely to let strategy emerge on the basis of opportunities identified by employees close to the market. Where new initiatives can muster enough support among colleagues, they are awarded more time and money to bring them to fruition. In this way, Semco can make the best possible use of the engagement and entrepreneurship of its employees. Nonetheless, Semco has strategic principles. For instance, it will not enter a business if it is not a highly complex operation. Semco believes that under very complex environments, its organizational competences to foster innovative solutions allow it to gain a particularly strong competitive advantage (*power of complexity*).

Summing up the Semco philosophy, Semler told the *Financial Times*:

> At Semco, the basic question we work on is: how do you get people to come to work on a gray Monday morning? This is the only parameter we really care about, which is a 100% motivation issue. Everything else – quality, profits, growth – will fall into place, if enough people are interested in coming to work on Monday morning. (*Power of leadership.*)

CONCLUSION AND SUMMARY

Carlos Ghosn at Nissan in Japan and Ricardo Semler at Semco in Brazil are both employing a wide range of control tools in the respective organizations that they are in charge of. At first sight, Carlos Ghosn's managerial style seems to be entirely top-down driven – but a closer look at the case reveals that he uses all four powers of control, and possibly the power of leadership least of them, to achieve control over Nissan. The Nissan case is so striking because it defies most of the standard management paradigms. Essentially everybody in the industry had written Nissan off, but given an effective cocktail of control, even a company so down on its luck as Nissan could be revitalized in a short time. Nor was Nissan just a matter of rigorous cost cutting, the second restructuring wave, the Nissan 180 plan, showed that Ghosn's control mix was as effective in creating innovation-driven growth, as it was in reducing costs.

The same could be said for Ricardo Semler. At first sight, Semco looks like a chaotic community with no planning or control. But closer inspection shows that there are many elements of competition, leadership and rules mechanisms that create control in the company – and make it flourish greatly even throughout a most strenuous business environment.

When studying these two executives one will notice how undogmatic both of them are in going about their management methods. It is not that they use control mechanisms arbitrarily, but neither are they blinded by singular approaches or cookie cutter recipes. Instead, both executives display a deep understanding of the mechanics of each of the four control mechanisms, and apply them pragmatically, especially concerning the fostering of innovation. With this flexibility to mix and match control mechanisms, they can maximize the advantages of each of the powers, and try to avoid their disadvantages.

Just as Ghosn and Semler did in their technology-driven companies, managers of technology will invariably encounter the need to manage innovation to keep on growing and finding new stimuli for value creation. In doing so, managers of technology will need to overcome the fundamental paradox of control and freedom: because the innovative process requires freedom, while the organization requires control. If one creates a well designed control system around and with the innovation system in the company, this can be achieved, and used for the company's competitive advantage. The four sources of power that can be used for these control systems are, first, the power of rules, second, the power of leadership, third, the power of competition and, fourth, the power of complexity.

Each of these four powers has distinctive advantages, which makes each of them attractive to be an important element in a total organizational control design system. By the same token, each of them also carries distinctive disadvantages. Thus, it cannot be expected that any one of the four sources of power for organizational control is inherently superior to the others – instead, the designer of the control system is challenged to find the right mixing and matching of all four sources of power for his particular organization. The chapter provides check lists of the critical elements that recognized thought leaders in the managerial sciences have found to be important attributes when using each of these four powers. But the chapter cannot provide the best recipe on which mix is most suitable for which circumstances – this will remain up to the skills and aptitude of the manager in face of the conditions at hand.

FURTHER READING

Of the sources above, the books by Ricardo Semler make particularly inspirational reading. Iansiti and Gratton put special emphasis on organization and technology management. The whole range of paradoxes in strategy management is most accessible in De Wit and Meyer's textbook on strategy.

REFERENCES

Beinhocker, E.D. (1997). Strategy at the Edge of Chaos. *The McKinsey Quarterly*, (1) 24–39.

Bennis, W.G. and Nanus, B. (1985). *Leaders: The Strategies for Taking Charge*. New York: Harper & Row.

Conger, J.A. (1999). Charismatic and Transformational Leadership in Organizations: An Insider's Perspective on these Developing Streams of Research. *The Leadership Quarterly*, 10 (2).

De Wit, B. and Meyer, R. (2004). *Strategy: Process, Content, Context*. Boston, MA: Harvard Business School.

Eisenhardt, K.M. and Brown, S.L. (1998). *Competing on the Edge: Strategy as Structured Chaos*. Boston, MA: Harvard Business School Press.

Gratton, L. (2004). *The Democratic Enterprise*. London: Pearson Education.

Hayek, F.A. (1976). *Individualismus und wirtschaftliche Ordnung*. 2. Auflage. Salzburg: Verlag Wolfgang Neugebauer.

Hayek, F.A. (1996). *Die verhängnisvolle Anmaßung: Die Irrtümer des Sozialismus*. Tübingen: Verlag J.C.B Mohr.

Iansiti, M. (1998). *Technology Integration: Making Critical Choices in a Dynamic World*. Boston, MA: Harvard Business School Press.

Jensen, M.C. (1994). The Nature of Man. *Journal of Applied Corporate Finance*, I7 (2).

Magee, D. (2003). *Turnaround – How Carlos Ghosn rescued Nissan*. New York: Harper-Collins Publishing.

Picot, A., Reichwald, R. and Wigand, R. (2003). *Die Grenzenlose Unternehmung: Information, Organisation und Management (The Boundaryfree Enterprise: I, O, M)*. 5th edition. Wiesbaden: Gabler Verlag.

Quinn, R.E. and Cameron, K.S. (1988). *Paradox and Transformation: Toward a Theory of Change in Organization and Management*. Cambridge, MA: Ballinger Publishing.

Semler, R. (1993). *Maverick: The Success Story behind the World's Most Unusual Workplace*. London: Century.

Semler, R. (2003). *The Seven Day Weekend*. London: Century.

Smith, A. (1976). *The Wealth of Nations*. Cannan's edition. Chicago, IL: University of Chicago Press.

Chapter 16

When failure is not an option

Managing complex technologies under intensifying interdependencies

Michel J.G. van Eeten, Emery Roe, Paul R. Schulman and Mark de Bruijne

OVERVIEW

Technology-intensive organizations are continually challenged to ensure the reliability of their complex technologies. In this respect, all organizations can learn from a special class of organizations for which failure is not an option: the so-called High Reliability Organizations (HROs). This chapter provides an overview of the research on HROs. This research, however, has exclusively focused on particular organizations which have command and control over their technical cores. Many technical systems, including electricity generation, telecommunications and other "critical infrastructures," are no longer the exclusive domain of single organizations. Reliability is more and more the outcome of networks of organizations. How do HROs ensure reliability within a network setting? This chapter presents the findings of an in-depth empirical case study of the California electricity system. It concludes by relating these findings to the management of technology.

Organizations that critically depend on technology are continually challenged to defeat Murphy's Law – things that can go wrong, will go wrong. At best, "things going wrong" means costly disruptions in service provision or production. The worst case is that the existence of the organization is threatened. Think of a bank whose financial systems are compromised; or of a large internet retailer like Amazon when its site is unreachable or its database with credit card information broken into; or think of a mobile phone operator whose network has substantially more disruptions than those of its competitors.

How can organizations manage their technologies so that they perform reliably? Most organizations rely on "trial and error learning" to make their technologies reliable: they do their best to design robust technologies – within the limits posed by market forces and cost-effectiveness – and adapt these technologies when faced with disruptions or poor reliability. But the "error" in "trial and error" actually means that failure *is* an acceptable option, even though it may be costly or painful. However, for an increasing number of organizations, failure is no longer an option. The more important the technology is to the organization, the more threatening the consequences of failure.

If trial and error is no longer an option, how can managers make technologies reliable? Here, all organizations can learn from a special class of organizations: the so-called High Reliability Organizations (HROs). These HROs manage highly complex and risky technologies. Typical examples of HROs are nuclear power plants, aircraft carriers, electricity grid operators and air traffic control. They have extremely low tolerances for failure because any failure can potentially have catastrophic consequences – such as nuclear meltdown.

First, we'll discuss the research on reliability and its findings on how organizations highly reliably manage their technology. Next, we discuss how HROs are responding to new developments in technology and markets, most notably the increasing interdependencies among organizations – technological and otherwise. This means that reliability is more and more the outcome of networks of organizations, rather than individual organizations. Here, we'll present recent empirical research in which we explored how the California electricity system operator deals with the new interdependencies in an unbundled and deregulated market. We'll conclude by briefly discussing the future implications of these findings.

TECHNOLOGY, RISK AND RELIABILITY

Interest continues to grow and intensify among organizational theorists about organizational reliability – the ability of organizations to manage hazardous technical systems safely and without serious error (LaPorte and Consolini, 1991; LaPorte, 1996; Von Meier, 1999; Schulman, 1993a; Perrow, 1999; Roberts, 1993; Sagan, 1993; Sanne, 2000; Weick *et al.*, 1999; Weick and Sutcliffe, 2001; Beamish, 2002; Evan and Manion, 2002). A number of reasons account for the stepped up interest.

First, society increasingly depends upon "high performance" but also highly hazardous technologies, ranging from nuclear weapons and nuclear power to large jet aircraft, medical technologies, complex electrical grids and telecommunication systems. Many of these technical systems impose relatively tight error tolerances upon operators and maintenance personnel, and the consequences of the errors can be disastrous, ramifying beyond the user to by-standers and society at large (Perrow, 1999). Second, concern for reliability among organization theorists has grown because of major high-profile accidents (so commonly known as to be recognizable from a single phrase) such as Exxon Valdez, Three Mile Island, Bhopal, Challenger, and the Tenerife air disaster. Many of the accidents illustrate all too vividly that technical design alone cannot guarantee safe and reliable performance (Weick, 1993; Vaughn, 1996). That is, these technologies are so complex and tightly coupled that their behavior is full of surprises. Finally, in the aftermath of September 11, there is a heightened interest in "critical infrastructures" and their reliability in the face of potential terrorist attack (National Research Council, 2002).

Despite the recent interest, the analytic roots of reliability research in organization theory are deeply set. Some of the earliest research appears in the analysis of quality control (QC). QC has been centered on assuring reliability in both organizational products and production processes. In its focus on identifying and averting failure in production, one quality control historian traces the beginnings of quality control to medieval guilds, with their emphasis on long training, formal testing of skills and careful inspection of task performance and products (Juran, 1986). Statistical QC began in the 1920s and helped isolate causal factors, and their precursors, in production errors (Duncan, 1986; Ott and Schilling, 1990).

A related approach to reliability research occurs in the area of human factors analysis. Led by psychologists (as opposed to the engineering roots of much QC research) human factors analysis focuses on the impact of physical designs and task requirements on human performance (Salvendy, 1997; Perrow, 1983; Norman, 2002). As general tendencies, QC seeks to achieve reliability by controlling worker behavior to match task requirements, while the human factors approach is directed toward securing reliability by sculpting strategic organizational and task variables to human requirements.

As important as both QC and human factors approaches to organizational reliability have been for many analysts and organizations, they are different from more recent reliability concerns. Recent research does not concern production reliability per se. For some organizations, errors, accidents and failures undermining safety affect much more than production. They may jeopardize the survival of the organization itself and its members, as well as significant numbers of people outside the organization.

The conventional approach to organizational reliability treats it as a marginal or fungible property whose costs can be traded off against other organizational values. Conventionally, "marginal reliability" has been considered to be an embedded variable – a probability coefficient attached to production estimations (Schulman, 2002). The new reliability analysis is quite different as it deals with error and failures that have far-reaching, often unacceptable implications for safety, not just inside but outside the organization as well. This is not reliability as a probabilistic property that can be traded off at the margin with other organizational values, but reliability directed toward a set of catastrophic events whose occurrence must, as nearly as possible, be completely precluded. In this respect the new organizational reliability analysis is about the high reliability achieved through precluding events – the high standards of performance that can be achieved against a set of unacceptable events. These differences between the marginal and precluded-event reliability are summarized in Table 16.1 below.

HIGH RELIABILITY THEORY

The beginning of the new perspective on organizational reliability can be traced to works by James Reason (1972) and Barry Turner (1978) that connected human and organizational factors as systematic producers of major technical failures. A key anchor point for the new approach, however, is Charles Perrow's *Normal Accidents* (1999). In this work Perrow added a new dimension to his earlier path-breaking work on technology and organizations (Perrow, 1979). Categorizing technologies along the dimensions of complexity and tight coupling, Perrow identified a specific class of technologies – complex and

Table 16.1 *Marginal and precluded-event ("high") reliability*

Variable	*Marginal reliability*	*Precluded event reliability*
Context	Efficiency	Social dread
Risk	Localized	Widely distributed
Calculation	Marginal (variable cost)	Non-fungible (fixed requirement)
Standards	Average or run of cases	Every last case
Learning	Trial and error learning	Formal learning with limited trial and error
Orientation	Retrospectively measured	Prospectively focused
Control	Probabilistic	Deterministic

tightly coupled – that are particularly problematic from the standpoint of organizational reliability. They pose the risk of "normal accidents" irrespective of what strategies organizations adopt for their management. These technologies are, in effect, accidents waiting to happen – they are capable of changing their conditions or states with a speed and interactivity that defies the understanding in real-time of operators or the anticipation of designers and planners. Further, the changing conditions threaten catastrophic consequences for users, managers and innocent third-parties alike. Perrow's framework set, in fact, a limiting condition for the organizational reliability of large technical systems. Perrow's analytic perspective has been echoed in a number of subsequent studies (Sagan, 1993; Evan and Manion, 2002).

Countering Perrow's argument has been the work of a group of researchers who have identified in case studies what they assert to be a set of "High Reliability Organizations" (HROs). These organizations (a nuclear aircraft carrier, nuclear power plant, and air traffic control centers) have established comparatively excellent performance records in managing technologies of high complexity and tight coupling (LaPorte and Consolini, 1991; Rochlin and von Meier, 1994; Roberts, 1993; Schulman, 1993a). They were found to be surviving in highly unforgiving political and regulatory niches with respect to reliability. They are able to do so, it was argued, because of organizational, managerial and cultural factors that buffer the organizations from the hazards of tightly coupled and complex technologies and, in effect, mitigate the risk of managing these systems. Among the factors observed by the HRO researchers are those summarized in Box 16.1.

These organizations begin with a clear specification of core events that simply must not happen (LaPorte, 1996). To this they add the specification through careful causal analysis of a set of precursor events or conditions that could lead to core events. These precluded and precursor events constitute the "envelope" of reliability within which these organizations seek to operate. They develop elaborate procedures to constrain behavior and task performance within the envelope. At the same time, the organizations feature, through a "culture of reliability," a widespread sensitivity and attentiveness toward previously unspecified conditions that might causally connect specified events. If careful formal specification underlies the identification of core and precursor events, here the organization avoids specific boundary criteria

BOX 16.1 PROPERTIES OF HIGH RELIABILITY ORGANIZATIONS (HROS)

- organizationally specified core events that must not occur
- established error priorities and trade-offs in support of precluding events
- identified set of precursor events or conditions that can funnel through specified chains of causation into precluded events
- established set of procedures that specify behaviors to guard against both precluded and precursor events
- maintained widespread sensitivity and attentiveness toward unanticipated, unspecified events or conditions that might also have a causal connection to precursor and precluded events
- pursued incompatible strategies simultaneously (e.g. buffering against paradoxes)
- established formal structure of roles, responsibilities and reporting relationships that can also be transformed under conditions of emergency or stress
- team approach to problem solving
- recognition that key features of strategy and structure are unstable and subject to decay: cycles of reinforcement
- external supports, constraints and regulations that allow for all of the above

for what constitutes a reliability or "safety" issue (Schulman, 1993a).

Additionally, HROs are characterized by the simultaneous pursuit of contradictory or paradoxical properties of reliability management (Rochlin, 1993). Error protection regimes that guard against one type of error (say an error of omission) are likely to make another type of error (errors of commission) more likely. HROs must clearly specify operational procedures and standards and yet protect ambiguity so they do not become insensitive or inattentive to the unexpected. They must pursue simultaneous strategies of anticipation and resilience (see Wildavsky, 1988 for an elaboration of this paradox). Another contradiction that must be managed is that which can arise between formal design principles and actual operational experience. HROs are able to buffer these paradoxes, having to reconcile both sides of the paradox if high reliability in averting catastrophic failure and securing widespread safety are to be achieved.

Another feature discovered in research on HROs is the ability to transform formal roles, reporting and authority relationships under emergency conditions or stress. Typically this means by-passing formal hierarchy and the development of lateral, less formal modes of communication and coordination (Roberts, 1990). Much of the work of the organizations is generally carried out in teams and there is a great emphasis throughout the organization on the cultivation of high levels of technical competence through personnel selection and training.

HROs also recognize that key organizational properties such as attention, close coordination and mutual trust across units that have to rely on one another are not constants and cannot be treated as givens. They are subject to decay and must be continually renewed to the high levels required in these organizations. Routines will numb mindfulness (Langer, 1989); shared understandings will erode. As noted in earlier research, it is not invariance but, rather, the attention to, and careful management of, fluctuations that helps define the HRO (Schulman, 1993a).

Finally, and perhaps most importantly, HROs exist in environments that share an intense aversion to the events they are trying to preclude. This means HROs are carefully watched and regulated, and are constrained from internal drift or changing their goals away from high reliability by the constraints imposed by the wider task environment. At the same time the environment supports the organization in treating reliability as non-fungible, that is, it generally insulates the organization from pressures to trade off reliability with other variables under close market competition. For example, ratepayers absorb the security and reliability costs of nuclear power plants, and all airlines are required to practice similar maintenance procedures under close regulation of the aviation authorities. This regulation and support allows HROs to incorporate redundancy in technical designs and to invest a great deal in anticipatory and contingency analysis.

HIGH RELIABILITY AND NORMAL ACCIDENTS

Whether the above features constitute a sufficient or even necessary set of conditions for preventing "normal accidents," i.e., precluding unacceptable events, is at present an unanswerable question. While the dispute between the normal accident theory and HRO research has continued (Perrow, 1994; LaPorte, 1994; Rijpma, 1997; Weick *et al.*, 1999), in its most extreme form the dispute centers around an assertion that is unfalsifiable. No amount of good performance can disprove Perrow's viewpoint concerning normal accidents because it can always be said that an organization is only as reliable as the first catastrophic failure that still lies ahead, not the many successful operations that lie behind. Along these lines, Perrow has insisted there have not been more serious nuclear accidents because "we have not given large plants . . . time to express themselves" (Perrow, 1999, p. 12) – that is, given enough time, the complexity of these plants will produce instances of erratic behavior that will prove to be disastrous.

Further, one variable that Perrow asserts to be an independent variable in his causal analysis – loose and tight coupling – has significant ambiguity surrounding its identification and understanding. Loose coupling means the system has multiple paths to achieve the desired outcomes, so a failure in one place need not disrupt the system. Tight coupling means there is no such slack and failures can quickly spread through the system causing other failures.

In Perrow's formal analysis of "tight coupling," he refers to the physical properties of technologies. But in later points in his analysis he applies the concept to social organizations themselves. There are reasons for wondering whether organizations are direct analogs of physical systems and tight coupling is equivalent in both contexts. This very point has, in fact, been the subject of historical debate in organization theory, leading to the shift from the "machine metaphor" of scientific management theory (Morgan, 1997).

In addition, it is sometimes difficult to distinguish tight coupling as a cause or a consequence of failure or accident. In July of 1993 a massive flooding occurred across mid-Western states in the US. During the flood, water flows overwhelmed dams and reached levels so high that suddenly a set of spillways, across several states, which had been considered independent state flood protection devices became tightly coupled relative to the impact of any one water diversion upon the others (Hey and Phillipi, 1994). The failure of physically separate dams and spillways to contain unprecedented water levels was really the independent variable that turned a loosely coupled set of elements into a tightly coupled system. At best we could ascribe tight coupling as a latent feature of the mid-Western spillways, a feature that follows upon a specific magnitude of failure.

On the other hand, the HRO research perspective has had its own conceptual and empirical difficulties. The research has centered on a small number of selective case studies at a single slice in time for each organization. This small number of cases does not constitute a proven argument that the features discovered in these organizations are truly necessary ones (Schulman, 1993b). Further, HRO research has in some respects asserted high reliability as a defining characteristic rather than a performance variable of its organizations. This leaves unanswered the question of which traits, if any, and in which amounts, can contribute to *higher* reliability (along a continuum) in organizations. Fortunately, more recent research is beginning to broaden the analytic focus on reliability from structure to process in organizations, especially the cognitive and sense-making skills and strategies of their members (Weick *et al.*, 1999; Sanne, 2000; Schulman *et al.*, 2004).

THE CHALLENGE OF NETWORKED RELIABILITY

In all the debate between normal accident and HRO approaches to reliability research one issue has been consistently ignored. Many technical systems for which we wish to attain the highest reliability in operation and management, specifically our "critical infrastructures," are not under the control of single organizations. In electricity transmission and distribution, water resource management, transportation, telecommunication, medicine and financial services, many critical services we depend on for reliable, error-free performance are derived from *networks* of organizations. In fact, reliability increasingly is, and has to be, a network property, not a consequence of the structure or behavior of a single organization.

The formation of networks and their increasing centrality in the understanding of high reliability is the focus of the remainder of this chapter. We illustrate with a specific and particularly apt case – electricity restructuring in California – why organizational networks pose particular challenges to reliability and to reliability theory in its present form. We frame the remainder of the chapter around a single research question: how do networks of organizations and units, many with competing, if not conflicting, goals and interests, provide highly reliable services in the absence of ongoing command and control and in the presence of rapidly changing task environments and technologies?

THE CASE OF CALIFORNIA'S ELECTRICITY NETWORK

From a reliability perspective, the California electricity system can be described in terms of three components: the reliability task environment (RTE) and within it, the regulatory reliability network (RRN) and the high reliability network (HRN). Together, the RTE, RRN and HRN constitute what we term the California electricity system. Our chapter is exclusively focused on the high reliability network created by California's electricity restructuring. For a more detailed description of this case study, see Roe *et al.* (2002) and Schulman *et al.* (2004). Simply put, the HRN organizes for electrical service provision and delivery and includes the organizations and units that have direct operational responsibilities for this provision and delivery.

In the California electricity system at the time of our study (2002), the HRN consisted of the control rooms and support staff of the California Independent System Operator (CAISO), the distribution utilities, the private generators along with scheduling coordinators (market traders and the Power Exchange (PX) and later the California Energy Resources Scheduling (CERS) division) and the adjacent control areas. The focal organization of interest to our research was CAISO. It manages the statewide high voltage grid that connects generators to industries and to the utilities that distribute the power to consumers. The core of California electricity restructuring was to unbundle the generation, transmission and distribution elements of the former vertically integrated utilities into the post-restructuring HRN.

The RRN sets the mandates and criteria for high reliability, i.e. it establishes and/or enforces standards that define reliability for units in the HRN. The standards include formal regulations under a government mandate and regulations developed and enforced by system participants themselves, such as WSCC (Western Systems Coordinating Council) standards. The RRN does not organize the provision of services directly. In the California electricity system, the RRN consists of the CPUC, Federal Energy Regulatory Commission (FERC), WSCC, North American Electricity Reliability Council (NERC), CEC and the Electricity Oversight Board (EOB), among others. While these agencies and units have reliability of electrical service as a mandate, they often also have other mandates to reconcile, e.g. cost to the electricity consumer.

The RTE sets the context for the reliability-related tasks of the RRN and HRN. The reliability task environment includes customers, voters, businesses, elected officials and the public. The customer has demands, expectations, and even certain contractual rights to receive service in a reliable fashion. A widespread notion in California is that electricity is not just any other commodity and that access to cheap, always-on electricity is an entitlement, if not a right. Clearly, the RTE, RRN and HRN are connected, as when the public's pervasive notion of electricity as a right influenced much of the intervention of the Governor's Office to have CERS fund the CAISO's provision of electricity during the California electricity crisis.

One last point before we turn to the body of the chapter. It has been argued that the chief feature of the California electricity crisis was the *unreliability* of electricity provision, such that it borders on the presumptuous for us to talk about a "high reliability network" in anything more than name only. Yet, notwithstanding the popular view of rolling blackouts sweeping California during its electricity crisis, blackouts were minimal – both when measured in hours and in megawatts (MW). In point of fact, rolling blackouts occurred on six days during 2001, accounting for no more than 30 hours. Load shedding ranged from 300 to 1000 MW in a state whose total daily load averaged in the upper 20,000 to lower 30,000 MW. The aggregate amount of megawatts that was actually shed during these rolling blackouts amounted to slightly more than 14,000 Megawatt-hours (MWh), the rough equivalent of 10.5 million homes being out of power for one hour in a state having some 11.5 million households – with business and other non-residential customers remaining unaffected. In short, the California electricity crisis had the effect of less than an hour's worth of outage for every household in the state in 2001. Why the lights by and large actually stayed on was due, we argue, to factors outlined in the rest of this chapter.

FRAMEWORK

The answer to our core research question – How does the California high reliability network maintain reliable electricity in real time? – is this: its focal organization, the CAISO, balances load and generation in real time by developing and maintaining a repertoire of responses and options in the face of unpredictable or uncontrollable system volatility, where the "system" in question is the California electricity system as defined above.

Our research leads us to focus on the match between, on one hand, the options and strategies of the HRN to achieve its reliability requirement (e.g. balancing load and generation, staying within thermal limits set for key paths) and, on the other hand, the unpredictable or uncontrollable threats to fulfilling the HRN reliability requirement. A match results from having at least one option sufficient to meet the requirement under given conditions. At any point, there is the possibility of a mismatch between the system variables that must be managed to achieve the reliability requirement and the options and strategies available for managing those variables. That match is not automatic and requires management – in this case high reliability management within the HRN.

The CAISO's reliability mandate is twofold: keeping the electricity flow always on and reliably protecting the grid from islanding or worse. Meeting the dual reliability mandate involves managing the options and strategies that coordinate actions of the independent generators, energy traders and the distribution utilities in the HRN. As the focal organization, the options the CAISO deploys are HRN-based or HRN-wide options, e.g. outage coordination is the responsibility of the CAISO, but involves the other partners in the HRN.

Consequently, the CAISO management can be categorized in terms of the variety of HRN-based options it, the CAISO, has available (high or low) and the volatility of the California electricity system (high or low), resulting in four performance conditions and modes as set out in Figure 16.1.

Volatility is the degree to which the focal organization, the CAISO, faces uncontrollable changes or unpredictable conditions that threaten the grid and service reliability

		System volatility	
		High	*Low*
Network option variety	*High*	Just-in-time performance	Just-in-case performance
	Low	Just-for-now performance	Just-this-way performance

Figure 16.1 *Performance conditions for CAISO*

of electricity supply, i.e. that threaten the task of balancing load and generation. Some days are of low volatility. A clear example of high volatility are those days where a large part of the forecasted load had not been scheduled through the day-ahead market desk, which means for the CAISO actual flows are unpredictable and congestion will have to be dealt with at the last minute. Volatility refers to any system-related instabilities, not to price movements alone.

Options variety is the amount of HRN resources, including strategies, available to the CAISO to respond to events in the system in order to keep load and generation balanced at any specific point in time. It can be approximated with conventional engineering parameters, including available operating reserves and other generation capacity and available transmission capacity. High option variety means, for instance, that the grid has more than the required regulatory resources available (there are high or wide margins), low options means the resources are below requirements and, ultimately, that very few resources are left (low or tight margins). These two dimensions together set the conditions under which the CAISO has to operate and demand different performance modes for ensuring reliability (i.e. for ensuring the balancing of load and generation): "just-in-case," "just-in-time," "just-for-now" and "just-this-way."

PERFORMANCE MODES

Each performance mode achieves the balancing of load and generation, but in very different ways because of the initial conditions both operators and engineers face in the control room.

"Just-in-case" performance mode for balancing load and generation

When options are high and volatility low, "just-in-case" performance is dominant because of high redundancy – that is, the CAISO still has multiple means available to respond to incidents. Generation reserves available to the CAISO are large, excess plant capacity exists at the generator level, and the distribution lines are working with ample backups, all much as forecasted with little or no volatility (again, unpredictability and/or uncontrollability). More formally, redundancy is a state where the number of different but effective options to balance load and generation is high relative to the market and technology requirements for this balance. There are, in brief, a number of different options and strategies to achieve the same balance. The state of high redundancy is best summed up as one of maximum equifinality, i.e. there are very many means to meet the reliability requirement, "just-in-case" they are needed. It is important, however, not to confuse this mode with the pre-restructuring condition where integrated utilities had "high reserves." CAISO reserves were around 7 percent at the best of times; prior to restructuring, the vertically integrated utilities may have had reserves of 13 percent or higher.

"Just-in-time" performance mode for balancing load and generation

When options and volatility are both high, "just-in-time" performance is dominant. Option variety to maintain load and generation remains high, but so is the volatility in system variables, such as due to a change in weather. High market volatility may be in the form of underscheduling or rapid price fluctuations leading to unexpected strategic behavior by market parties, while higher grid volatility may be in the form of contingencies such as sagging transmission lines during unexpected hot weather. This performance condition demands real-time flexibility, that is, the ability to utilize and develop different options and strategies quickly in order to balance load and generation. Operators in the control room are in constant communication with each other and others in the HRN, options are reviewed and updated continually, and informal communications are much more frequent. Flexibility in real time is the state where the control room operators and engineers are so focused on meeting the reliability requirement and the options to do so that more often than not they "customize" the match between them, i.e. the options are just enough and "just-in-time."

More formally, the state of real-time flexibility is best summed up as adaptive equifinality: there are effective alternative options, which are developed or assembled as required to meet the reliability requirement. Substitutability of options and strategies is high for "just-in-time" performance, where the increased volatile network behavior is matched by the flexibility in options and strategies for keeping performance within formal reliability tolerances and bandwidths.

"Just-for-now" performance for balancing load and generation

When option variety is low but volatility is high, "just-for-now" performance is dominant. It is the most unstable performance mode of the four and the one control room operators and engineers want most to avoid or get out of as soon as possible. Options to maintain load and generation have become visibly fewer and increasingly insufficient to what is needed in order to balance load and generation. This state could result from various reasons related to the behavior of the electricity system. Unexpected outages can occur, load may increase to the physical limits of transmission capacity; and the use of some options can preclude or exhaust other options, e.g. using stored hydro capacity now rather than later. Under these performance conditions, unpredictability or uncontrollability has increased (i.e., volatility has increased), with the variety of effective options and strategies diminished or less available. For instance, a Stage 1 or 2 emergency has been declared by the CAISO and a senior CAISO official goes outside of official channels and calls his counterpart at a private generator, who agrees to keep the unit online, "just-for-now."

More formally, "just-for-now" performance is a state best summed up as one of maximum potential for "deviance amplification": even small deviations in elements of the market, technology or other factors in the system can ramify widely throughout the system. Marginal changes can have maximum impact in threatening the reliability requirement, i.e. the loss of a low-megawatt generator can tip the system into blackouts. From the standpoint of reliability, this state is untenable over an extended period of time. Here people have no delusions that they are in full control. They understand how vulnerable the network is, how limited the options are and precarious the balance, they are keeping communications lines open to monitor the state of the network, and they are busily engaged in developing options and strategies to move out of this state. They are not panicking and, indeed, by prior design, they still retain the crucial option to reconfigure the electricity system itself, by declaring a Stage 3.

"Just-this-way" performance for balancing load and generation

When options variety and volatility are both low, "just-this-way" performance is dominant. This performance condition occurs in the California electricity system as a short-term "emergency" solution. In an electricity crisis, the option is to tamp down volatility directly with the hammer of crisis controls and forced network reconfigurations. The ultimate instrument of crisis management strategy is acknowledged to be the Stage 3 declaration, which requires interruption of firm load in order to bring back the balance of load and generation from the brink of "just-for-now" performance.

More formally, "just-this-way" performance is a state best summed up as one of zero equifinality: whatever flexibility could be squeezed through the remaining option and strategies is forgone on behalf of maximum control of a single system variable, in this case load. The Stage 3 declaration has become both a necessary and sufficient condition for balancing load and generation, in this case reducing load directly. You are left with only one way, or no way.

For the purposes of the chapter, we term "just-in-time" and "just-for-now" performance modes of balancing load and generation under high system volatility (i.e. the left side of the Figure 16.1 typology) to be, "real-time reliability." The differences between performance modes, along with others discussed more fully in the chapter, are summarized in Table 16.2.

When we returned to the CAISO in mid-2002 to present the results of our research – this being a year after our control room observations during the peak of the 2001 electricity crisis – we were informed that well over 85 percent of control room activity was still in the real-time reliability performance modes of "just-in-time" and "just-for-now." Yes, price volatility was no longer the problem it had been in April 2001, but now it was a continuous stream of "something else" arising out of the electricity restructuring.

Our research findings and subsequent presentations to staff at the CAISO and CPUC lead us to believe that "high system volatility" is here to stay for the foreseeable future as a result of the California restructuring. The reasoning for this is detailed in the second section of the chapter, to which we now turn. Simply put, there are very important factors not only pushing the HRN to real-time reliability, but also pulling control room operators and engineers there.

IMPLICATIONS FOR THE FUTURE OF TECHNOLOGY MANAGEMENT

Our findings at CAISO are not unique to this organization. In other systems, we have also encountered similar patterns in dealing with the increased interdependencies that these organizations face (e.g. van Eeten and Roe, 2002). The salience of these findings is further highlighted when we look at a number of trends that will only intensify inter-organizational interdependencies. Deregulation in all kinds of markets – e.g. air traffic, electricity and telecommunications – has led to the unbundling of utilities and breakup of other organizations operating the system. Increasing governmental requirements – e.g. environmental policy, consumer protection, etc. – have brought new organizations and mandates into the management of large technical systems. Furthermore, technological innovations have enabled increasing participation of third-party service providers, the outsourcing of critical technological components, and the rapid deployment of new services. And last, but not least, terrorist

Table 16.2 *Features of high reliability performance modes*

	Performance mode			
	Just-in-case	*Just-in-time*	*Just-for-now*	*Just-this-way*
Volatility	low	high	high	low
Option variety	high	high	low	low
Principal feature	high redundancy	real-time flexibility	maximum potential for amplified deviance	command and control
Equifinality	maximum equifinality	adaptive equifinality	low equifinality	zero equifinality
Operational risks	risk of inattention and complacency	risk of misjudgment because of time and system constraints	risk of exhausted options and lack of maneuverability (most untenable mode)	risk of control failure over what needs to be controlled
Variables of attention	structural variables (e.g. operating reserves)	escalating variables (e.g. cascading accidents)	triggering variables (e.g. a single push over the edge)	control variables (e.g. enforcing load shedding requirements)
Information strategy	vigilant watchfulness	keeping the bubble	localized firefighting	compliance monitoring
Lateral communication	little lateral communication during routine operations	rich, lateral communication for complex system operations in real-time	lateral communication around focused issues and events	little lateral communication, during fixed protocol (closest to command and control)
Rules and procedures	performing according to wide-ranging established rules and procedures	performing in and outside analysis; many situations not covered by procedures	performing reactively, waiting for something to happen, i.e. "I'm all tapped out"	performing to very specific set of detailed procedures
Orientation toward Area Control Error	having control	keeping control	losing control	forcing command and control

threats have brought to light unexpected links between different technological systems and have posed new vulnerabilities.

These and other developments have a common denominator: highly reliable services are more and more the product of networks of organizations, rather than individual organizations. For managers, this raises a key challenge: how can they ensure the reliability of their systems and services, while at the same time they become more and more dependent on other organizations in the network. The typical managerial response of command and control is increasingly inadequate and impossible. In light of this development toward networked reliability, the research at CAISO points to a number of important implications for the future management of technology.

First, and perhaps foremost, we expect that organizations concerned with reliability will partially shift organizational resources from long-term planning to real-time management. Conventionally, reliability was ensured by carefully designing, planning and building the organization's technological systems. The day-to-day operational management of these systems was meant to deal with the few remaining surprises and disturbances. However, organizations are more and more exposed to volatility in their environment. This seriously undermines their ability to plan and develop robust systems that are intrinsically reliable – not least because critical parts of those systems may now be outsourced. Therefore, we have seen organizational attention and resources being shifted from product development and system planning to the real-time operations that manage the systems from day to day. Practically, this may mean larger control centers with a more diverse set of professionals. It also brings out the rather different, and at times conflicting, roles for engineers and operators (see Von Meier, 1999). The shifts we have described imply and require that power is shifted from the former to the latter group of professionals.

Related to this development, we have witnessed other shifts. When planning gives way to real-time management, we should also expect a shifting emphasis from design to improvisation. As systems grow more complex, their behavior will present the organization with more surprises. "The chief manifestation of complexity is surprise," in the words of Demchak (1991). Surprises are, by definition, not covered by the systems design, and therefore require improvisation by operators if reliability is not to suffer.

A correlated shift is that from anticipation to resilience (see Wildavsky, 1988). Anticipation as a risk management strategy relies heavily on the ability to foresee future disturbances. If increased system complexity makes it more difficult to foresee all risks and deal with them through careful planning and design, then systems will need to be more resilient: that is, their ability to bounce back from disturbances becomes more important, since their ability to prevent disturbances is being undermined. This also implies a shift of attention from analysis to operational experience. The more experienced operators are, the larger their repertoire to correctly diagnose surprises when they occur.

In sum: we argue that if the management of technology is to meet the challenge of networked reliability, it needs to develop the capabilities to support the shifts from planning to real-time management, from reliable design to reliable improvisation, from anticipation to resilience, and from analysis to experience. Organizational performance, and in some cases even survival, depends on successfully navigating these shifts.

FURTHER READING

Perrow's classic study of *Normal Accidents* (1984/1999) is still highly relevant for anyone who wants to understand the challenges that are posed by complex technological systems. Weick and Sutcliffe (2001) have written a wonderful summary of the lessons of HROs for "normal" organizations. For detailed empirical analyses of how organizations respond to the challenge of networked reliability in large-scale water and electricity systems, see van Eeten and Roe (2002) and Roe *et al.* (2002).

REFERENCES

Beamish, T.D. (2002). *Silent Spill.* Cambridge, MA: MIT Press.

Demchak, C.C. (1991). *Military Organizations, Complex Machines: Modernization in the U.S. Armed Services.* Ithaca, NY: Cornell University Press.

Duncan, A.J. (1986). *Quality Control and Industrial Statistics.* Homewood, IL: Irwin.

Evan, W.M. and Manion, M. (2002). *Minding the Machines: Preventing Technological Disasters.* Saddle River, NJ: Prentice Hall.

Hey, D.L. and Philippi, N.S. (1994). "Reinventing Flood Control Strategy." Wetlands Initiative.

Juran, J.M. (1986). The Quality Trilogy: A Universal Approach to Managing for Quality. *Quality Progress*, 19 (8), 19–24.

Langer, E. (1989). *Mindfulness.* New York: Addison-Wesley.

LaPorte, T. (1994). A Strawman Speaks Up: Comments on Limits of Safety. *Journal of Contingencies and Crisis Management*, 2, 207–211.

LaPorte, T. (1996). High Reliability Organizations: Unlikely, Demanding and At Risk. *Journal of Contingencies and Crisis Management*, 4, 60–71.

LaPorte, T. and Consolini, P. (1991). Working in Practice but not in Theory: Theoretical Challenges of High Reliability Organizations. *Public Administration Research and Theory*, 1, 19–47.

Meier, A. von (1999). Occupational Cultures as a Challenge to Technological Innovation. *IEEE Transactions on Engineering Management*, 46(1), 104–114.

Morgan, G. (1997). *Images of Organization.* New York: Sage.

National Research Council. (2002). *Making The Nation Safer: The Role of Science and Technology in Countering Terrorism.* Washington, DC: National Academy Press.

Norman, D. (2002). *The Design of Everyday Things.* New York: Basic Books.

Ott, E.R. and Schilling, E.G. (1990). *Process Quality Control.* New York: McGraw-Hill.

Perrow, C. (1979). *Complex Organizations; A Critical Essay.* New York: Wadsworth.

Perrow, C. (1983) The Organizational Context of Human Factors Engineering. *Administrative Science Quarterly*, 28, 521–541.

Perrow, C. (1994) Review of S.D. Sagan, Limits of Safety. *Journal of Contingencies and Crisis Management*, 2, 212–220.

Perrow, C. (1999) [1984] *Normal Accidents.* Princeton, NJ: Princeton University Press.

Reason, J. (1972). *Human Error.* Cambridge: Cambridge University Press.

Rijpma, J.A. (1997). Complexity, Tight Coupling and Reliability. *Journal of Contingencies and Crisis Management*, 5, 15–23.

Roberts, K. (1990). Some Characteristics of One Type of High Reliability Organization. *Organization Science*, 1, 160–176.

Roberts, K. (1993). *New Challenges to Understanding Organizations.* New York: Macmillan.

Rochlin, G.I. (1993). Defining High Reliability Organizations in Practice. In K. Roberts (ed.), *New Challenges To Understanding Organizations*: 11–32. New York: Macmillan.

Rochlin, G.I. and Meier, A. von (1994). Nuclear Power Operations; A Cross Cultural Perspective. *Annual Review of Energy and Environment*, 19, 153–187.

Roe, E., Eeten, M.J.G. van, Schulman, P.R. and Bruijne, M. de (2002). *Real-Time Reliability: Provision of Electricity under Adverse Performance Conditions Arising from California's Electricity Restructuring and Crisis.* A report prepared for the California Energy Commission, Lawrence Berkeley National Laboratory, and the Electrical Power Research Institute. San Francisco, CA: California Energy Commission.

Sagan, S. (1993). *The Limits of Safety.* Princeton, NJ: Princeton University Press.

Salvendy, G. (1997). *Handbook of Human Factors and Ergonomics.* New York: Wiley.

Sanne, J.M. (2000). *Creating Safety in Air Traffic Control.* Lund, Sweden: Arkiv Forlag.

Schulman, P.R. (1993a). The Negotiated Order of Organizational Reliability. *Administration and Society*, 25, 353–372.

Schulman, P.R. (1993b). The Analysis of High Reliability Organizations: A Comparative Framework. In K. Roberts (ed.), *New Challenges To Understanding Organizations*: 33–54. New York: Macmillan.

Schulman, P.R. (2002). Medical Errors: How Reliable is Reliability Theory? In M.M. Rosenthal and K.M. Sutcliffe (eds), *Medical Error*: 200–216. San Francisco, CA: Jossey Bass.

Schulman, P.R., Roe, E., Eeten, M.J.G. van and Bruijne, M. de (2004). High Reliability and the Management of Critical Infrastructures. *Journal of Contingencies and Crisis Management*, 12 (1), 14–28.

Turner, B.M. (1978). *Man-Made Disasters.* London: Wykeham.

Van Eeten, M.J.G. and Roe, E. (2002). *Ecology, Engineering and Management: Reconciling Ecosystem Rehabilitation and Service Reliability.* New York: Oxford University Press.

Vaughn, D. (1996). *The Challenger Launch Decision.* Chicago, IL: University of Chicago Press.

Weick, K.E. (1993). The Vulnerable System: An Analysis of the Tenerife Air Disaster. In K. Roberts (ed.), *New Challenges to Understanding Organizations*: 173–197. New York: Macmillan.

Weick, K.E. and Sutcliffe, K.M. (2001). *Managing the Unexpected.* San Francisco, CA: Jossey Bass.

Weick, K.E., Sutcliffe, K.M. and Obstfeld, D. (1999). Organizing for High Reliability. *Research in Organizational Behavior*, 21, 81–123.

Wildavsky, A. (1988). *Searching For Safety.* New Brunswick, NJ: Transaction Books.

Chapter 17

Managing performance in firms

Hans de Bruijn

OVERVIEW

This chapter deals with performance management in primarily privately owned high tech firms. First, the advantages and disadvantages of performance management will be outlined. Then a number of conditions, under which the disadvantages will emerge will be discussed. More interesting than this inventory is the question under what conditions the positive or negative effects will dominate. This issue will be further discussed and the chapter will conclude with a number of recommendations for the proper use of performance management in companies.

INTRODUCTION

In 2004, Shell, the Anglo-Dutch oil company, suddenly announced that it had seriously overestimated its proven oil and gas reserves. The reserves, as stated in the accounts, should be reduced by more than 20 percent because they no longer satisfied the criteria of the SEC. The consequences for Shell were very serious: top executives were forced to quit and Shell was fined by the SEC and the SFA, Britain's stock-exchange watchdog. US attorneys launched an investigation into a number of Shell's managers and former managers.

One major cause of overestimating the reserves was the system of performance management that Shell used. The bonuses of a number of important players in the organization were linked to the volume of the proved reserves. The bigger these reserves were, the bigger were the bonuses. According to Roel Murris, Shell's former Exploration Director: "Then you encourage people to present too rosy a picture" (De Graaf and Meeus, 2004).

It is not the first time a performance management system has had such a negative effect. One year earlier, Royal Ahold, the parent of US Foodservice (USF) among other companies, earned a bad name. USF was found to have misrepresented purchase discounts, enabling it to inflate its figures and meet head office expectations. The pressure to meet targets eventually led to a major accounting scandal and the departure of top executives (Smit, 2004). To show his audience that accounting scandals occur not only in the US, SEC manager Paul Atkins placed Royal Ahold in a row of European companies such as Vivendi, Swissair and Robert Maxwell.

Apparently, performance management systems can have a perverse effect. For what reason? This chapter shows that the use of performance management stems from a major dilemma. On the one hand, a company's management needs sound and reliable information about the performance of the various parts of the organization. The mere fact that an enterprise is accountable to its shareholders requires it. On the other hand, the primary process in many companies will turn out to be so complex that it is difficult to reduce it to a few performance figures. Should companies do so, nonetheless, they may suffer major adverse effects. This chapter focuses on this tension between the need to account for one's performance in figures on the one hand, and the complexity of the primary process on the other.

THE ADVANTAGES AND DISADVANTAGES OF PERFORMANCE MANAGEMENT

Performance management may be defined as measuring and quantifying the performance of groups or individuals as objectively as possible (see also Chapter 3). Performance management takes place where this objectively measured and quantified performance is used as a management tool. It is used to make performance agreements, for example, or to award bonuses or as a benchmark for groups or individuals in a company (see Harvard Business Review, 1988; Behn and Kant, 1999; De Bruijn, 2002; Johnson, 1991; Klein and Carter, 1988; Smith, 2001, for overviews). In the remainder of this chapter, I will refer to "actors" within the organization, by which I mean individuals, groups and organizational actors such as divisions or units.

Like everything else, a system of performance management has both advantages and disadvantages (review of advantages and disadvantages based upon De Bruijn, 2002). Of course, the first major advantage is that performance management is an incentive to perform. Actors are set a clear target and know what is expected of them: to reach the target. Performance management thus provides a company's business process with a clear focus and has a mobilizing function: every effort is made to reach the target. Performance management can also be an incentive to formulate goals as ambitiously as possible ("goal stretching"), which may continually improve performance. Research shows that this incentive is effective in many cases, both in the public and in the private sector.

The second advantage is that performance management provides managers with the most relevant information about the activities performed in the capillaries of the organization at the most operational level. Technology-intensive enterprises tend to have a complex and opaque primary process. Highly educated and/or specialized engineers do work that requires specific expertise and experience. This makes the task for the management of such enterprises difficult. Not all managers have the same in-depth expertise as the engineers in the operation. By definition, managers also have a smaller span of control and are therefore unable to enter into the details of the operation. Performance management solves this problem: it offers managers information about the performance of the organization's operational levels. The information reflects the essence of the performance without requiring in-depth knowledge of the operation. In addition, the information is highly compact, which matches a manager's limited span of control.

The third advantage is that performance management improves transparency in a company. It makes clear who contributes what to the desired performance. This is important management information, which can be used to make decisions about strategic, organizational and operational issues. Performance management can thus *rationalize* decision-making: activities that do not clearly contribute to an organization's results are suspect.

However, performance management also has a number of major disadvantages. First, performance management is an incentive for strategic behavior or "gaming the numbers" (Osborne and Gaebler, 1992). In every organization of some size and complexity there are plenty of possibilities for the actors on the operational levels to fiddle the figures. The events at Shell and Royal Ahold are examples of this: what is delivered is output on paper, which has nothing to do with reality. The literature contains numerous other examples of organizations where only the output on paper is raised: schools, the FBI, municipalities. At best, this output is meaningless, and at worst, it is misleading and fraudulent.

Second, performance management may cause fixation on quantitative performance, sometimes called targetitis or myopia (Pollitt, 2003). Such a fixation is risky:

- It may harm *entrepreneurship* in the organization. An organization should be able to take advantage of unforeseen opportunities. In a dynamic environment, such opportunities always occur and entrepreneurship is largely the art of taking advantage of them. Overemphasizing the achieving of pre-arranged performance may blind an organization to such opportunities. Research by Iaquito among Japanese companies illustrates this. Companies focusing too long on a particular set of indicators develop a one-sided orientation and, as a result, eventually perform worse (Iaquito, 1999).
- It may harm *professionalism* in the organization. An example of this is a rat catcher who is judged by the number of rats caught and therefore only tries to catch as many rats as possible. As a result, he no longer asks himself whether he might be catching too many of them, leaving the rats that survive so much physical space that they have more and stronger young rats.
- It may harm *system responsibility*. Performance management may lead to a strong focus on nothing but the quantified performance that has to be delivered. An organization has many other tasks that are crucial to its functioning, but that cannot be expressed in quantitative indicators and that may affect its own performance. Examples of these are long-term interests, strategic interests or cooperation with other actors if they create synergy for the company as a whole. An example of this is an IT consultancy, comprising several divisions. Each division tries to reach its annual targets for turnover and billable hours, and the system of performance management creates strong rivalry between the divisions. This rivalry may prevent the divisions from cooperating, since cooperation allows the other divisions to look behind the scenes and possibly copy advice models or snatch customers. Moreover, if realizing the annual targets takes up all the energy,

> not enough time may be freed up for making new advice models. This is not without risk, because the existing models will be obsolete at some point.

Third, performance management may optimize input. This is also called "creaming" or "cherry picking." Many schools that are judged by output show this behavior: they only admit the better students or, if this is not allowed for legal reasons, they use a "counseling out" strategy, advising weak students to choose a different school. Input optimizing means that a school compromises on its ambitions. Good performance is delivered by admitting good students; if the school also admits bad students, the drive for good performance is far more ambitious.

As a result, information about quantitative performance may *pervert*: the information fails to present a correct picture of the performance: the extent of gaming the numbers, targetitis or input optimization is unclear, because good quantitative performance may mean that an actor has, for example, neglected its professionalism or system responsibility.

Performance management, therefore, has advantages as well as disadvantages. The question is now under what conditions these advantages or disadvantages will manifest themselves. I will address this issue in the following sections.

THE RISK OF TYPE 2 RESULTS

In the first place, the nature of the performance is important. Table 17.1 distinguishes between type 1 performance and type 2 performance (see also De Bruijn, 2002).

Type 1 performance is amenable to a system of performance management; type 2 performance is so to a lesser extent. Technology-intensive enterprises tend to deliver type 2 performance. The following should make this clear.

Unambiguous versus ambiguous

An unambiguous product is easy to measure and quantify objectively, whereas an ambiguous product is not. A "proved reserve" is an ambiguous product that cannot be determined unambiguously. According to the SEC:

> Proved oil and gas reserves are the estimated quantities of crude oil, natural gas, and natural gas liquids which geological and engineering data demonstrate with reasonable certainty to be recoverable in future years from known reservoirs under existing economic and operating conditions.
>
> (SEC, 2001)

This definition leaves ample room for interpretation ("estimated" quantities, "reason-

Table 17.1 *Two types of performance*

Type 1 performance	*Type 2 performance*
Unambiguous product	Ambiguous product
Causal relationship actor's action and result	Non-causal relationship
Divisible, unbundled	Indivisible, bundled
Aligned criteria	Competing criteria
Output = outcome	Outcome > output

able" certainty) by which a "proved reserve" cannot be measured objectively. Moreover, a "proved reserve" is difficult to quantify. Although there are technologies to establish the volume, uncertainty remains high, in many cases higher than by a factor of 2. A clearer picture will not emerge until the fields are actually exploited.

The consequence for performance management will be clear. When the performance for ambiguous products is quantified, there is considerable room for gaming the numbers, which actors can use when accounting for their performance.

Causal versus non-causal

In type 1 performance, there is a causal relationship between an actor's action and that actor's results. In type 2 results, there is no such causal relationship and an action–result pattern of thought is far too simple. There may be several reasons for this:

- Actions of third parties may partly determine the result. The results of a school partly depend on the efforts of the parents. The school can only influence the behavior of these parents to a limited extent.
- The causality between action and result is unknown. Take an engineering consultancy firm that uses the volume of the commissions it wins as a performance indicator. Who can claim the performance if a customer contacts the consultancy by telephone and gives a commission? The person who happens to answer the telephone? The person who advised this customer earlier?

Here, too, if performance is measured by results that cannot unambiguously be reduced to the activities of one actor, it presents a distorted picture of reality, because there is plenty of room for credit claiming and so for gaming the numbers. The link made between activity and result is a social construction and the result of, for example, negotiations between actors (e.g. consultants mutually agree who is allowed to report what performance).

Divisible versus bundled

The next question concerns the extent to which various achievements are intertwined and therefore influence each other. If they are intertwined, one performance is difficult to isolate from another, which hampers performance management.

A "bonanza," a big oil or gas find, tends to be a matter of good luck, and requires a great deal of expertise and perseverance. This makes it difficult for oil companies to decide how much they will bid for an oilfield: the value of such an oilfield cannot be established unambiguously and ex ante. In addition, there is keen competition. In such a situation, many oil companies use the strategy of the ubiquitous overbid: they overbid to make sure they will secure an oil field. This makes many oil fields loss-making operations, but these losses are compensated because there are always a few bonanzas among them.

This makes the performance of isolated oilfields difficult to measure. Performance should always be regarded in the context of the larger whole. If the same profit requirement is set for each oilfield, as in the case of Shell, it will have a perverting effect. Achievements are bundled, so a quantitative requirement for each performance will clash with professional practice, in which bonanzas play an important part. Performance management may thus affect, for example, professionalism and entrepreneurship in an organization (de Graaf and Meeus 2004).

Aligned criteria versus competing criteria

Almost every product that a company makes has to satisfy several criteria simultaneously: it has to meet economic criteria, quality criteria, legal criteria and safety criteria. It has to meet both short-term and long-term criteria, etc. Some of the products have to meet competing criteria: a strong accent on the one value may affect the other value. If there are competing criteria, the risk is that performance management will focus on one or just a few criteria and that the other, or the others, will be neglected. Research shows that, in many cases, fund managers' bonuses are based on short-term performance (i.e. one year or less) of their funds (Morning Star European Fund Trends Survey, September 2004). According to the researchers, this is "worrying," because criteria for short-term and long-term performance may clash and many clients are interested in the performance over a number of years. Performance management may thus lead to targetitis (seeking short-term success) and harm the long-term effect. Here, too, performance management will harm professionalism and entrepreneurship within an organization because it hampers a trade-off between short-term and long-term.

Output = outcome or outcome > output

Outcome is the ultimately intended effect, for example the safety level (police), the educational level (school) or the degree of continuity (company). Output is the direct and, in many cases, clearly measurable effect: the clear-up rate (police), the test score (school) or the volume of the profit or loss (company).

The distance between output and outcome may be either greater or smaller: do more arrests improve safety? Does a good test score mean that a school delivers students with a high educational level? Or does it mean that it trains its students particularly in taking tests? Do high profits enhance a company's continuity or have long-term investments been scrapped to boost annual shareholder value? The bigger the distance is between output and outcome, the more problematic is performance management, because high output need not improve outcome. Performance management may cause an actor to opt for achieving output, even though the contribution of this output to the outcome is limited and drives out professional and entrepreneurial considerations.

The conclusion is that the more the results of an organization have the features of a type 2 result, the more limited is the significance of quantitative performance. Consequently, the risks of perverted information (information that does not represent reality because of gaming playing, input optimization etc.), in type 2 results are greater than in type 1 results.

The characteristics of type 2 performance are applicable to an MOT-context. An example may clarify this. The product of the auto industry seems to be type 1. But on closer consideration, the product also has features of a type 2 performance. There can be tensions between the esthetics of a design and safety requirements, or between a car's environmental performance and it's power (competing criteria). The performance even has some ambiguous elements. Does the management of the industry want to sell as many cars as possible or is it also interested in creating an image of the car and building long-term relationships with its clients? An image or long-term relationship is far less amenable to performance management than the number of cars sold.

THE RISK OF THE LAW OF DECREASING EFFECTIVENESS

Another question is how an organization uses performance management. What is the managerial function of performance management? We can distinguish a number of functions:

- *Information* Performance management provides insight into the quantitative performance of actors. This quantitative information is used by management, but other information sources are also used, particularly qualitative ones. This allocates a modest role to performance management: there is loose coupling between the quantitative performance and the judgment.
- *Information plus judgment* Performance management provides insight into quantitative performance and this information is an important basis for a judgment: there is tight coupling between the quantitative performance and the judgment. On the basis of the quantitative data, management judges that a performance is either good or bad, that performance has either improved or worsened, is either better or worse than that of others, etc.
- *Information plus judgment plus financial consequences* This means that financial consequences are attached to the output figures. There is tight coupling between the output figure and, for example, a personal bonus, extra funds or the hiring and firing policy.

If products are type 1 results, the tight coupling between performance measurement and financial consequences is no problem at all. The tighter this coupling is, the stronger are the incentives to perform. The actors in the organization will start behaving in accordance with the incentives posed by performance management. They will make every effort to reach the output target. In type 2 results, however, another

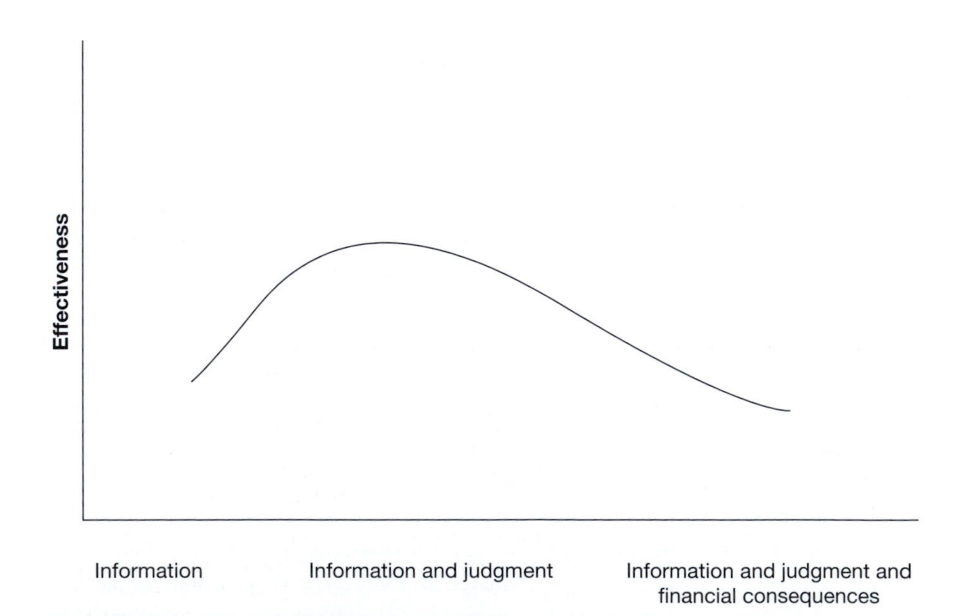

Figure 17.1 *The Law of Decreasing Effectiveness*

mechanism operates. The pressure of meeting the performance norms may lead to gaming the numbers, targetitis or input optimization. Every effort is made to meet the output norm, even though it adversely affects entrepreneurship, professionalism and system responsibility or requires strategic behavior, because the costs may be very high if the quantitative performance is not achieved. This is called the Law of Decreasing Effectiveness: the tighter the coupling is, the stronger are the incentives for perverting the information. The manager who provides powerful steering by performance management thus causes the adverse effects (De Bruijn, 2002).

The conclusion is that the more performance is judged by quantified type 2 performance, the stronger are the incentives for perverting the information. This is exactly what happens at Shell. External regulator Anton Barendrecht warns in three successive annual reports that the reliability of the estimated reserves will come under pressure by linking bonuses to reserves. Dutch research journalists have received confirmation from several sources that the country managers within Shell have rebuked Barendrecht several times, telling him to "keep his paws off the bonuses" (De Graaf and Meeus, 2004). Meanwhile, many people in the organization believed that Shell was doing very well.

THE RISK OF COLLECTIVE BLINDNESS

Suppose a system of performance management has the adverse effects. "Gaming the numbers" will be played, information will be inflated, input will be optimized or entrepreneurship and system responsibility will be neglected on a large scale. Why does nobody intervene? Why can a situation such as the one at Shell or at Royal Ahold persist for so long? Table 17.2 holds the answer to this question.

The management of an organization may pay attention to particular subjects: it is either interested in them or it is not (Davenport and Beck, 2001). The greater the interest it takes in them, the more information it wants to receive about them. In addition, a management may, or may not, hold strong views about particular subjects (i.e. preferences). If the management of an organization introduces a system of performance management, it implies that it pays considerable attention to particular subjects (e.g. oil reserves) and that it also has clear preferences (e.g. oil reserves should be as high as possible).

On an imaginary spectrum, one extreme is a management without attention and without preferences, the other extreme is a management that pays great attention and has strong preferences.

Table 17.2 *Information flows between management and actors for type 2 results*

	Top management: No attention, no preferences	*Top management: Strong attention, strong preferences*
Actors in the organization: low failure costs	1 No information supply to management	2 Balanced information supply to management
Actors in the organization: high failure costs	3 Incentive for over-egging information to draw the attention of management	4 Incentive for over-egging information to satisfy management and prevent intervention by management

The actors in the organization are faced with a system of performance management. If this system has many functions (information supply plus judgment plus reward) and the performance cannot be achieved, failure costs will be high. This is not so if a system has limited functions.

The table shows four patterns in the information flows in an organization.

- If a management pays no attention nor has any preferences and the failure costs are low (quadrant 1), there are no incentives to inform the management. "Low failure costs" means that an under-performance has relatively limited consequences.
- If there is little attention and there are no preferences, but the failure costs are high (quadrant 3), there are incentives to over-egg information so as to draw the attention of the management, for example to change the system of performance management.
- If there is great attention and there are strong preferences, but the failure costs are low (quadrant 2), there are either limited incentives or no incentives at all for over-egging information. The information supply is likely to be well-balanced.

A major risk is imminent in quadrant 4. Suppose the management of an organization pays great attention and holds strong views and the failure costs of an actor are high. This may spark a very dangerous dynamic:

- The management pays attention to particular issues, has its preferences and is therefore highly sensitive to information on these issues.
- The actors confirm the preferences by perverting information: they supply the information that confirms the management's preferences, but which, for example, is based on gaming the numbers or targetitis. Roel Murris, Shell's former Exploration Director: "If you demand success of your subordinates, you'll get it. Especially if it involves a financial or career bonus" (De Graaf and Meeus, 2004).
- This results in a peaceful balance between management and actors. The management is satisfied because the targets it fixes are apparently achieved. The actors are satisfied because they are protected from interventions by the management or are even rewarded by the management. The perverted information has become established in the organization. Collective blindness is the result.

The conclusion is that a perverted system of performance measurement can become established in an organization, if the organization has insufficient checks and balances and the system is therefore insufficiently dynamic.

THREE PRINCIPLES FOR PERFORMANCE MANAGEMENT

The question that now presents itself is how to deal with a system of performance management, because despite all the criticism leveled at systems of performance management, they continue to be necessary. They solve the problem of the limited expertise and the limited span of control of managers in technology-intensive enterprises and have major advantages. Suppose there is type 2 performance – most products in large, complex organizations will be of a type 2 nature – and that there is a system of performance management: how can it be arranged so as to prevent the adverse effects? How can there be a system of performance management that contains incentives for performance on the one hand while moderating the incentives for the negative behavior as much as possible on the other hand? To answer this question, I present three principles for performance management in type 2 results.

Principle 1: no direct coupling between output and high financial rewards

This rule of the game logically follows from the Law of Decreasing Effectiveness: there must be no direct link between output and a high financial reward. Direct coupling means that the fixing of the reward is a self-executing process: if an actor achieves the quantified performance, the reward agreed has to be paid. This is very risky for type 2 results: the higher the reward, the stronger is the incentive for perverse behavior.

One risk of this rule of the game is the loss of predictability of performance management, because an actor that delivers a performance is not sure whether it will earn a reward. It may cause the loss of one of the important advantages of performance management: incentivizing performance. How should this risk be addressed?

- Use a mixed system for type 2 results: the reward is directly linked partly to the quantified performance, partly to the quantified performance plus an additional qualitative judgment. This is a disincentive for the adverse effects: those who only focus on reaching the production target and neglect other aspects risk missing their additional rewards. In concrete terms, a bonus should never depend completely on a proven reserve. If an actor states that a proven reserve has increased greatly, a qualitative judgment is passed as well: how did this actor deal with concepts such as "estimated" quantities, "reasonable" certainty? Were they estimated over-optimistically or was a conservative estimate made?
- Agree a number of rules of the games to make the award of this additional reward more predictable. For example, an organization might very well indicate beforehand what qualitative aspects of the performance it regards as important and will include in its judgment (also see principle 2).

Principle 2: in addition to paying attention to quantities, always pay attention to underlying processes

The essence of the above explanations for the negative effects is that steering is only based on figures, although by their very nature figures present a distorted picture of reality. This is also the essence of the remedy: organizations should shape their system of performance management in such a way that other, qualitative, information also has its place in this system. The literature mentions the need for a process approach in addition to a product approach. In a product approach, the key words are nouns: profits from an oilfield, arrests, student pass rates. In a process approach, the key words are verbs: to prospect for and exploit oilfields, to solve crimes, to teach. What matters is the way the performance came about. Far more than a product approach, a process approach leaves room for the ambiguity of the performance when the performance is judged.

- Room for the rat catcher to explain that he has caught fewer rats than last year

because catching more rats would only result in a stronger population.
- Room for an actor that exploits oilfields to explain that the bonanza elsewhere compensates poor performance.

This idea may be worked out in practice by involving an interview protocol in the judgment, covering the main qualitative aspects of the performance. The interview protocol may be built up round the potential adverse effects of performance management and the characteristics of type 2 results. Examples of questions are:

- How did the actor's input come about? How does the performance delivered relate to the organization's commitment to entrepreneurship and professionalism?
- What activities were undertaken that, although they do not produce quantitative performance, fall within an actor's system responsibility?
- How was the ambiguity of the performance dealt with?
- How was the trade-off between competing values made? How was it prevented from working out one-sidedly to the advantage of the quantitative indicators?
- How does the output achieved relate to the outcome desired? How is "good output, bad outcome" prevented?

Such an interview protocol may enhance the predictability of the judgment (see principle 1). A qualitative approach may confirm the picture presented by the quantitative data. It may also unmask it: although the quantitative performance looks fine, the reason may be that the actor has compromised too much on entrepreneurship.

Principle 3: dynamize performance management

Where performance and reward are linked directly, the role of management is merely administrative: it checks whether a performance has been delivered and, if so, it pays the reward. Adding qualitative aspects to the system of performance management makes performance management a lively activity. In a game of interaction, management and actor form a rich picture of the performance. This game requires a knowledgeable manager, who has many learning opportunities during this game.

The information supplied by the various actors may be used to make mutual comparisons and then to develop the system of performance management further. This makes the system dynamic.

- New qualitative information may lead to changes in the quantitative performance norms. A railway company faced with the tension between departing on time and allowing time to change trains may introduce an extra performance indicator: the time to change trains may also be measured.
- New qualitative information may lead to changes in or tightening of the interview protocol. A manager who learns that catching too many rats may result in an extra strong new population can ask the rat catcher under what conditions the phenomenon occurs. If it occurs mainly in dry springs, (which limit the rats' habitat and therefore cause considerable natural weakening of the population), this fact may be used to tighten the interview protocol.

As a result of this dynamic, a system of performance management continues to develop and becomes a learning system, and the risk of collective blindness will decrease. Moreover, the system of performance management develops in interaction between managers and actors. This is important, because too much dynamic is undesirable. It makes a system insufficiently predictable, which may cause it to lose its incentives for performance. Dynamic resulting from interaction between managers and actors means that both of them influence the system of performance management, because an actor who feels that changes to the system destroy the incentive for performance can put this forward in the interaction with the manager. Interaction thus helps to strike a balance between stability and dynamic. This is what performance management is all about: striking balances between the many dilemmas surrounding performance management.

Finally, the trends described in this book – bundling and unbundling of chains, privatization, decentralization, etc. – imply that companies have to meet more and more standards. Technology-intensive companies have to operate in a less and less predictable context. This makes considerable demands on managers in such companies. If fewer and fewer issues are self-evident, they have to make decisions on more and more matters. They can do so only if they have the correct information. Systems of performance management provide this information, but they also pose the risks set out in this book. Managers should realize these risks to safeguard the quality of the information supply.

FURTHER READING

Bacal, R. (1998). *Performance Management.* New York: McGraw-Hill Education.

Daniels, Aubrey C., Daniels, James E. (2004). *Performance Management: Changing Behavior that Drives Organizational Effectiveness.* New York: McGraw-Hill Education.

REFERENCES

Behn, Robert D. and Kant, Peter A. (1999). Strategies for Avoiding the Pitfalls of Performance Contracting. *Public Productivity and Management Review,* 22 (4), 470–489.

Bruijn, Hans de (2002). *Managing Performance in the Public Sector.* London: Routledge.

Davenport, Thomas H. and Beck, John C. (2001). *The Attention Economy.* Boston, MA: MIT Press 2.

Graaf, Heleen de and Meeus, Tom Jan (2004). Dossier Shell. In *NRC Handelsblad,* see also www.nrc.nl.

Harvard Business Review (1998). *Harvard Business Review on Measuring Corporate Performance.* Boston, MA.

Iaquito, Anthony L. (1999). Can Winners be Losers? The Case of the Deming Prize for Quality and Performance among Large Japanese Manufacturing Firms. *Managerial Auditing Journal,* 14 (1/2), 28–35.

Johnson, Thomas H. (1991). *Relevance Lost. The Rise and Fall of Management Accounting.* Boston, MA: Harvard Business School Press.

Klein, R. and Carter, N. (1988). Performance Measurement: a Review of Concepts and Issues. D. Beeton (ed.) *Performance Measurement. Getting the Concepts Right.* London: Public Finance Foundation.

Morning Star European Fund Trends Survey (September 2004).

Osborne, David and Gaebler, Ted (1992). *Reinventing Government.* Reading: Penguin.

Pollitt, Christopher (2003). *The Essential Public Manager.* Maidenhead: Open University Press.

SEC (2001). Financial Accounting and Reporting for Oil and Gas Producing Activities Pursuant to the Federal Securities Laws and the Energy Policy and Conservation Act of 1975. Washington, DC.

Smit, J. (2004). *Het Drama Ahold.* Amsterdam: Balans.

Smith, M., Fiedler, B., Braun, B. and Kestel, J. (2001). Structure Versus Appraisal in the Audit Process: a Test of Kinney's Classification. *Managerial Auditing Journal*, 16 (1), 40–49.

Chapter 18

Management dilemmas and strategies in practice

Willemijn M. Dicke

OVERVIEW

Managers are confronted with many developments that arise from both inside and outside the organization. In responding to these developments, managers come to realize that some trends demand conflicting responses: an action that provides an adequate answer to one development may have an adverse effect on a different part of the organization. We asked six senior managers who deal with issues of technology and innovation on a daily basis to describe the dilemmas that they face in their work and their strategies for overcoming these dilemmas. The aim of this chapter is to show the practice of technology and innovation management: what do managers of technology actually do with all of the theories and insights that have been presented in the former chapters?

INTRODUCTION

Managers are confronted with many developments that arise from both inside and outside the organization. They seek to respond adequately to all of these developments, whether they involve globalization, the development of new technologies, new legal requirements affecting the financial management of the production line, or demographic changes that result in the need for different products. In responding to these developments, managers come to realize that some trends demand conflicting responses: an action that provides an adequate answer to one development may have an adverse effect on a different part of the organization.

We asked managers to use their own words to describe the dilemmas that they face in their daily work and their strategies for dealing with these dilemmas. Six senior managers working in typical MOT contexts tell their stories: an innovation manager (Siemens), a global business manager CO_2 (Shell), a senior strategist (Gasunie), a manager of a network operating centre (KPN Mobile), a recruitment manager for Research and Development (Unilever) and an open source developer.

The interviews cluster around three themes. The first two interviews deal with dilemmas regarding innovation (Siemens and Shell); the second theme concerns how to deal with increased uncertainty (Gasunie and KPN). The third theme is about the design of the firm: how can multiple, often conflicting, requirements be incorporated into one organizational design (Unilever and open source software development). The interviews show that the issues raised in earlier chapters are reflected in the day-to-day practices of these managers. The boxes contain theoretical reflections and explicit references to previous chapters.

The aim of this chapter is to show the practice of technology and innovation management: what do managers of technology actually *do* with all of the theories and insights that have been presented in the former chapters? Let us now turn to the managers. These are their stories.

INTERVIEW WITH HANS HELLENDOORN, INNOVATION MANAGER, SIEMENS NEDERLAND N.V.

Siemens Nederland is part of Siemens International, a company that is active in more than 190 countries. This multinational is specialized in electrical engineering and electronics.

Siemens Nederland offers products, systems and services in the fields of energy, living, working, mobility, education, communication, safety, health and industry. Innovation is the spearhead of their commercial policy.

Hans Hellendoorn is the innovation manager for Siemens in the Netherlands. He is responsible for the generation of new ideas, the process of innovation and the successful implementation of the innovations. His tasks involve both the organizational and the technological aspects of the innovation process. Hellendoorn faces a number of dilemmas. We first discuss the dilemma arising from Siemens in relation to its wider context. Second, we address dilemmas of internal organization.

What if societal preferences change overnight?

For the commercial areas of Siemens (e.g. mobility, health, education) the government is an important stakeholder. This means that Siemens cannot plan and implement innovations in isolation. Instead, the company must cooperate with the stakeholders. While cooperation is a precondition for successful innovation, it makes the innovation process more complex. Hellendoorn illustrates his point with the example of innovation in road pricing.

At present, Dutch car owners pay a flat tax, independent of the frequency or time of day in which they make use of the highway. As a means of fighting traffic jams, the Netherlands Ministry of Transport and other stakeholders developed the idea of road pricing. In this concept, fees are determined as a function of both the frequency and the time of day that motorists use the highway: highway tariffs during rush hour are higher than those during off-peak hours.

All stakeholders had long been convinced that road pricing would provide a sustainable solution to mobility. This encouraged Siemens Nederland to develop and refine the techniques and systems needed for road pricing. After years of research, the company was able to present a road-pricing system that worked and was ready to be tested in a pilot study. Almost overnight, however, the atmosphere changed completely. Regardless of how sophisticated road pricing had become, and regardless of how many academics and policy makers were in favor of the system, politicians decided against the road-pricing experiment. Years of research were nullified.

This is one example of how societal preferences can change overnight. How does Siemens deal with this level of uncertainty? One possible response, of course, is to keep the innovation on the shelf. Should the societal preferences change once again, a new window of opportunity could arise. Keeping innovations on the shelf, however, is not without risks: competitors might run off with the invention, or the employees that were the innovators of this system may leave the company. The departure of such people is always accompanied, at least in part, by the departure of their knowledge. Although "storing" an innovation may sometimes be necessary, it is not an outcome for which to aim. On the contrary, the general guideline within Siemens is to seek return on investment within a period of two years for all innovations.

Strategies for dealing with uncertainty

Siemens Nederland has sought for more proactive ways to deal with uncertainty arising from shifts in societal preferences and has developed two strategies. The first is a method for keeping abreast of societal preferences. Siemens Nederland derives its focus for innovation from the governmental program, which is presented once every four years. This program is thought to reflect societal preferences at a given point in time.

Another strategy is to spread the risk. It is important that companies do not focus solely on one area or on one innovation. In addition to the governmental program, Siemens selects a few other areas of importance that are not addressed by the program.

This illustration provides a clear example of the relation between technology and society (Chapter 6). Success in innovation and success in the diffusion of innovation (Chapter 7) require close ties with society. Hellendoorn has institutionalized this link by taking the governmental program as the starting point for his company's innovations. In addition, storage of knowledge innovations further relates to the chapter on Knowledge Management (Chapter 14).

From product innovation to integral solutions

A decade ago, managers of railway stations may have demanded more sophisticated monitoring cameras in order to increase safety at their stations. At present, such a manager is more likely to ask a supplier to provide a safe railway station, and not just a few cameras. For this reason, most innovations at Siemens seek to provide integral solutions. Siemens does not propose to deliver a set of cameras; instead, it offers "a safe railway station solution" that includes a full security staff, a call center and all other relevant components. While technological innovation is still an important part of these innovations, the combination of different links in the value chain is gaining increasing attention in the innovation process.

The shift from product innovation towards integral solutions in the innovation process creates new organizational challenges for innovation managers. In order to explicate this challenge, we provide a brief sketch of how innovation is organized at Siemens Nederland.

A separate R&D department is set up for the international holding. Radical technological innovations are developed within this central department. At Siemens Nederland, innovations are developed and implemented by employees within each product division. There is no separate R&D department at the subsidiary in the Netherlands. Innovations are developed in a project organization, meaning that the innovation manager "borrows" employees from a product division for certain innovation projects.

The current structure has both advantages and disadvantages. One advantage is that the employees working on a certain innovation are aware of the needs and demands of the market. In this way, Siemens seeks to avoid situations in which specialists put their effort in the development of a perfect technology that nobody wants. The risk of such a project organization, however, is that people are very much focused on their own specialized product divisions.

To follow up on the example of innovations regarding integral safety: an excellent engineer from a product division is likely to focus on refining the technology of a camera, but is not likely to think of new integral concepts for "providing safety," which implies the consideration of call centers, security people and communication solutions. In other words, it is often difficult for specialized engineers to "think outside the box." It is precisely this creative thinking across different segments of the value chain that is needed.

How do innovation managers deal with this dilemma? Hellendoorn is developing incentives to stimulate cross-sectoral and cross-segment thinking, and these incentives apply not only to innovators. For example, in the past, account managers had targets. As a result, they would seek to sell as many cameras as possible. This made perfect sense in the former situation. In the new situation, however, account managers should be rewarded for selling the integral concept of safety, or for involving new stakeholders that enable Siemens to provide more integral services. This situation requires a different performance management system.

The theories explained in Chapter 10 (inter-organizational decision-making) and in Chapter 17 (managing performance in firms) are relevant for innovation. Apparently, Siemens Nederland is of the opinion that innovation is not something that can be done by "innovation specialists." It must be conducted with specialists from the field who know their markets and the developments in those markets. In order to innovate, various disciplines within the organization are needed. This requires considerable intra- and inter-organizational decision making (as opposed to the idea that innovation takes place within the confined space of an R&D laboratory) and monitoring of the developments in society (see Chapter 6). Moreover, innovation requires changes in performance management (Chapter 17). Instead of relying solely on measurable outputs, such systems should provide rewards for less tangible elements, such as networking and involving the stakeholders. This illustration also relates to Chapter 3, with regard to pay and rewards (performance management).

INTERVIEW WITH STIJN SANTEN, GLOBAL BUSINESS MANAGER CO_2, SHELL GLOBAL SOLUTIONS[1]

Shell Global Solutions International (SGSi) is a 100 percent daughter company of the Royal Dutch Shell Group. SGSi is a consultancy company that offers its services to subsidiaries of Shell, as well as to other companies in the energy sector. Stijn Santen is the founder of a new business within SGSi that sells a portfolio of novel services under the heading of carbon dioxide (CO_2) emissions management. The goal of the emissions management consultancy is to help companies compete effectively in business environments in which there are legislative or other incentives for curbing carbon dioxide emissions.

We begin by describing the innovation process in a large company: how can an innovator convince peers, line managers and senior management that an idea is path breaking and worthy of the investment of Shell resources? New challenges arise once the organization has been convinced: how can a company sell a product to customers that have not – yet – asked for it?

Birth of an innovation . . . and the long road to implementation

In the late 1990s, Santen was working as a chemical engineer at a Shell plant in Moerdijk, the Netherlands. This specific Shell plant emits pure CO_2 as a by-product of its operations. One day, representatives of a Swiss company came to the Shell plant and asked whether they could use the Shell plant's carbon dioxide emissions. They were planning to build a plant in close proximity to the Shell facility. In other words, this company wanted to use the Shell plant's pure CO_2 emissions as a raw material for their own production process. The CO_2 gas had a value for the Swiss company; it therefore appeared to Santen that the emissions could have a value for Shell as well.

Santen worked on the contract negotiation and on the subsequent project to enable the sale and transport of carbon dioxide to the neighboring factory, which was under construction. The delivery and sales of CO_2 from Moerdijk to the customer began February 1999.

Santen was convinced that more factories could benefit from this idea. To his surprise, the concept of commercializing carbon dioxide emissions had not yet been introduced at a central level. He decided to develop this idea into a globally profitable business for Shell, based on sales of CO_2 gas from Shell sites to all potential industrial CO_2 buyers. The core idea of the concept was to sell CO_2 under long-term contracts, with asset financing provided by the customer. In this concept, SGSi would act as a broker. The business model was based on revenue sharing.

A long process

Shell has an internal review process, "Gamechanger," for assessing and selecting innovative ideas that have been generated by employees and that are eligible for funding. This business process is used to stimulate innovation and remove such potential barriers as lack of funding. At the time, the process was rather formal, and it was organized as a kind of "beauty contest." The assessment was organized in different rounds. Each round was a milestone, in which the Gamechanger panel returned a ruling of either "go" or "no go" for additional funding.

In June of 2000, Santen decided to submit his idea for commercializing carbon dioxide emissions to the Gamechanger internal review process. He survived the first round, and he had to work out his plan in more detail for the next round in December 2000, which would be followed by many subsequent rounds.

One factor that complicated the acceptance of his idea was that, although annual reports since 1999 had mentioned CO_2 emissions as a greenhouse issue, it had not yet been internalized in the core of Shell's corporate strategy. It was therefore an important task to convince senior management and the stakeholders that the strategy could also incorporate opportunities such as CO_2. Santen devoted considerable effort not only to improving his plan for each subsequent Gamechanger round, but also to convincing his peers, the line managers and his superiors. A great stimulus was that Santen gained the support of the Gamechanger process and Shell's CEO. He personally encouraged Santen to pursue his plan. At the end, the decision was made to commercialize the opportunity within SGSi.

Several rounds of the challenge sessions followed. Finally, Santen was rewarded with a "go"; he could begin implementation of the CO_2 sales within SGSi.

And another innovation

During the period in which Santen was implementing the CO_2 sales business within SGSi, CO_2 became increasingly important in international policy debates. The international community was becoming increasingly aware that CO_2 emissions have an important adverse impact on the environment – the "Greenhouse effect." In the wake of the United Nations Framework Convention on Climate Change in Kyoto, Japan on December 11, 1997, the situation changed. An initial statement was made that CO_2 emissions should be curbed in the future, and CO_2 continued to appear in the newspaper headlines on an almost daily basis (see Box 18.1).

Santen was convinced that legislation would soon follow the initial agreements on CO_2.

He used the impetus of the aftermath of Kyoto to develop yet another innovation: integral emissions management. The envisaged company would sell integral solutions for carbon dioxide management, which would require technical, operational, policy and trading expertise for compliance needs. The business could also realize opportunities resulting from legislation that Santen expected to result from Kyoto.

In November 2002, Santen introduced the concept of "emissions management" to Shell. Although connected to the first innovation, it was set up quite differently. It involved a strategic consultancy targeted toward industrial customers that would have to comply with the EU legislation and the European Trading Systems concerning CO_2. This legislation limits the amount of CO_2 that companies can release into the atmosphere. According to this legislation, companies must detail their actual emissions in relation to their allowances in annual reports. Companies exceeding the imposed limits could face financial penalties.

Santen's idea was that the consultancy would help these companies in the careful consideration of all of the technical and commercial options open to them, thus enabling them to construct a cost effective response to the legislation. The range of available solutions included CO_2 sales (the first innovation), as well as other options (e.g. using the

BOX 18.1 ARTICLE 3 OF THE KYOTO PROTOCOL

The Parties included in Annex I shall, individually or jointly, ensure that their aggregate anthropogenic carbon dioxide equivalent emissions of the greenhouse gases listed in Annex A do not exceed their assigned amounts, calculated pursuant to their quantified emission limitation and reduction commitments inscribed in Annex B and in accordance with the provisions of this Article, with a view to reducing their overall emissions of such gases by at least 5 per cent below 1990 levels in the commitment period 2008 to 2012.

flexible market mechanisms in the Kyoto Protocol: energy management and trading of CO_2 emissions allowances). For this endeavor, Santen proposed a consultancy business model, followed, if required, by implementation based on revenue sharing. As before, a long and painstaking process followed, in which Santen had to convince his peers and his superiors.

For an innovator, the fairytale can stop here: he managed to convince his organization to introduce two innovations. The story does not stop here, however. Santen faced a number of new challenges in the implementation of his ideas.

Too innovative?

Implementation of the innovations posed two different kinds of challenges. In the case of CO_2 sales, the challenge was to balance the interests of the Shell customer and the external customer. Second, Shell had never before seen CO_2 as a potential product. This was a genuine paradigm shift for them. Santen solved these problems by using genuine network management and stakeholder involvement.

A different kind of challenge arose with regard to emissions management. During implementation, it became clear that companies were not aware that they might have a problem, and clients, of course do not ask for solutions to problems of which they are not aware. Santen approached the market in a way that was different from what usually occurs in the traditional producer–client situation. He looked for stakeholders, and not solely for clients. He put great effort into trying to raise awareness among companies that CO_2 emissions were a problem that had to be solved, and that it brought benefits to companies that did. It was a painstaking process. After much talking, Santen found a company that was willing to be the pilot case. At this stage, the aim was not to sell a standard service. Instead, the aim was to develop a solution in close cooperation with the stakeholder in order to show Shell and the world that the solution worked.

Now the story does look like a fairy tale. In 2004, CO_2 emissions management is a full-fledged and endorsed business within SGSi, with a network of business relations throughout Shell and many large companies – a happy ending, at last.

Dilemmas . . . and strategies to overcome them

What does Santen's story have to say about innovation in a multinational such as Shell? In the first place, it shows that a strict perception of corporate strategy may impede genuine innovation. Vision is needed to provide direction to a company, and the vision was there. At the same time, however, the implementation of that vision (the strategy), which may reflect historical knowledge and perceptions rather than the actions that are needed for the future, requires priority setting with existing business interests. In hindsight, it is easy to see the importance of CO_2. After all, the Kyoto protocol is currently in existence because Russia finally ratified it (in 2004), thereby achieving the critical mass necessary for the document to go into effect. At the time that Santen submitted his idea to the Shell innovation community, however, this awareness was only simmering.

Had Shell adopted the corporate strategy rigorously, the two innovations would never have made it through the internal review process. The Shell Gamechanger process supported the nurturing of the idea and gave the resulting business opportuity a strategic priority. Guidance by strategy is good; it provides robustness and coherence to a company. At the same time, however, a company should search for ways to enable flexibility and adaptation to prepare itself to meet future demands. In this case, world events (Kyoto) and the support of the CEO and the Gamechanger process opened up the company for flexibility. The CEO showed leadership in encouraging Santen to pursue his plans.

Second, the story shows a tension between the characteristics needed to develop innovations into a successful product and the culture of a technological multinational, such as Shell. Let us first assess the technological part: in order for innovations to flourish, creativity is needed. Subsequently a business process and culture is required to develop the innovative ideas into solid business opportunities and to implement them in the main business. The latter requires a raise of strategic profile. In the Shell case this flexibility was shown by the Gamechanger process.

Another tension arises in the next phase of the innovation process. Entrepreneurs are needed in order to develop ideas into successful products. A large company such as Shell, however, does not attract genuine entrepreneurs. It attracts excellent scientists, engineers and analytical people. This is reflected in the reward system. While bonuses and targets stimulate incremental improvements in current products and services (e.g. sales volume), these reward systems are insufficient to reward risk-taking and entrepreneurial behavior. Santen persisted, not because of Shell's reward system, but because of his internal drive. He was so convinced that the innovation would be successful that he pushed forward, sometimes seemingly against all odds.

A third point is related to the former and concerns the implementation of the innovation. What if the innovation is so new that clients do not ask for it? How can an innovator sell a new product? Santen's answer was stakeholder management. He searched for partners to develop the idea jointly. New reward systems are needed in this phase as well. Existing reward systems may stimulate managers to become better in what they are already doing. A company that encourages innovation should reward managers who engage in new activities. They should be rewarded for their attempts to convince clients and create new markets. In other words, the company should create and reward leaders by using entrepreneurs as role models within the organization.

Intermission

So far, we have heard from two managers (Siemens and Shell) on innovation. A few general remarks can be made. In the first place, both managers discuss the importance of an adequate performance management system (Chapter 17). The current systems tend to be rigorous in rewarding measurable achievements instead of such less tangible accomplishments as networking and stakeholder management. The latter is indispensable for generating support for innovation within the organization and, later, for the diffusion of the innovation within the market. It also relates to the chapter on finance management (Chapter 4). Furthermore, while scientists and analytical work are indispensable for innovation, the successful implementation of an innovation requires more. It starts with the origin of the innovation; there must be a strong link between technology and society (Chapter 6) in order to ensure that the diffusion of the innovation will happen in the end (Chapter 7). Furthermore, the Shell case illustrated that considerable effort is put into both intra-organizational lobbying and stakeholder management (Chapter 10).

The interviews also show that strategy is important (Chapter 11), but that it is just as important for leaders to be able to deviate from the corporate strategy when necessary. The strategy should not live a life of its own. There will always be a tension between flexibility and control in the management of innovation (Chapter 15). On the one hand, there are strict procedures for innovation; on the other hand, however, entrepreneurs in the organization need trust and confidence, even if they do not formally meet all of the requirements. In the Shell case, this flexibility was shown by the CEO and the Gamechanger process.

INTERVIEW WITH MARTIEN VISSER, MANAGER COMMERCIAL STAFF/SENIOR STRATEGIST, GASUNIE TRADE AND SUPPLY

Dealing with uncertainty

Gasunie buys, transports and sells natural gas and promotes the safe, effective and innovative use of this source of energy. The company operates both on the national and international markets. Trade and Supply is one of the four companies within the holding. It buys and sells gas in the national and international market and supplies related services. The principal task of Gasunie Trade and Supply is to maximize the value of the national gas resources for the benefit of the government of the Netherlands and, to a much lesser extent, for the two private shareholders of Gasunie: ExxonMobil and Shell.

Among other tasks, Martien Visser is responsible for working on the strategy of Gasunie Trade and Supply. This is not an easy task, as the environment has changed drastically over the last few years. Like most utility sectors in the world, the gas sector is subject to liberalization policies. How does this development change the job of a strategist?

Increase of import . . . but how much?

Two trends have increased the European import of gas. The first is that the demand for gas keeps going up while the European supply is falling. A second fact is that a European agreement to decrease CO_2 emissions is making coal less suitable as a raw material for producing energy. At the same time, nuclear energy appears to remain too expensive, even aside from possible environmental objections. Because of these trends, the European import of gas is increasing, and it will continue to increase.

This situation is most relevant for the planning of Gasunie, which is one of the four or five major companies that sell gas on the European market. Gasunie needs to know how much gas its customers would like to buy from them in the (near) future. At this stage, however, the magnitude of future demand remains unclear. One important factor contributing to this uncertainty is the ultimate decision of the European states: will they stop producing nuclear energy and energy out of coal altogether, thus setting off an enormous rise in the demand for gas? Alternatively, will there only be a slight increase? A further complication is that governments may change their policies regarding nuclear or coal significantly after the elections.

The pipelines for transporting gas are already experiencing heavy use. The only answer to the increase in demand and the decrease in European supply is to construct new pipelines to enable Europe to import more gas. The planning and construction of a new pipeline would take at least five and maybe ten years.

The first dilemma is as follows: uncertainty about the rise and fall of demand is not in line with the long-term planning needed for the exploitation of natural gas. How can companies solve this mismatch between supply and demand?

Characteristics of a liberalized market versus requirements of the gas market

Liberalization and privatization are hot topics for utility sectors in general, and the gas sector is no exception. The European gas sector is in transition as well. The main characteristics of a liberalized market include the following: many competitors leading to "perfect competition" and a splitting up of infrastructure and trade operations. The gas sector, however, requires long-term planning and companies of sufficient size to carry the risks associated with the large, new projects required by Europe. For instance, the currently planned pipeline between Russia (Petersburg) and Northwestern Europe will cost around €8 billion and will not be in production before 2012. Who knows what the rapidly changing energy market will look like in 2012?

That there is a conflict between the liberalized market (fragmentation) and the requirements of the gas sector (big companies with great purchasing power) is obvious.

Dealing with dilemmas: planning, contracts and champions

What are Visser's strategies for dealing with the dilemmas of increasing uncertainty in a market that requires long-time planning and fragmentation in a market that requires big players with purchase power?

To deal with the first dilemma, Gasunie has refined its planning methods, building in mechanisms to create more flexibility. In addition to long-term planning, Gasunie has increasingly developed short-term planning. With regard to the fragmentation, the importance of contracts has increased. Such contracts aim to minimize the total risk of a project and to bring the final risks of the various contract partners into better balance. Given the considerable level of political uncertainty, which the common economic risk-mitigating measures cannot reasonably address, the governments also have a role to play in the contract game. This political stakeholder tends to complicate contract negotiations tremendously. With regard to the problem of fragmentation, other countries (e.g. France and Germany) have appointed national "champions." This means that at least one company within the country has the purchase power required by the gas sector at its disposal. This strategy, however, is not in the hands of the strategists at Gasunie, but in the hands of national policy makers. Even with the arrival of "champions," the strategists at Gasunie will continue to face these dilemmas.

Conclusion

The general trends toward liberalization and privatization that were described in Chapter 1 appear to be having a major impact on the gas industry. The response imposed by the EU and the national governments (liberalization of the market and more players of less significant size in order to break existing monopolies) is not in line with the requirements derived from the technical characteristics of the market. A strategist must deal with this tension, in part by using the strategies outlined in Chapter 16 to increase the flexibility. At the same time, however, the application of methods of cost and financial accounting (Chapter 4) and forecasting (Chapter 9) have become more rigid and thorough. The interview also shows that corporate strategy (Chapter 11) is entirely embedded in the interplay between technology and society (Chapter 6); without the decision of the EU and the national governments to liberalize the gas markets, Visser's job at Gasunie would have been completely different from his job today.

INTERVIEW WITH MICHIEL VALK, MANAGER OF THE NETWORK OPERATING CENTRE, KPN MOBIEL NL

Uncertainty and mobile telephony services: standardization or improvisation?

KPN Mobiel NL is the official name of the mobile branch of the Netherlands' incumbent telecom provider. In this book we will refer to this company as KPN Mobile. The com-pany works from within a high-rise office block, not unlike any other company. At least, that is the appearance a visitor is likely to have before opening the door of the Network Operating Centre (NOC). This is no run-of-the-mill workplace; it is more like a war room.

The first feature to demand a visitor's attention is the video wall, on which various images are projected. A number of screens display graphs that are the size of the average adult. In addition, the room contains a map of the Netherlands with flashing light bulbs of various sizes, which could easily be construed as a warning that enemy maneuvers have been spotted. In the meantime, the news runs non-stop.

This is the Network Operating Centre of KPN Mobile. KPN is the largest telecom provider in the Netherlands. The NOC closely monitors the physical infrastructure (e.g. the masts and connections) day and night; it coordinates repair jobs, corrects disturbances and measures end-to-end quality. Fifty people work at the NOC: twenty perform monitoring in shifts, twenty work during the day, and ten are responsible for management and support.

Ever-increasing complexity

Valk's job has changed dramatically over the years. In the past, services were relatively easy to monitor. Disturbances could be easily traced and remedied, as there was only one chain. Schematically, the voice telephony chain looked like this (part a), but the advent of such new services such as I-mode, Voicemail and Blackberry, however, has changed it (part b):

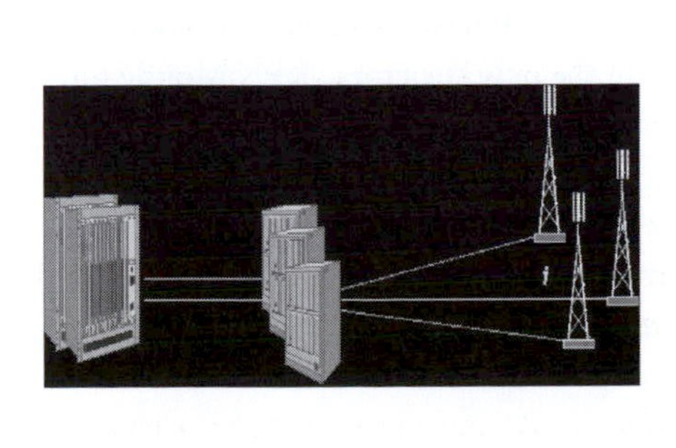

(a)

(b)

Figure 18.1 *Base station, control*

To complicate things further, various external companies are involved in the management of certain platforms in the chain, each with its own software, processes and technological systems. As a result, this long chain also consists of innumerable components. The software and hardware must mesh seamlessly with each other in order to deliver a reliable, efficient and effective service – but sometimes things go wrong.

This example thus offers a real-life case of the theory discussed in "Managing complex technologies under intensifying interdependencies" (Chapter 16). What is the real-life response to this dilemma?

The answer to complexity: obligatory uniformity?

The most popular solution for increased complexity in the environment is standardization. Initially, the best strategy for KPN appeared to be the standardization of processes and technology of all the companies in the value chain. This soon proved impossible in practice, however. Many companies had already optimized their working methods and processes to suit their own scales. Moreover, companies often outsource some of their activities to sub-contractors, who use different technologies. Even if a company can oblige the main operators in the chain to work with "their" technology, therefore, different procedures and technologies will always remain at the level of the sub-contractors.

Specialization and improvisation

The intertwining of sectors and services and the increase in the length of the chains have jettisoned uniform technology into "Cloud Cuckoo Land." The ex ante strategy ("use all the same technology") proved unworkable. How can reliable services be guaranteed under these new conditions? The answer of KPN Mobile is to invest in the improvisational ability of the staff. It has proved impossible, however, to train employees to trace and solve *all* disturbances in *all* services. The chains are too long, and the services are simply too diverse. It is hoped that a new structure at the NOC will improve the employees' improvisational skills through training and coaching.

Second, it was decided that NOC employees would specialize more, as the increasingly complex chains require more detailed knowledge to analyze disturbances correctly. On January 1, 2005, the services will be split into "Voice" and "Data." The new levels of specialization will be achieved by investing in knowledge management, among other things. Knowledge will then become more easily accessible to the personnel.

Standardization to *enable* improvisation

On the one hand, specialization and improvisation are strategies for coping with the increasing complexity caused by sector interaction, the rise in the number of companies, and links between different technologies. Improvisation means giving more scope to the personnel. On the other hand, a trend is emerging that actually calls for standardization. In the new structure, KPN Mobile must communicate with many more companies in the event of a disturbance. When trying to trace the disturbance, they must cooperate with system integrators, hosting parties and similar entities. Standard definitions and working methods could facilitate communication.

Whereas improvisation requires more scope for the employees, standardization imposes limitations. A question that arises, therefore, involves how to find the right way to manage increasing complexity?

Improvisation, specialization and standardization

Henry Ford's answer to this problem at the end of the nineteenth century was revolutionary in its simplicity: uniformity and standardization. Only black Model T Fords rolled off the assembly line at his factory. This kind of strategy no longer works, simply because there is no central control to enforce uniformity. Moreover, customers want all sorts of extra services and are no longer happy with just black Model T Fords. Neither the partners of KPN nor their sub-contractors will allow anyone to dictate to them the kind of technology that they must use. Improvisation is the only workable strategy.

As in the case of innovation, improvisation does not necessarily imply *laissez-faire*. In fact, improvisation is helped by standardization, especially when applied to working methods and definitions. Standardization is needed in order to improve internal communication and knowledge management and, most importantly, to facilitate communication between KPN and the other companies.

Valk was faced with finding an answer to increased uncertainty. In this case, improvisation is not opposed to standardization. On the contrary, specializing and standardizing at the same time actually pave the way for improvisation (see also Chapter 16). This choice has consequences for the design of KPN Mobile (see Chapter 2): improvisation enables inter-organizational collaboration, which, in turn, calls for more inter-organizational decision making (Chapter 10 and Chapter 17 on knowledge management).

INTERVIEW WITH MARGIEN MUTTER, RECRUITMENT MANAGER FOR UNILEVER RESEARCH AND DEVELOPMENT

Specialized skills and generalized competences: wanting the impossible?

Unilever is one of the world's leading suppliers of fast-moving consumer goods. Some of Unilever brands are world famous, such as the food brands Knorr, Becel, Lipton, Iglo, SlimFast and Bertolli, and such personal care brands as Dove, Lux and Rexona, to name only a few. Other brands have a special role in just one or a few countries. The term “multi-national” aptly applies to Unilever; the company employs 234,000 people in approximately 100 countries worldwide.

Innovation is essential for fast-moving consumer goods, and this interest is reflected in the size of the R&D department in Unilever: in 2003, Unilever spent €1,065 million on research and development, which is 2.5 percent of their turnover. The R&D department in Vlaardingen, the Netherlands, employs 1,000 researchers (worldwide, there are 3,000 researchers in the central R&D laboratories and 3,000 in regional innovation centers and global technology centers).

Margien Mutter is the Recruitment Manager for Research and Development in the Netherlands. In her daily job, she faces two important dilemmas. The first involves the tension between two different kinds of requirements that Unilever has developed for its employees: requirements regarding skills and requirements concerning competences. Second, the tasks and position of R&D within Unilever has changed over the years, resulting in competing demands with regard to the qualities that new R&D employees should possess.

How can a recruitment manager contribute to finding new employees for Unilever who have the scientific skills and who, at the same time, meet additional and sometimes competing requirements with regard to corporate competences?

Scientific skills versus corporate competences

Highly specialized scientific skills are very important in R&D. In the Unilever laboratories, researchers work on problems that require highly specialized expertise. Without scientific thoroughness, breakthrough innovations in detergents, in food and personal care would not have been possible. For years, scientific qualifications constituted the first and foremost criterion for an R&D candidate. Given the specialized nature of the work, this qualification can only be judged by the candidate's peers and superiors. It is too specialized to be tested by such general staff functionaries as recruiters or HR officers.

In the early 1990s, Unilever formulated a list of competences that all employees should possess, including the researchers at R&D. This set of competences changes every five to seven years to reflect the changing demands of Unilever and its employees. In 2001, Unilever introduced a new set of competences, as part of its Path to Growth Strategy. These competences included "change catalyst, empowering others, breakthrough thinking, team commitment, organizational awareness, seizing the future, passion for growth, holding people accountable, developing self and others."

In theory, the organizational competences and scientific skills can be combined or even strengthen each other. In practice, there is a tension between the characteristics of a scientific researcher and the Unilever competences. After all, research requires specialized employees who are motivated to spend years behind microscopes or in chemical plants to work meticulously on an innovation, analyzing test result after test result. They have had their training in universities, in which research results are the only criterion for success.

In the new situation, this same person with research skills and specialized expertise is required to possess such generic management qualities as "empowering others, holding people accountable and showing organizational awareness." The trouble is that specialists are rarely good generalists, and vice versa.

How does Mutter solve this dilemma? Mutter has chosen to raise awareness among line managers concerning the Unilever competences in the selection procedure, as these managers are pivotal in the selection process. The suitability of applicants for such highly specialized jobs is difficult for an outsider to judge. Evaluating the suitability of a candidate requires expertise in the content. In terms of expertise, peers and line managers are obviously the best qualified to select their future colleagues. Because they are all senior researchers themselves, however, line managers are inclined to select according to scientific skills rather than ensuring that the Unilever competences are fulfilled.

For this reason, R&D recruitment has developed tools that R&D line managers can use to test candidates. The testing of competences has, thus, become an integral part of the procedure, which is led by the candidate's future superior. Of course, this can only be realized if the line managers within R&D are convinced of the value of the Unilever competences. This is a long process, which needs continuous attention from the recruitment manager.

The process described above is part of Mutter's broader effort to create more sensitivity for competences among line managers. This awareness can be stimulated by adding "selection of persons with both scientific skills and generic competences" to the personal goals of the managers. Such a change would hold line managers accountable for hiring staff that meet the challenges that Unilever considers important. Her experience is that managers with such targets usually choose higher quality recruits than do managers who have no such targets.

Researcher versus entrepreneur

Mutter is convinced that competences are important for R&D employees, and that their importance will increase in the future. The reason is that the position of R&D within Unilever has changed. In the past, R&D resembled a "university" within the multinational, where dedicated and highly specialized researchers could work for years on a project, which may or may not result in a profitable innovation for Unilever. Recently, however, Unilever has been holding the R&D division increasingly accountable for the investments. There is a stricter regime concerning the number of innovations that should be realized within a given time frame. In other words, R&D must show results in order to be eligible for future funding.

This development has led to a shift from a university type of approach towards "product development" and "technology development." Managers within R&D must be aware of marketing, cost and productivity in order to legitimize future investments in their project. This means that they must sell their concepts to both their peers and their superiors.

Convincing one's own organization requires skills that are different – or at least in addition to – those that are needed to carry out fundamental scientific research. Managers must be aware of the markets; they must cooperate with other departments within Unilever in order to turn their innovations into successful implementation in the market. This measure calls for broadening the scope and range of skills that researchers are required to possess. Employees in R&D face the challenge of combining qualities that are traditionally not combined: commercial skills and scientific thoroughness.

For Mutter, this presents a new challenge: how to find candidates who are not only excellent researchers, but also good entrepreneurs. This challenge is met by recruiting in unorthodox and creative ways instead of using the usual channels. One of the ways is to invite promising students personally for open-house days at Unilever and using the networks of managers to recruit new employees. As before, however, Mutter's most important strategy is the long-term approach. Together with line managers, she seeks to raise awareness that entrepreneurship is an important selection criterion and will increasingly be so, in order to guarantee future funding for R&D activities within Unilever.

Conclusion

Recruitment is never an easy task, and specific issues occur in the case of advanced technology, as Chapter 3 described. The Unilever interview illustrates this point. It is hard to combine requirements that result from the design of the firm (Chapter 2) and the corporate strategy (Chapter 11) that ensures both innovation and diffusion (Chapters 7 and 8). Mutter's answer to this tension is to make recruitment an integral part of the primary process of the company, instead of an isolated activity in the organization. She does so for two reasons. First, since recruitment requires in-depth knowledge of the expertise of the future R&D employees, it can only be done by peer review. Second, entrepreneurship is thought to be vital for Unilever. This competence should not only be tested by a separate recruitment officer; it should be dispersed throughout the organization. For these reasons, recruitment has become an ongoing activity.

INTERVIEW WITH DIRK-WILLEM VAN GULIK, PRESIDENT OF THE APACHE SOFTWARE FOUNDATION, OPEN SOURCE DEVELOPER AND CTO OF A SOFTWARE SOLUTIONS COMPANY

Dirk-Willem van Gulik is an open source developer. This means that he develops software, but there is more to source development than simply writing programs and fixing errors in existing code. Van Gulik is part of a community – an open source community – within which open source software is developed. Most of the developers are paid, either directly or indirectly, to work on open source, or their work in the open source community relates directly to the demands of their consulting services or products. Programmers who work for product companies (both large and small), consultants for "the big 5," consultants for small companies, system integrators and value-added resellers are typical participants in open source communities.

Many open source developers never see each other in real life; they meet virtually on the Internet. Nonetheless, they are able to work collectively to develop highly complex software that is very successful, according to some measures. The Apache community is a well-known example of an open source community, which consists of a large number of sub-projects, including the Apache http Webserver. According to the Gardner group, some ninety percent of the world's Internet and web systems relies in some way on Apache.

The Apache community is formalized in the Apache Software Foundation (ASF), which is a non-profit (502c3) corporation. Its members develop new, technically not-for-profit, software. The Apache Software Foundation was formed primarily to supply hardware, communication and business infrastructure in order to provide a foundation for open, collaborative software development projects. The Foundation also plays a legal role as a holder and steward of copyrights, trademarks and similar artifacts. It is important to note that the software is not sold, but it is offered freely to society and made available through the Internet.

Standardization or tailor-made?

When ordering software, a company seeks a tailor-made solution for a specific problem or situation. The software should not be too specific, however, as the company must also be able to communicate and exchange with other companies, and with other systems within the company. This situation thus creates an incentive for standardization in software. There is yet another incentive, however: standardization is cost-efficient. After all, increasing the scale can decrease the costs of developing software. Another reason for standardization or neutrality is that no IT manager wishes to become too dependent on a single vendor. The aversion to dependency stems largely from two issues: stability and access. Stability is an issue, as a vendor may suddenly cease to exist, end the sale of a product or suddenly decide to change the product radically, making it no longer suitable for the IT manager's purposes. Access is an issue, as a vendor may decide to raise the price, enforce costly upgrades or support contracts, for vendors know that clients face significant costs should they decide to change suppliers. In the industry, this strategy is known as "lock-in."

In the past, IBM was known, and feared by CFOs, for its highly effective lock-in; today, Microsoft offers a prime example. When a company orders a product from such suppliers as IBM, Getronics or Oracle, it seeks software that is both specific and neutral. How can this be achieved?

Open source as the answer?

Open source software can provide a "neutral," cost-efficient, lowest layer of software that can enable communication with other systems. Companies can make the software more specific in the layers that are based on the lower layers. Typically, the higher or "back-ends" are highly specific, not only according to vendor, but often varying from customer to customer as well.

Open source software offers another advantage. The software is freely available to everyone who is interested in it. Users can use, test and refine the software, allowing it to mature under a more realistic and wider range of real-world requirements than does software that is conceived within a marketing department. Open source software therefore tends to fit the real-world problems better and with less modification. This point is illustrated with an example from the company within which van Gulik works. As a general guideline, the number of days that his company estimates for the system integration of a non-open source product is roughly double that required to integrate an open source product.

Another advantage of open source software is that it enables better communication between programs. This advantage arises purely from the fact that both parties use the same open source product. This has led to a "network effect," which has caused open source protocol stacks to dominate http, smtp, DNS, xml and similar systems. Finally, the blueprints for open source software are, for the most part, available to the customer. This means that any competent practitioner in the relevant field is able to maintain, refine, tailor or develop the software further. As a result, it becomes possible and economically viable for a customer to make a purely business-driven decision concerning whether to get an outsider to make the fix needed to continue to use the software should the vendor cease to exist, refuse or delay a fix, or raise the price above a certain level. In short, the public availability of these blueprints makes vendor dependency and lock-in less of an issue.

Open source developers are proud to say that their products are cheaper and can communicate with other programs; some of the world's best developers have worked on it, and it has matured in a more realistic environment, resulting in better solutions and minimizing the lock-in issue.

Dilemma for the company

Open source software appears to offer a solution to the dilemma arising out of the tension between neutrality and specificity. Companies that use open source software, however, face new dilemmas. One problem is that the rationale of part of the open source community conflicts with business models. A group of "evangelists" within the community are open source developers by conviction. They feel strongly about the ideology of open source software, namely that all software should be open and freely accessible to all. This ideology is based on the opinion that if developers have used the code to develop software, they should give the new product or the fixes in the old code, back to the community. According to these evangelists, the entire code should be published on the Internet. This desire conflicts with demands made on the software by the companies, especially with regard to secrecy and copyrights. Although most open source licenses do not interact or conflict with business models, some commonly used license models do. Even if this is not the case, as when there are no legal conflicts, there may still be a risk that the company could be "branded" as one that "cheats" or "makes money at the expense of some poor open source developer who never got a dime."

The dilemma in practice

In practice, however, practically everyone makes use of open software, according to van Gulik. The basic layers of Internet software are all based on open source; examples include the software platform behind Google, the Apache HTTP web server, the Sendmail SMTP mail software and the BIND DNS (Internet Address book) software. The last example illustrates how one company deals with open source in practice. BIND is open source software, but it is written by a company under contract with a consortium of vendors. The company pays full-time developers, and is paid, in turn, by the consortium of approximately twenty large companies, all of whom need this DNS address book software.

In order to make sure that no one cheats, and in order to make sure that the Internet as a whole remains a viable market, these companies do not mind the fact that the software is open source. On the contrary, they realize that this is a way of protecting their investments.

The same holds true for programs developed by such companies as SAP, IBM, HP and Oracle. Each of these is built on significant layers of open source (e.g. the Apache Web Server, the Xerces XML parsers and the Axis SOAP protocol stack). Companies who use open source software can choose what to do with their products. One option is to give the code – with fixes – back to the community, to the extent that copyright and other restrictions allow. This option has both advantages and disadvantages. On the one hand, giving the code back to the community requires less work for integrating and tracking the company's own changes, and it allows improvements and deeper integration by others with similar needs. Furthermore, it provides a natural generalization of the addition, making it easier for the company to use it for other customers. Finally, the choice to return the revised code to the open source community strengthens and improves the open source product.

On the other hand, returning the code to the community results in lost benefits; the new product is less unique. A company can choose to keep the newly developed software for themselves. While this option does generate a competitive advantage, it also increases the amount of work necessary to track the new changes, and it makes the new product quite specific for the customer, which means that there is no generalization.

A company does not always have a free choice in all circumstances. For example, companies who build onto GPL Licensed code are obliged by law to give it back to the community. This is not the case for the more commonly used open source software, which falls under the "BSD style license," and companies can keep their improvements for themselves.

Standardization versus specialization is an old, perhaps the oldest, issue in the management of technology (see also Chapter 2). Van Gulik's illustrations of the use of open source software make clear that it mitigates the original standardization–specialization dilemma. In some cases, open source actively enables a long-desired combination of standardization and specialization. New dilemmas have arisen simultaneously, however. MOT managers will always have something to manage, even in open source contexts.

CONCLUDING REMARKS: MATCHING THE SOCIAL AND THE TECHNOLOGICAL

The management of technology is a demanding job. The four preceding sections in this book have shown that even one discipline (e.g. human resources management, innovation, manufacturing or marketing) is already a difficult job in itself. These interviews with managers of technology have demonstrated that the situation becomes even more complicated when all of these various disciplines must be combined. What is good for innovation may be less desirable from the perspective of efficiency; what is good for managing human resources, may not be the best policy with regard to product development. A manager must balance all of these different, often conflicting, demands.

If we aggregate the interviews, a common theme emerges. Many of the dilemmas facing the managers arise out of a need to integrate the social and the technological. For example, the managers at Siemens, Shell and Unilever mentioned that the social dimension has been gaining increasing importance in the innovation process. They referred to the involvement of stakeholders in society, and to the necessity of persuading peers and superiors within the organization as a precondition for a successful process of innovation and dissemination. These are not just words; all of the managers brought up concrete and specific measures with which they aimed to integrate the technological and the social. Examples are the grouping of innovation around the themes described in the national governmental program (Siemens), bringing about a shift in the reward system to encourage intra-organizational entrepreneurship (Siemens and Shell), embedding the formerly isolated position of R&D more firmly within the organization (Unilever and Siemens) and involving end users from the start of the innovation (Shell and Siemens).

Integrating the social and the technological is also a theme in the interviews with the managers at Gasunie and KPN Mobile. The main concern of these managers is how to deal with uncertainty. Gasunie's uncertainty stems mainly from the liberalization of the market that resulted from policy decisions of the EU and the national governments. In order to be able to operate effectively and efficiently in the future, Gasunie must engage actively in national and international debates on liberalization. It would not suffice simply to respond to developments in the market; they must operate pro-actively. As before, this requires stakeholder management. For KPN Mobile, matching the technological with the social forms the core dilemma. They face increased uncertainty resulting from technological developments and changes in the product-service chain. Although they are supported by technological aids, their main response to these conditions of increased uncertainty is to enhance the ability of their employees to improvise.

The open source developer presented a dilemma of a different kind. His main concern was the tension between standardization and specialization. His account of open source software showed that new technologies can provide answers to deep-rooted management dilemmas. The good news is that open source software has reconciled differences that have been considered irreconcilable since the advent of management theory. New dilemmas have arisen at the same time, however.

NOTE

1 Mr Santen has left Shell to start up his own business. The opinions expressed in this interview are those of Mr Santen personally.

Index

Page references to Boxes or Tables are in *italic* print.